T0317761

FINANCIAL SIGNAL PROCESSING AND MACHINE LEARNING

FINANCIAL SIGNAL PROCESSING AND MACHINE LEARNING

Edited by

Ali N. Akansu

New Jersey Institute of Technology, USA

Sanjeev R. Kulkarni

Princeton University, USA

Dmitry Malioutov

IBM T.J. Watson Research Center, USA

IEEE PRESS

WILEY

This edition first published 2016
© 2016 John Wiley & Sons, Ltd
First Edition published in 2016

Registered office
John Wiley & Sons Ltd, The Atrium, Southern Gate, Chichester, West Sussex, PO19 8SQ, United Kingdom

For details of our global editorial offices, for customer services and for information about how to apply for permission to reuse the copyright material in this book please see our website at www.wiley.com.

The right of the author to be identified as the author of this work has been asserted in accordance with the Copyright, Designs and Patents Act 1988.

All rights reserved. No part of this publication may be reproduced, stored in a retrieval system, or transmitted, in any form or by any means, electronic, mechanical, photocopying, recording or otherwise, except as permitted by the UK Copyright, Designs and Patents Act 1988, without the prior permission of the publisher.

Wiley also publishes its books in a variety of electronic formats. Some content that appears in print may not be available in electronic books.

Designations used by companies to distinguish their products are often claimed as trademarks. All brand names and product names used in this book are trade names, service marks, trademarks or registered trademarks of their respective owners. The publisher is not associated with any product or vendor mentioned in this book

Limit of Liability/Disclaimer of Warranty: While the publisher and author have used their best efforts in preparing this book, they make no representations or warranties with respect to the accuracy or completeness of the contents of this book and specifically disclaim any implied warranties of merchantability or fitness for a particular purpose. It is sold on the understanding that the publisher is not engaged in rendering professional services and neither the publisher nor the author shall be liable for damages arising herefrom. If professional advice or other expert assistance is required, the services of a competent professional should be sought.

Library of Congress Cataloging-in-Publication Data applied for

ISBN: 9781118745670

A catalogue record for this book is available from the British Library.

Set in 10/12pt, TimesLTStd by SPi Global, Chennai, India.

1 2016

Contents

List of Contributors

Ali N. Akansu, New Jersey Institute of Technology, USA

Marco Cuturi, Kyoto University, Japan

Alexandre d'Aspremont, CNRS - Ecole Normale supérieure, France

Christine De Mol, Université Libre de Bruxelles, Belgium

Jianqing Fan, Princeton University, USA

Jun-ya Gotoh, Chuo University, Japan

Nicholas A. James, Cornell University, USA

Prabhanjan Kambadur, Bloomberg L.P., USA

Alexander Kreinin, Risk Analytics, IBM, Canada

Sanjeev R. Kulkarni, Princeton University, USA

Yuan Liao, University of Maryland, USA

Han Liu, Princeton University, USA

Matthew Lorig, University of Washington, USA

Aurélie C. Lozano, IBM T.J. Watson Research Center, USA

Ronny Luss, IBM T.J. Watson Research Center, USA

Dmitry Malioutov, IBM T.J. Watson Research Center, USA

David S. Matteson, Cornell University, USA

William B. Nicholson, Cornell University, USA

Ronnie Sircar, Princeton University, USA

Akiko Takeda, The University of Tokyo, Japan

Mustafa U. Torun, New Jersey Institute of Technology, USA

Stan Uryasev, University of Florida, USA

Onur Yilmaz, New Jersey Institute of Technology, USA

Preface

This edited volume collects and unifies a number of recent advances in the signal-processing and machine-learning literature with significant applications in financial risk and portfolio management. The topics in the volume include characterizing statistical dependence and correlation in high dimensions, constructing effective and robust risk measures, and using these notions of risk in portfolio optimization and rebalancing through the lens of convex optimization. It also presents signal-processing approaches to model return, momentum, and mean reversion, including both theoretical and implementation aspects. Modern finance has become global and highly interconnected. Hence, these topics are of great importance in portfolio management and trading, where the financial industry is forced to deal with large and diverse portfolios in a variety of asset classes. The investment universe now includes tens of thousands of international equities and corporate bonds, and a wide variety of other interest rate and derivative products-often with limited, sparse, and noisy market data.

Using traditional risk measures and return forecasting (such as historical sample covariance and sample means in Markowitz theory) in high-dimensional settings is fraught with peril for portfolio optimization, as widely recognized by practitioners. Tools from high-dimensional statistics, such as factor models, eigen-analysis, and various forms of regularization that are widely used in real-time risk measurement of massive portfolios and for designing a variety of trading strategies including statistical arbitrage, are highlighted in the book. The dramatic improvements in computational power and special-purpose hardware such as field programmable gate arrays (FPGAs) and graphics processing units (GPUs) along with low-latency data communications facilitate the realization of these sophisticated financial algorithms that not long ago were "hard to implement."

The book covers a number of topics that have been popular recently in machine learning and signal processing to solve problems with large portfolios. In particular, the connections between the portfolio theory and sparse learning and compressed sensing, robust optimization, non-Gaussian data-driven risk measures, graphical models, causal analysis through temporal-causal modeling, and large-scale copula-based approaches are highlighted in the book.

Although some of these techniques already have been used in finance and reported in journals and conferences of different disciplines, this book attempts to give a unified treatment from a common mathematical perspective of high-dimensional statistics and convex optimization. Traditionally, the academic quantitative finance community did not have much overlap with the signal and information-processing communities. However, the fields are seeing more interaction, and this trend is accelerating due to the paradigm in the financial sector which has embraced state-of-the-art, high-performance computing and signal-processing technologies. Thus, engineers play an important role in this financial ecosystem. The goal of this edited volume is to help to bridge the divide, and to highlight machine learning and signal processing as disciplines that may help drive innovations in quantitative finance and electronic trading, including high-frequency trading.

The reader is assumed to have graduate-level knowledge in linear algebra, probability, and statistics, and an appreciation for the key concepts in optimization. Each chapter provides a list of references for readers who would like to pursue the topic in more depth. The book, complemented with a primer in financial engineering, may serve as the main textbook for a graduate course in financial signal processing.

We would like to thank all the authors who contributed to this volume as well as all of the anonymous reviewers who provided valuable feedback on the chapters in this book. We also gratefully acknowledge the editors and staff at Wiley for their efforts in bringing this project to fruition.

1

Overview
Financial Signal Processing and Machine Learning

Ali N. Akansu[1], Sanjeev R. Kulkarni[2], and Dmitry Malioutov[3]
[1]*New Jersey Institute of Technology, USA*
[2]*Princeton University, USA*
[3]*IBM T.J. Watson Research Center, USA*

1.1 Introduction

In the last decade, we have seen dramatic growth in applications for signal-processing and machine-learning techniques in many enterprise and industrial settings. Advertising, real estate, healthcare, e-commerce, and many other industries have been radically transformed by new processes and practices relying on collecting and analyzing data about operations, customers, competitors, new opportunities, and other aspects of business. The financial industry has been one of the early adopters, with a long history of applying sophisticated methods and models to analyze relevant data and make intelligent decisions – ranging from the quadratic programming formulation in Markowitz portfolio selection (Markowitz, 1952), factor analysis for equity modeling (Fama and French, 1993), stochastic differential equations for option pricing (Black and Scholes, 1973), stochastic volatility models in risk management (Engle, 1982; Hull and White, 1987), reinforcement learning for optimal trade execution (Bertsimas and Lo, 1998), and many other examples. While there is a great deal of overlap among techniques in machine learning, signal processing and financial econometrics, historically, there has been rather limited awareness and slow permeation of new ideas among these areas of research. For example, the ideas of stochastic volatility and copula modeling, which are quite central in financial econometrics, are less known in the signal-processing literature, and the concepts of sparse modeling and optimization that have had a transformative impact on signal processing and statistics have only started to propagate slowly into financial

Financial Signal Processing and Machine Learning, First Edition.
Edited by Ali N. Akansu, Sanjeev R. Kulkarni and Dmitry Malioutov.
© 2016 John Wiley & Sons, Ltd. Published 2016 by John Wiley & Sons, Ltd.

applications. The aim of this book is to raise awareness of possible synergies and interactions among these disciplines, present some recent developments in signal processing and machine learning with applications in finance, and also facilitate interested experts in signal processing to learn more about applications and tools that have been developed and widely used by the financial community.

We start this chapter with a brief summary of basic concepts in finance and risk management that appear throughout the rest of the book. We present the underlying technical themes, including sparse learning, convex optimization, and non-Gaussian modeling, followed by brief overviews of the chapters in the book. Finally, we mention a number of highly relevant topics that have not been included in the volume due to lack of space.

1.2 A Bird's-Eye View of Finance

The financial ecosystem and markets have been transformed with the advent of new technologies where almost any financial product can be traded in the globally interconnected cyberspace of financial exchanges by anyone, anywhere, and anytime. This systemic change has placed real-time data acquisition and handling, low-latency communications technologies and services, and high-performance processing and automated decision making at the core of such complex systems. The industry has already coined the term *big data finance*, and it is interesting to see that technology is leading the financial industry as it has been in other sectors like e-commerce, internet multimedia, and wireless communications. In contrast, the knowledge base and exposure of the engineering community to the financial sector and its relevant activity have been quite limited. Recently, there have been an increasing number of publications by the engineering community in the finance literature, including *A Primer for Financial Engineering* (Akansu and Torun, 2015) and research contributions like Akansu *et al.*, (2012) and Pollak *et al.*, (2011). This volume facilitates that trend, and it is composed of chapter contributions on selected topics written by prominent researchers in quantitative finance and financial engineering.

We start by sketching a very broad-stroke view of the field of finance, its objectives, and its participants to put the chapters into context for readers with engineering expertise. Finance broadly deals with all aspects of money management, including borrowing and lending, transfer of money across continents, investment and price discovery, and asset and liability management by governments, corporations, and individuals. We focus specifically on trading where the main participants may be roughly classified into hedgers, investors, speculators, and market makers (and other intermediaries). Despite their different goals, all participants try to balance the two basic objectives in trading: to maximize future expected rewards (returns) and to minimize the risk of potential losses.

Naturally, one desires to buy a product cheap and sell it at a higher price in order to achieve the ultimate goal of profiting from this trading activity. Therefore, the expected return of an investment over any holding time (horizon) is one of the two fundamental performance metrics of a trade. The complementary metric is its variation, often measured as the standard deviation over a time window, and called investment risk or market risk.[1] Return and risk are two typically conflicting but interwoven measures, and risk-normalized return (Sharpe ratio)

[1] There are other types of risk, including credit risk, liquidity risk, model risk, and systemic risk, that may also need to be considered by market participants.

finds its common use in many areas of finance. Portfolio optimization involves balancing risk and reward to achieve investment objectives by optimally combining multiple financial instruments into a portfolio. The critical ingredient in forming portfolios is to characterize the statistical dependence between prices of various financial instruments in the portfolio. The celebrated Markowitz portfolio formulation (Markowitz, 1952) was the first principled mathematical framework to balance risk and reward based on the covariance matrix (also known as the variance-covariance or VCV matrix in finance) of returns (or log-returns) of financial instruments as a measure of statistical dependence. Portfolio management is a rich and active field, and many other formulations have been proposed, including risk parity portfolios (Roncalli, 2013), Black–Litterman portfolios (Black and Litterman, 1992), log-optimal portfolios (Cover and Ordentlich, 1996), and conditional value at risk (cVaR) and coherent risk measures for portfolios (Rockafellar and Uryasev, 2000) that address various aspects ranging from the difficulty of estimating the risk and return for large portfolios to the non-Gaussian nature of financial time series, and to more complex utility functions of investors.

The recognition of a price inefficiency is one of the crucial pieces of information to trade that product. If the price is deemed to be low based on some analysis (e.g. fundamental or statistical), an investor would like to buy it with the expectation that the price will go up in time. Similarly, one would shortsell it (borrow the product from a lender with some fee and sell it at the current market price) when its price is forecast to be higher than what it should be. Then, the investor would later buy to cover it (buy from the market and return the borrowed product back to the lender) when the price goes down. This set of transactions is the building block of any sophisticated financial trading activity. The main challenge is to identify price inefficiencies, also called *alpha* of a product, and swiftly act upon it for the purpose of making a profit from the trade. The efficient market hypothesis (EMH) stipulates that the market instantaneously aggregates and reflects all of the relevant information to price various securities; hence, it is impossible to beat the market. However, violations of the EMH assumptions abound: unequal availability of information, access to high-speed infrastructure, and various frictions and regulations in the market have fostered a vast and thriving trading industry.

Fundamental investors find alpha (i.e., predict the expected return) based on their knowledge of enterprise strategy, competitive advantage, aptitude of its leadership, economic and political developments, and future outlook. Traders often find inefficiencies that arise due to the complexity of market operations. Inefficiencies come from various sources such as market regulations, complexity of exchange operations, varying latency, private sources of information, and complex statistical considerations. An *arbitrage* is a typically short-lived market anomaly where the same financial instrument can be bought at one venue (exchange) for a lower price than it can be simultaneously sold at another venue. Relative value strategies recognize that similar instruments can exhibit significant (unjustified) price differences. Statistical trading strategies, including statistical arbitrage, find patterns and correlations in historical trading data using machine-learning methods and tools like factor models, and attempt to exploit them hoping that these relations will persist in the future. Some market inefficiencies arise due to unequal access to information, or the speed of dissemination of this information. The various sources of market inefficiencies give rise to trading strategies at different frequencies, from high-frequency traders who hold their positions on the order of milliseconds, to midfrequency trading that ranges from intraday (holding no overnight position) to a span of a few days, and to long-term trading ranging from a few weeks to years. High-frequency trading requires state-of-the-art computing, network communications, and trading infrastructure: a large number of trades are made where each position is held for a very short time period and typically produces a small return with very little risk. Longer term strategies are less dependent on latency and sophisticated technology, but individual positions are typically held for a longer time horizon and can pose substantial risk.

1.2.1 Trading and Exchanges

There is a vast array of financial instruments ranging from stocks and bonds to a variety of more sophisticated products like futures, exchange-traded funds (ETFs), swaps, collateralized debt obligations (CDOs), and exotic options (Hull, 2011). Each product is structured to serve certain needs of the investment community. Portfolio managers create investment portfolios for their clients based on the risk appetite and desired return. Since prices, expected returns, and even correlations of products in financial markets naturally fluctuate, it is the portfolio manager's task to measure the performance of a portfolio and maintain (rebalance) it in order to deliver the expected return.

The market for a security is formed by its buyers (bidding) and sellers (asking) with defined price and order types that describe the conditions for trades to happen. Such markets for various financial instruments are created and maintained by exchanges (e.g., the New York Stock Exchange, NASDAQ, London Stock Exchange, and Chicago Mercantile Exchange), and they must be compliant with existing trading rules and regulations. Other venues where trading occurs include dark pools, and over-the-counter or interbank trading. An order book is like a look-up table populated by the desired price and quantity (volume) information of traders willing to trade a financial instrument. It is created and maintained by an exchange. Certain securities may be simultaneously traded at multiple exchanges. It is a common practice that an exchange assigns one or several market makers for each security in order to maintain the robustness of its market.

The health (or liquidity) of an order book for a particular financial product is related to the bid–ask spread, which is defined as the difference between the lowest price of sell orders and the highest price of buy orders. A robust order book has a low bid–ask spread supported with large quantities at many price levels on both sides of the book. This implies that there are many buyers and sellers with high aggregated volumes on both sides of the book for that product. Buying and selling such an instrument at any time are easy, and it is classified as a high-liquidity (liquid) product in the market. Trades for a security happen whenever a buyer–seller match happens and their orders are filled by the exchange(s). Trades of a product create synchronous price and volume signals and are viewed as discrete time with irregular sampling intervals due to the random arrival times of orders at the market. Exchanges charge traders commissions (a transaction cost) for their matching and fulfillment services. Market-makers are offered some privileges in exchange for their market-making responsibilities to always maintain a two-sided order book.

The intricacies of exchange operations, order books, and microscale price formation is the study of market microstructure (Harris, 2002; O'Hara, 1995). Even defining the price for a security becomes rather complicated, with irregular time intervals characterized by the random arrivals of limit and market orders, multiple definitions of prices (highest bid price, lowest ask price, midmarket price, quantity-weighted prices, etc.), and the price movements occurring at discrete price levels (ticks). This kind of fine granularity is required for designing high-frequency trading strategies. Lower frequency strategies may view prices as regular discrete-time time series (daily or hourly) with a definition of price that abstracts away the details of market microstructure and instead considers some notion of aggregate transaction costs. Portfolio allocation strategies usually operate at this low-frequency granularity with prices viewed as real-valued stochastic processes.

1.2.2 Technical Themes in the Book

Although the scope of financial signal processing and machine learning is very wide, in this book, we have chosen to focus on a well-selected set of topics revolving around the concepts of high-dimensional covariance estimation, applications of sparse learning in risk management and statistical arbitrage, and non-Gaussian and heavy-tailed measures of dependence.[2]

A unifying challenge for many applications of signal processing and machine learning is the high-dimensional nature of the data, and the need to exploit the inherent structure in those data. The field of finance is, of course, no exception; there, thousands of domestic equities and tens of thousands of international equities, tens of thousands of bonds, and even more options contracts with various strikes and expirations provide a very rich source of data. Modeling the dependence among these instruments is especially challenging, as the number of pairwise relationships (e.g., correlations) is quadratic in the number of instruments. Simple traditional tools like the sample covariance estimate are not applicable in high-dimensional settings where the number of data points is small or comparable to the dimension of the space (El Karoui, 2013). A variety of approaches have been devised to tackle this challenge – ranging from simple dimensionality reduction techniques like principal component analysis and factor analysis, to Markov random fields (or sparse covariance selection models), and several others. They rely on exploiting additional structure in the data (sparsity or low-rank, or Markov structure) in order to reduce the sheer number of parameters in covariance estimation. Chapter 1.3.5 provides a comprehensive overview of high-dimensional covariance estimation. Chapter 1.3.4 derives an explicit eigen-analysis for the covariance matrices of AR processes, and investigates their sparsity.

The sparse modeling paradigm that has been highly influential in signal processing is based on the premise that in many settings with a large number of variables, only a small subset of these variables are active or important. The dimensionality of the problem can thus be reduced by focusing on these variables. The challenge is, of course, that the identity of these key variables may not be known, and the crux of the problem involves identifying this subset. The discovery of efficient approaches based on convex relaxations and greedy methods with theoretical guarantees has opened an explosive interest in theory and applications of these methods in various disciplines spanning from compressed sensing to computational biology (Chen *et al.*, 1998; Mallat and Zhang, 1993; Tibshirani, 1996). We explore a few exciting applications of sparse modeling in finance. Chapter 1.3.1 presents sparse Markowitz portfolios where, in addition to balancing risk and expected returns, a new objective is imposed requiring the portfolio to be sparse. The sparse Markowitz framework has a number of benefits, including better statistical out-of-sample performance, better control of transaction costs, and allowing portfolio managers and traders to focus on a small subset of financial instruments. Chapter 1.3.2 introduces a formulation to find sparse eigenvectors (and generalized eigenvectors) that can be used to design sparse mean-reverting portfolios, with applications

[2] We refer the readers to a number of other important topics at the end of this chapter that we could not fit into the book.

to statistical arbitrage strategies. In Chapter 1.3.3, another variation of sparsity, the so-called group sparsity, is used in the context of causal modeling of high-dimensional time series. In group sparsity, the variables belong to a number of groups, where only a small number of groups is selected to be active, while the variables within the groups need not be sparse. In the context of temporal causal modeling, the lagged variables at different lags are used as a group to discover influences among the time series.

Another dominating theme in the book is the focus on non-Gaussian, non-stationary and heavy-tailed distributions, which are critical for realistic modeling of financial data. The measure of risk based on variance (or standard deviation) that relies on the covariance matrix among the financial instruments has been widely used in finance due to its theoretical elegance and computational tractability. There is a significant interest in developing computational and modeling approaches for more flexible risk measures. A very potent alternative is the cVaR, which measures the expected loss below a certain quantile of the loss distribution (Rockafellar and Uryasev, 2000). It provides a very practical alternative to the value at risk (VaR) measure, which is simply the quantile of the loss distribution. VaR has a number of problems such as lack of coherence, and it is very difficult to optimize in portfolio settings. Both of these shortcomings are addressed by the cVaR formulation. cVaR is indeed coherent, and can be optimized by convex optimization (namely, linear programming). Chapter 1.3.9 describes the very intriguing close connections between the cVaR measure of risk and support vector regression in machine learning, which allows the authors to establish out-of-sample results for cVaR portfolio selection based on statistical learning theory. Chapter 1.3.9 provides an overview of a number of regression formulations with applications in finance that rely on different loss functions, including quantile regression and the cVaR metric as a loss measure.

The issue of characterizing statistical dependence and the inadequacy of jointly Gaussian models has been of central interest in finance. A number of approaches based on elliptical distributions, robust measures of correlation and tail dependence, and the copula-modeling framework have been introduced in the financial econometrics literature as potential solutions (McNeil *et al.*, 2015). Chapter 1.3.7 provides a thorough overview of these ideas. Modeling correlated events (e.g., defaults or jumps) requires an entirely different set of tools. An approach based on correlated Poisson processes is presented in Chapter 1.3.8. Another critical aspect of modeling financial data is the handling of non-stationarity. Chapter 1.3.6 describes the problem of modeling the non-stationarity in volatility (i.e. stochastic volatility). An alternative framework based on autoregressive conditional heteroskedasticity models (ARCH and GARCH) is described in Chapter 1.3.7.

1.3 Overview of the Chapters

1.3.1 Chapter 2: "Sparse Markowitz Portfolios" by Christine De Mol

Sparse Markowitz portfolios impose an additional requirement of sparsity to the objectives of risk and expected return in traditional Markowitz portfolios. The chapter starts with an overview of the Markowitz portfolio formulation and describes its fragility in high-dimensional settings. The author argues that sparsity of the portfolio can alleviate many of the shortcomings, and presents an optimization formulation based on convex relaxations. Other related problems, including sparse portfolio rebalancing and combining multiple forecasts, are also introduced in the chapter.

1.3.2 Chapter 3: "Mean-Reverting Portfolios: Tradeoffs between Sparsity and Volatility" by Marco Cuturi and Alexandre d'Aspremont

Statistical arbitrage strategies attempt to find portfolios that exhibit mean reversion. A common econometric tool to find mean reverting portfolios is based on co-integration. The authors argue that sparsity and high volatility are other crucial considerations for statistical arbitrage, and describe a formulation to balance these objectives using semidefinite programming (SDP) relaxations.

1.3.3 Chapter 4: "Temporal Causal Modeling" by Prabhanjan Kambadur, Aurélie C. Lozano, and Ronny Luss

This chapter revisits the old maxim that correlation is not causation, and extends the definition of Granger causality to high-dimensional multivariate time series by defining graphical Granger causality as a tool for temporal causal modeling (TCM). After discussing computational and statistical issues, the authors extend TCM to robust quantile loss functions and consider regime changes using a Markov switching framework.

1.3.4 Chapter 5: "Explicit Kernel and Sparsity of Eigen Subspace for the AR(1) Process" by Mustafa U. Torun, Onur Yilmaz and Ali N. Akansu

The closed-form kernel expressions for the eigenvectors and eigenvalues of the AR(1) discrete process are derived in this chapter. The sparsity of its eigen subspace is investigated. Then, a new method based on rate-distortion theory to find a sparse subspace is introduced. Its superior performance over a few well-known sparsity methods is shown for the AR(1) source as well as for the empirical correlation matrix of stock returns in the NASDAQ-100 index.

1.3.5 Chapter 6: "Approaches to High-Dimensional Covariance and Precision Matrix Estimation" by Jianqing Fan, Yuan Liao, and Han Liu

Covariance estimation presents significant challenges in high-dimensional settings. The authors provide an overview of a variety of powerful approaches for covariance estimation based on approximate factor models, sparse covariance, and sparse precision matrix models. Applications to large-scale portfolio management and testing mean-variance efficiency are considered.

1.3.6 Chapter 7: "Stochastic Volatility: Modeling and Asymptotic Approaches to Option Pricing and Portfolio Selection" by Matthew Lorig and Ronnie Sircar

The dynamic and uncertain nature of market volatility is one of the important incarnations of nonstationarity in financial time series. This chapter starts by reviewing the Black–Scholes formulation and the notion of implied volatility, and discusses local and stochastic models of volatility and their asymptotic analysis. The authors discuss implications of stochastic volatility models for option pricing and investment strategies.

1.3.7 Chapter 8: "Statistical Measures of Dependence for Financial Data" by David S. Matteson, Nicholas A. James, and William B. Nicholson

Idealized models such as jointly Gaussian distributions are rarely appropriate for real financial time series. This chapter describes a variety of more realistic statistical models to capture cross-sectional and temporal dependence in financial time series. Starting with robust measures of correlation and autocorrelation, the authors move on to describe scalar and vector models for serial correlation and heteroscedasticity, and then introduce copula models, tail dependence, and multivariate copula models based on vines.

1.3.8 Chapter 9: "Correlated Poisson Processes and Their Applications in Financial Modeling" by Alexander Kreinin

Jump-diffusion processes have been popular among practitioners as models for equity derivatives and other financial instruments. Modeling the dependence of jump-diffusion processes is considerably more challenging than that of jointly Gaussian diffusion models where the positive-definiteness of the covariance matrix is the only requirement. This chapter introduces a framework for modeling correlated Poisson processes that relies on extreme joint distributions and backward simulation, and discusses its application to financial risk management.

1.3.9 Chapter 10: "CVaR Minimizations in Support Vector Machines" by Junya Gotoh and Akiko Takeda

This chapter establishes intriguing connections between the literature on cVaR optimization in finance, and the support vector machine formulation for regularized empirical risk minimization from the machine-learning literature. Among other insights, this connection allows the establishment of out-of-sample bounds on cVaR risk forecasts. The authors further discuss robust extensions of the cVaR formulation.

1.3.10 Chapter 11: "Regression Models in Risk Management" by Stan Uryasev

Regression models are one of the most widely used tools in quantitative finance. This chapter presents a general framework for linear regression based on minimizing a rich class of error measures for regression residuals subject to constraints on regression coefficients. The discussion starts with least squares linear regression, and includes many important variants such as median regression, quantile regression, mixed quantile regression, and robust regression as special cases. A number of applications are considered such as financial index tracking, sparse signal reconstruction, mutual fund return-based style classification, and mortgage pipeline hedging, among others.

1.4 Other Topics in Financial Signal Processing and Machine Learning

We have left out a number of very interesting topics that all could fit very well within the scope of this book. Here, we briefly provide the reader some pointers for further study.

In practice, the expected returns and the covariance matrices used in portfolio strategies are typically estimated based on recent windows of historical data and, hence, pose significant uncertainty. It behooves a careful portfolio manager to be cognizant of the sensitivity of portfolio allocation strategies to these estimation errors. The field of robust portfolio optimization attempts to characterize this sensitivity and propose strategies that are more stable with respect to modeling errors (Goldfarb and Iyengar, 2003).

The study of market microstructure and the development of high-frequency trading strategies and aggressive directional and market-making strategies rely on short-term predictions of prices and market activity. A recent overview in Kearns and Nevmyvaka (2013) describes many of the issues involved.

Managers of large portfolios such as pension funds and mutual funds often need to execute very large trades that cannot be traded instantaneously in the market without causing a dramatic market impact. The field of optimal order execution studies how to split a large order into a sequence of carefully timed small orders in order to minimize the market impact but still execute the order in a timely manner (Almgren and Chriss, 2001; Bertsimas and Lo, 1998). The solutions for such a problem involve ideas from stochastic optimal control.

Various financial instruments exhibit specific structures that require dedicated mathematical models. For example, fixed income instruments depend on the movements of various interest-rate curves at different ratings (Brigo and Mercurio, 2007), options prices depend on volatility surfaces (Gatheral, 2011), and foreign exchange rates are traded via a graph of currency pairs. Stocks do not have such a rich mathematical structure, but they can be modeled by their industry, style, and other common characteristics. This gives rise to fundamental or statistical factor models (Darolles *et al.*, 2013).

A critical driver for market activity is the release of news, reflecting developments in the industry, economic, and political sectors that affect the price of a security. Traditionally, traders act upon this information after reading an article and evaluating its significance and impact on their portfolio. With the availability of large amounts of information online, the advent of natural language processing, and the need for rapid decision making, many financial institutions have already started to explore automated decision-making and trading strategies based on computer interpretation of relevant news (Bollen *et al.*, 2011; Luss and d'Aspremont, 2008) ranging from simple sentiment analysis to deeper semantic analysis and entity extraction.

References

Akansu, A.N., Kulkarni, S.R., Avellaneda, M.M. and Barron, A.R. (2012). Special issue on signal processing methods in finance and electronic trading. *IEEE Journal of Selected Topics in Signal Processing*, 6(4).

Akansu, A.N. and Torun, M. (2015). *A primer for financial engineering: financial signal processing and electronic trading*. New York: Academic-Elsevier.

Almgren, R. and Chriss, N. (2001). Optimal execution of portfolio transactions. *Journal of Risk*, 3, pp. 5–40.
Bertsimas, D. and Lo, A.W. (1998). Optimal control of execution costs. *Journal of Financial Markets*, 1(1), pp. 1–50.
Black, F. and Litterman, R. (1992). Global portfolio optimization. *Financial Analysts Journal*, 48(5), pp. 28–43.
Black, F. and Scholes, M. (1973). The pricing of options and corporate liabilities. *Journal of Political Economy*, 81(3), p. 637.
Bollen, J., Mao, H. and Zeng, X. (2011). Twitter mood predicts the stock market. *Journal of Computational Science*, 2(1), pp. 1–8.
Brigo, D. and Mercurio, F. (2007). *Interest Rate Models – Theory and Practice: With Smile, Inflation and Credit*. Berlin: Springer Science & Business Media.
Chen, S., Donoho, D. and Saunders, M. (1998). Atomic decomposition by basis pursuit. *SIAM Journal on Scientific Computing*, 20(1), pp. 33–61.
Cover, T. and Ordentlich, E. (1996). Universal portfolios with side information. *IEEE Transactions on Information Theory*, 42(2), pp. 348–363.
Darolles, S., Duvaut, P. and Jay, E. (2013). *Multi-factor Models and Signal Processing Techniques: Application to Quantitative Finance*. Hoboken, NJ: John Wiley & Sons.
El Karoui, N. (2013). On the realized risk of high-dimensional Markowitz portfolios. *SIAM Journal on Financial Mathematics*, 4(1), 737–783.
Engle, R. (1982). Autoregressive conditional heteroscedasticity with estimates of the variance of United Kingdom inflation. *Econometrica: Journal of the Econometric Society*, 50(4), pp. 987–1007.
Fama, E. and French, K. (1993). Common risk factors in the returns on stocks and bonds. *Journal of Financial Economics*, 33(1), pp. 3–56.
Gatheral, J. (2011). *The Volatility Surface: A Practitioner's Guide*. Hoboken, NJ: John Wiley & Sons.
Goldfarb, D. and Iyengar, G. (2003). Robust portfolio selection problems. *Mathematics of Operations Research*, 28(1), pp. 1–38.
Harris, L. (2002). *Trading and Exchanges: Market Microstructure for Practitioners*. Oxford: Oxford University Press.
Hull, J. (2011). *Options, Futures, and Other Derivatives*. Upper Saddle River, NJ: Pearson.
Hull, J. and White, A. (1987). The pricing of options on assets with stochastic volatilities. *The Journal of Finance*, 42(2), 281–300.
Kearns, M. and Nevmyvaka, Y. (2013). Machine learning for market microstructure and high frequency trading. In *High-frequency trading – New realities for traders, markets and regulators* (ed. O'Hara, M., de Prado, M.L. and Easley, D.). London: Risk Books, pp. 91–124.
Luss, R. and d'Aspremont, A. (2008). Support vector machine classification with indefinite kernels. In *Advances in neural information processing systems* 20 (ed. Platt, J., Koller, D., Singer, Y. and Roweis, S.). Cambridge, MA, MIT Press, pp. 953–960.
Mallat, S.G. and Zhang, Z. (1993). Matching pursuits with time-frequency dictionaries. *IEEE Transactions on Signal Processing*, 41(12), 3397–3415.
Markowitz, H. (1952). Portfolio selection. *The Journal of Finance*, 7(1), 77–91.
McNeil, A.J., Frey, R. and Embrechts, P. (2015). *Quantitative risk management: concepts, techniques and tools*. Princeton, NJ: Princeton University Press.
O'Hara, M. (1995). *Market Microstructure Theory*. Cambridge, MA: Blackwell.
Pollak, I., Avellaneda, M.M., Bacry, E., Cont, R. and Kulkarni, S.R. (2011). Special issue on signal processing for financial applications. *IEEE Signal Processing Magazine*, 28(5).
Rockafellar, R. and Uryasev, S. (2000). Optimization of conditional value-at-risk. *Journal of Risk*, 2, 21–42.
Roncalli, T. (2013). Introduction to risk parity and budgeting. Boca Raton, FL: CRC Press.
Tibshirani, R. (1996). Regression shrinkage and selection via the lasso. *Journal of the Royal Statistical Society: Series B (Methodological)*, 58(1), 267–288.

2

Sparse Markowitz Portfolios

Christine De Mol
Université Libre de Bruxelles, Belgium

2.1 Markowitz Portfolios

Modern portfolio theory originated from the work of Markowitz (1952), who insisted on the fact that returns should be balanced with risk and established the theoretical basis for portfolio optimization according to this principle. The portfolios are to be composed from a universe of N securities with returns at time t given by $r_{i,t}$, $i = 1, \dots, N$, and assumed to be stationary. We denote by $\mathbf{E}[\mathbf{r}_t] = \boldsymbol{\mu}$ the $N \times 1$ vector of the expected returns of the different assets, and by $\mathbf{E}[(\mathbf{r}_t - \boldsymbol{\mu})(\mathbf{r}_t - \boldsymbol{\mu})^\top] = \boldsymbol{C}$ the covariance matrix of the returns ($\boldsymbol{\mu}^\top$ is the transpose of $\boldsymbol{\mu}$).

A portfolio is characterized by a $N \times 1$ vector of weights $\mathbf{w} = (w_1, \dots, w_N)^\top$, where w_i is the amount of capital to be invested in asset number i. Traditionally, it is assumed that a fixed capital, normalized to one, is available and should be fully invested. Hence the weights are required to sum to one: $\sum_{i=1}^N w_i = 1$, or else $\mathbf{w}^\top \mathbf{1}_N = 1$, where $\mathbf{1}_N$ denotes the $N \times 1$ vector with all entries equal to 1. For a given portfolio $\mathbf{w}$, the expected return is then equal to $\mathbf{w}^\top \boldsymbol{\mu}$, whereas its variance, which serves as a measure of risk, is given by $\mathbf{w}^\top \boldsymbol{C} \mathbf{w}$. Following Markowitz, the standard paradigm in portfolio optimization is to find a portfolio that has minimal variance for a given expected return $\rho = \mathbf{w}^\top \boldsymbol{\mu}$. More precisely, one seeks $\mathbf{w}_*$ such that:

$$\begin{aligned} \mathbf{w}_* &= \arg\min_{\mathbf{w}} \mathbf{w}^\top \boldsymbol{C} \mathbf{w} \\ \text{s. t. } \mathbf{w}^\top \boldsymbol{\mu} &= \rho \\ \mathbf{w}^\top \mathbf{1}_N &= 1. \end{aligned} \tag{2.1}$$

The constraint that the weights should sum to one can be dropped when including also in the portfolio a risk-free asset, with fixed return r_0, in which one invests a fraction w_0 of the unit capital, so that

$$w_0 + \mathbf{w}^\top \mathbf{1}_N = 1. \tag{2.2}$$

Financial Signal Processing and Machine Learning, First Edition.
Edited by Ali N. Akansu, Sanjeev R. Kulkarni and Dmitry Malioutov.
© 2016 John Wiley & Sons, Ltd. Published 2016 by John Wiley & Sons, Ltd.

The return of the combined portfolio is then given by

$$w_0 r_0 + \mathbf{w}^\top \mathbf{r}_t = r_0 + \mathbf{w}^\top (\mathbf{r}_t - r_0 \mathbf{1}_N). \tag{2.3}$$

Hence we can reason in terms of "excess return" of this portfolio, which is given by $\mathbf{w}^\top \widetilde{\mathbf{r}}_t$ where the "excess returns" are defined as $\widetilde{\mathbf{r}}_t = \mathbf{r}_t - r_0 \mathbf{1}_N$. The "excess expected returns" are then $\widetilde{\boldsymbol{\mu}} = \mathbf{E}[\widetilde{\mathbf{r}}_t] = \mathbf{E}[\mathbf{r}_t] - r_0 \mathbf{1}_N = \boldsymbol{\mu} - r_0 \mathbf{1}_N$. The Markowitz optimal portfolio weights in this setting are solving

$$\widetilde{\mathbf{w}}_* = \arg\min_{\mathbf{w}} \mathbf{w}^\top \boldsymbol{C} \mathbf{w} \tag{2.4}$$
$$\text{s. t. } \mathbf{w}^\top \widetilde{\boldsymbol{\mu}} = \widetilde{\rho}$$

with the same covariance matrix as in (2.1) since the return of the risk-free asset is purely deterministic instead of stochastic. The weight corresponding to the risk-free asset is adjusted as $\widetilde{w}_{*,0} = 1 - \widetilde{\mathbf{w}}_*^\top \mathbf{1}_N$ (and is not included in the weight vector $\widetilde{\mathbf{w}}_*$). Introducing a Lagrange parameter and fixing it in order to satisfy the linear constraint, one easily sees that

$$\widetilde{\mathbf{w}}_* = \frac{\widetilde{\rho}}{\widetilde{\boldsymbol{\mu}}^\top \boldsymbol{C}^{-1} \widetilde{\boldsymbol{\mu}}} \boldsymbol{C}^{-1} \widetilde{\boldsymbol{\mu}} \tag{2.5}$$

assuming that $\boldsymbol{C}$ is strictly positive definite so that its inverse exists. This means that, whatever the value of the excess target return $\widetilde{\rho}$, the weights of the optimal portfolio are proportional to $\boldsymbol{C}^{-1}\widetilde{\boldsymbol{\mu}}$. The corresponding variance is given by

$$\widetilde{\sigma}^2 = \widetilde{\mathbf{w}}_*^\top \boldsymbol{C} \widetilde{\mathbf{w}}_* = \frac{\widetilde{\rho}^2}{\widetilde{\boldsymbol{\mu}}^\top \boldsymbol{C}^{-1} \widetilde{\boldsymbol{\mu}}} \tag{2.6}$$

which implies that, when varying $\widetilde{\rho}$, the optimal portfolios lie on a straight line in the plane $(\widetilde{\sigma}, \widetilde{\rho})$, called the *capital market line* or *efficient frontier*, the slope of which is referred to as the Sharpe ratio:

$$S = \frac{\widetilde{\rho}}{\widetilde{\sigma}} = \sqrt{\widetilde{\boldsymbol{\mu}}^\top \boldsymbol{C}^{-1} \widetilde{\boldsymbol{\mu}}}. \tag{2.7}$$

We also see that all efficient portfolios (i.e, those lying on the efficient frontier) can be obtained by combining linearly the portfolio containing only the risk-free asset, with weight $\widetilde{w}_{*,0} = 1$, and any other efficient portfolio, with weights $\widetilde{\mathbf{w}}_*$. The weights of the efficient portfolio, which contains only risky assets, are then derived by renormalization as $\widetilde{\mathbf{w}}_* / \widetilde{\mathbf{w}}_*^\top \mathbf{1}_N$, with of course $\widetilde{w}_{*,0} = 0$. This phenomenon is often referred to as Tobin's two-fund separation theorem. The portfolios on the frontier to the right of this last portfolio require a short position on the risk-free asset $\widetilde{w}_{*,0} < 0$, meaning that money is borrowed at the risk-free rate to buy risky assets.

Notice that in the absence of a risk-free asset, the efficient frontier composed by the optimal portfolios satisfying (2.1), with weights required to sum to one, is slightly more complicated: it is a parabola in the variance – return plane (σ^2, ρ) that becomes a "Markowitz bullet" in the plane (σ, ρ). By introducing two Lagrange parameters for the two linear constraints, one can derive the expression of the optimal weights, which are a linear combination of $\boldsymbol{C}^{-1}\boldsymbol{\mu}$ and $\boldsymbol{C}^{-1}\mathbf{1}_N$, generalizing Tobin's theorem in the sense that any portfolio on the efficient frontier can be expressed as a linear combination of two arbitrary ones on the same frontier.

The Markowitz portfolio optimization problem can also be reformulated as a regression problem, as noted by Brodie *et al.* (2009). Indeed, we have $\boldsymbol{C} = \mathbf{E}[\mathbf{r}_t \mathbf{r}_t^{\mathsf{T}}] - \boldsymbol{\mu}\boldsymbol{\mu}^{\mathsf{T}}$, so that the minimization problem (2.1) is equivalent to

$$\begin{aligned} \mathbf{w}_* = \arg \min_{\mathbf{w}} \mathbf{E}[|\rho - \mathbf{w}^{\mathsf{T}} \mathbf{r}_t|^2] \\ \text{s. t. } \mathbf{w}^{\mathsf{T}} \boldsymbol{\mu} = \rho \\ \mathbf{w}^{\mathsf{T}} \mathbf{1}_N = 1. \end{aligned} \tag{2.8}$$

Let us remark that when using excess returns, there is no need to implement the constraints since the minimization of $\mathbf{E}[|\widetilde{\rho} - \widetilde{\mathbf{w}}^{\mathsf{T}} \widetilde{\mathbf{r}}_t|^2]$ (for any constant $\widetilde{\rho}$) is easily shown to deliver weights proportional to $\boldsymbol{C}^{-1}\widetilde{\boldsymbol{\mu}}$, which by renormalization correspond to a portfolio on the capital market line.

In practice, for empirical implementations, one needs to estimate the returns as well as the covariance matrix and to plug in the resulting estimates in all the expressions above. Usually, expectations are replaced by sample averages (i.e., for the returns by $\widehat{\boldsymbol{\mu}} = \frac{1}{T}\sum_{t=1}^{T} \mathbf{r}_t$ and for the covariance matrix by $\widehat{\boldsymbol{C}} = \frac{1}{T}\sum_{t=1}^{T} [\mathbf{r}_t \mathbf{r}_t^{\mathsf{T}}] - \widehat{\boldsymbol{\mu}}\widehat{\boldsymbol{\mu}}^{\mathsf{T}}$).

For the regression formulation, we define $\boldsymbol{R}$ to be the $T \times N$ matrix of which row t is given by $\mathbf{r}_t^{\mathsf{T}}$, namely $\boldsymbol{R}_{t,i} = (\mathbf{r}_t)_i = r_{i,t}$. The optimization problem (2.8) is then replaced by

$$\begin{aligned} \widehat{\mathbf{w}} = \arg \min_{\mathbf{w}} \frac{1}{T} \|\rho \mathbf{1}_T - \boldsymbol{R}\mathbf{w}\|_2^2 \\ \text{s. t. } \mathbf{w}^{\mathsf{T}} \widehat{\boldsymbol{\mu}} = \rho \\ \mathbf{w}^{\mathsf{T}} \mathbf{1}_N = 1, \end{aligned} \tag{2.9}$$

where $\|\mathbf{a}\|_2^2$ denotes the squared Euclidean norm $\sum_{t=1}^{T} \mathbf{a}_t^2$ of the vector $\mathbf{a}$ in $\mathbb{R}^T$.

There are many possible variations in the formulation of the Markowitz portfolio optimization problem, but they are not essential for the message we want to convey. Moreover, although lots of papers in the literature on portfolio theory have explored other risk measures, for example more robust ones, we will only consider here the traditional framework where risk is measured by the variance. For a broader picture, see for example the books by Campbell *et al.* (1997) and Ruppert (2004).

2.2 Portfolio Optimization as an Inverse Problem: The Need for Regularization

Despite its elegance, it is well known that the Markowitz theory has to face several difficulties when implemented in practice, as soon as the number of assets N in the portfolio gets large. There has been extensive effort in recent years to explain the origin of such difficulties and to propose remedies. Interestingly, DeMiguel *et al.* (2009a) have assessed several optimization procedures proposed in the literature and shown that, surprisingly, they do not clearly outperform the "naive" (also called "Talmudic") strategy, which consists in attributing equal weights, namely $1/N$, to all assets in the portfolio. The fact that this naive strategy is hard to beat—and therefore constitutes a tough benchmark – is sometimes referred to as the $1/N$ *puzzle*.

A natural explanation for these difficulties comes in mind when noticing, as done by Brodie *et al.* (2009), that the determination of the optimal weights solving problem (2.1) or (2.4) can be viewed as an inverse problem, requiring the inversion of the covariance matrix $\boldsymbol{C}$ or, in practice, of its estimate $\widehat{\boldsymbol{C}}$. In the presence of collinearity between the returns, this matrix is most likely to be "ill-conditioned." The same is true for the regression formulation (2.9) where it is the matrix $\boldsymbol{R}^{\mathrm{T}}\boldsymbol{R}$ which has to be inverted. Let us recall that the condition number of a matrix is defined as the ratio of the largest to the smallest of its singular values (or eigenvalues when it is symmetric). If this ratio is small, the matrix can be easily inverted, and the corresponding weights can be computed numerically in a stable way. However, when the condition number gets large, the usual numerical inversion procedures will deliver unstable results, due to the amplification of small errors (e.g., rounding errors would be enough) in the eigendirections correponding to the smallest singular or eigenvalues. Since, typically, asset returns tend to be highly correlated, the condition number will be large, leading to numerically unstable, hence unreliable, estimates of the weight vector $\mathbf{w}$. As a consequence, some of the computed weights can take very large values, including large negative values corresponding to short positions.

Contrary to what is often claimed in the literature, let us stress the fact that improving the estimation of the returns and of the covariance matrix will not really solve the problem. Indeed, in inverting a true (population) but large covariance matrix, we would have to face the same kind of ill-conditioning as with empirical estimates, except for very special models such as the identity matrix or a well-conditioned diagonal matrix. Such models, however, cannot be expected to be very realistic.

A standard way to deal with inverse problems in the presence of ill-conditioning of the matrix to be inverted is provided by so-called regularization methods. The idea is to include additional constraints on the solution of the inverse problem (here, the weight vector) that will prevent the error amplification due to ill-conditioning and hence allow one to obtain meaningful, stable estimates of the weights. These constraints are expected, as far as possible, to represent prior knowledge about the solution of the problem under consideration. Alternatively, one can add a penalty to the objective function. It is this strategy that we will adopt here, noticing that most often, equivalence results with a constrained formulation can be established as long as we deal with convex optimization problems. For more details about regularization techniques for inverse problems, we refer to the book by Bertero and Boccacci (1998).

A classical procedure for stabilizing least-squares problems is to use a quadratic penalty, the simplest instance being the squared ℓ_2 norm of the weight vector: $\|\mathbf{w}\|_2^2 = \sum_{i=1}^{N} |\mathbf{w}_i|^2$. It goes under the name of Tikhonov regularization in inverse problem theory and of ridge regression in statistics. Such a penalty can be added to regularize any of the optimization problems considered in Section 2.1. For example, using a risk-free asset, let us consider problem (2.4) and replace it by

$$\widetilde{\mathbf{w}}_{ridge} = \arg\min_{\mathbf{w}}[\mathbf{w}^{\mathrm{T}}\boldsymbol{C}\mathbf{w} + \lambda\|\mathbf{w}\|_2^2] \quad \text{s. t. } \mathbf{w}^{\mathrm{T}}\widetilde{\boldsymbol{\mu}} = \widetilde{\rho} \tag{2.10}$$

where λ is a positive parameter, called the regularization parameter, allowing one to tune the balance between the variance term and the penalty. Using a Lagrange parameter and fixing its value to satisfy the linear constraint, we get the explicit solution

$$\widetilde{\mathbf{w}}_{ridge} = \frac{\widetilde{\rho}}{\widetilde{\boldsymbol{\mu}}^{\mathrm{T}}(\boldsymbol{C}+\lambda\boldsymbol{I})^{-1}\widetilde{\boldsymbol{\mu}}}(\boldsymbol{C}+\lambda\boldsymbol{I})^{-1}\widetilde{\boldsymbol{\mu}} \tag{2.11}$$

where $\boldsymbol{I}$ denotes the $N \times N$ identity matrix. Hence, the weights of the "ridge" optimal portfolio are proportional to $(\boldsymbol{C} + \lambda \boldsymbol{I})^{-1}\widetilde{\boldsymbol{\mu}}$, whatever the value of the excess target return $\widetilde{\rho}$. The corresponding variance is given by

$$\widetilde{\sigma}^2 = \widetilde{\mathbf{w}}_{ridge}^{\mathsf{T}} \boldsymbol{C}\, \widetilde{\mathbf{w}}_{ridge} = \frac{\widetilde{\rho}^2}{(\widetilde{\boldsymbol{\mu}}^{\mathsf{T}}(\boldsymbol{C} + \lambda \boldsymbol{I})^{-1}\widetilde{\boldsymbol{\mu}})^2}\, \widetilde{\boldsymbol{\mu}}^{\mathsf{T}}(\boldsymbol{C} + \lambda \boldsymbol{I})^{-1}\boldsymbol{C}(\boldsymbol{C} + \lambda \boldsymbol{I})^{-1}\widetilde{\boldsymbol{\mu}} \tag{2.12}$$

which implies that, when λ is fixed, $\widetilde{\sigma}$ is again proportional to $\widetilde{\rho}$ and that the efficient ridge portfolios also lie on a straight line in the plane $(\widetilde{\sigma}, \widetilde{\rho})$, generalizing Tobin's theorem to this setting. Notice that its slope, the Sharpe ratio, does depend on the value of the regularization parameter λ.

Another standard regularization procedure, called *truncated singular value decomposition*, (TSVD), consists of diagonalizing the covariance matrix and using for the inversion only the subspace spanned by the eigenvectors corresponding to the largest eigenvalues (e.g., the K largest). This is also referred to as reduced-rank or principal-components regression. and it corresponds to replacing in the formulas (2.11, 2.12) the regularized inverse $(\boldsymbol{C} + \lambda \boldsymbol{I})^{-1}$ by $\boldsymbol{V}_K \boldsymbol{D}_K^{-1} \boldsymbol{V}_K^{\mathsf{T}}$, where $\boldsymbol{D}_K$ is the diagonal matrix containing the K largest eigenvalues d_k^2 of $\boldsymbol{C}$ and $\boldsymbol{V}_K$ is the $N \times K$ matrix containing the corresponding orthonormalized eigenvectors. Whereas this method implements a sharp (binary) cutoff on the eigenvalue spectrum of the covariance matrix, notice that ridge regression involves instead a smoother filtering of this spectrum where the eigenvalues d_k^2 (positive since $\boldsymbol{C}$ is positive definite) are replaced by $d_k^2 + \lambda$ or, equivalently, in the inversion process, $1/d_k^2$ is replaced by $1/(d_k^2 + \lambda) = \phi_\lambda(d_k^2)/d_k^2$, where $\phi_\lambda(d_k^2) = d_k^2/(d_k^2 + \lambda)$ is a filtering, attenuation, or "shrinkage" factor, comprised between 0 and 1, allowing one to control the instabilities generated by division by the smallest eigenvalues. More general types of filtering factors can be used to regularize the problem. We refer the reader, for example, to the paper by De Mol *et al.* (2008) for a discussion of the link between principal components and ridge regression in the context of forecasting of high-dimensional time series, and to the paper by Carrasco and Noumon (2012) for a broader analysis of linear regularization methods, including an iterative method called Landweber's iteration, in the context of portfolio theory.

Regularized versions of the problems (2.1) and (2.9) can be defined and solved in a similar way as for (2.4). Tikhonov's regularization method has also been applied to the estimation of the covariance matrix by Park and O'Leary (2010). Let us remark that there are many other methods, proposed in the literature to stabilize the construction of Markowitz portfolios, which can be viewed as a form of explicit or implicit regularization, including Bayesian techniques as used for example in the so-called Black–Litterman model. However, they are usually more complicated, and reviewing them would go beyond the scope of this chapter.

2.3 Sparse Portfolios

As discussed in Section 2.2, regularization methods such as rigde regression or TSVD allow one to define and compute stable weights for Markowitz portfolios. The resulting vector of regularized weights generically has all its entries different from zero, even if there may be a lot of small values. This would oblige the investor to buy a certain amount of each security, which is not necessarily a convenient strategy for small investors. Brodie *et al.* (2009) have proposed to use instead a regularization based on a penalty that enforces sparsity of the weight vector, namely the presence of many zero entries in that vector, corresponding to assets that will not be included in the portfolio. More precisely, they introduce in the optimization problem, formulated as (2.9), a penalty on the ℓ_1 norm of the vector of weights $\mathbf{w}$, defined by $\|\mathbf{w}\|_1 = \sum_{i=1}^{N} |w_i|$. This problem then becomes

$$\mathbf{w}_{sparse} = \arg\min_{\mathbf{w}}[\|\rho \mathbf{1}_T - \boldsymbol{R}\mathbf{w}\|_2^2 + \tau\|\mathbf{w}\|_1] \tag{2.13}$$

$$\text{s. t. } \mathbf{w}^{\mathrm{T}}\hat{\boldsymbol{\mu}} = \rho$$

$$\mathbf{w}^{\mathrm{T}}\mathbf{1}_N = 1,$$

where the regularization parameter is denoted by τ. Note that the factor $1/T$ from (2.9) has been absorbed in the parameter τ. When removing the constraints, a problem of this kind is referred to as lasso regression, after Tibshirani (1996). Lasso, an acronym for least absolute shrinkage and selection operator, helps by reminding that it allows for variable (here, asset) selection since it favors the recovery of sparse vectors $\mathbf{w}$ (i.e., vectors containing many zero entries, the position of which, however, is not known in advance). This sparsifying effect is also widely used nowadays in signal and image processing (see, e.g., the review paper by Chen *et al.* (2001) and the references therein).

As argued by Brodie *et al.* (2009), besides its sparsity-enforcing properties, the ℓ_1-norm penalty offers the advantage of being a good model for the transaction costs incurred to compose the portfolio, costs that are not at all taken into account in the Markowitz original framework. Indeed, these can be assumed to be roughly proportional, for a given asset, to the amount of the transaction, whether buying or short-selling, and hence to the absolute value of the portfolio weight w_i. There may be an additional fixed fee, however, which would then be proportional to the number K of assets to include in the portfolio (i.e., proportional to the cardinality of the portfolio, or the number of its nonzero entries, sometimes also called by abuse of language the ℓ_0 "norm" ($\|\mathbf{w}\|_0$) of the weight vector $\mathbf{w}$). Usually, however, such fees can be neglected. Let us remark, moreover, that implementing a cardinality penalty or constraint would render the portfolio optimization problem very cumbersome (i.e., nonconvex and of combinatorial complexity). It has become a standard practice to use the ℓ_1 norm $\|\mathbf{w}\|_1$ as a "convex relaxation" for $\|\mathbf{w}\|_0$. Under appropriate assumptions, there even exist some theoretical guarantees that both penalties will actually deliver the same answer (see, e.g., the book on compressive sensing by Foucart and Rauhut (2013) and the references therein).

Let us remark that, in problem (2.13), it is actually the amount of "shorting" that is regulated; indeed, because of the constraint that the weights should add to one, the objective function can be rewritten as

$$\|\rho \mathbf{1}_T - \boldsymbol{R}\mathbf{w}\|_2^2 + 2\tau \sum_{i \text{ with } w_i<0} |w_i| + \tau, \tag{2.14}$$

in which the last term, being constant, is of course irrelevant for determining the solution. In this setting, we see that the ℓ_1-norm penalty is equivalent to a penalty on the negative weights (i.e., on the short positions), only. In the limit of very large values of the regularization parameter τ, we get, as a special case, a portfolio with only positive weights (i.e., no short positions). Such no-short optimal portfolios had been considered previously in the financial literature by Jagannathan and Ma (2003) and were known for their good performances, but, surprisingly, their sparse character had gone unnoticed. As shown by Brodie *et al.* (2009), these no-short portfolios, obtained for the largest values of τ, are typically also the sparsest in the family defined by (2.13). When decreasing τ beyond some point, negative weights start to appear, but the ℓ_1-norm penalty allows one to control their size and to ensure numerical stability of the portfolio weights. The regularizing properties of the ℓ_1-norm penalty (or constraint) for high-dimensional regression problems in the presence of collinearity is well known since the paper by Tibshirani (1996), and the fact that the lasso strategy yields a proper regularization method (as is the quadratic Tikhonov regularization method) even in an infinite-dimensional framework has been established by Daubechies *et al.* (2004). Notice that these results were derived in an unconstrained setting, but the presence of additional linear constraints can only reinforce the regularization effect. A paper by Rosenbaum and Tsybakov (2010) investigates the effect of errors on the matrix of the returns.

Compared to more classical linear regularization techniques (e.g., by means of a ℓ_2-norm penalty), the lasso approach not only presents advantages as described above but also has some drawbacks. A first problem is that the ℓ_1-norm penalty enforces a nonlinear shrinkage of the portfolio weights that renders the determination of the efficient frontier much more difficult than in the unpenalized case or in the case of ridge regression. For any given value of τ, such frontier ought to be computed point by point by solving (2.13) for different values of the target return ρ. Another difficulty is that, though still convex, the optimization problem (2.13) is more challenging and, in particular, does not admit a closed-form solution. There are several possibilities to solve numerically the resulting quadratic program. Brodie *et al.* (2009) used the homotopy method developed by Osborne *et al.* (2000a, 2000b), also known as the least-angle regression (LARS) algorithm by Efron *et al.* (2004). This algorithm proceeds by decreasing the value of τ progressively from very large values, exploiting the fact that the dependence of the optimal weight on τ is piecewise linear. It is very fast if the number of active assets (nonzero weights) is small. Because of the two additional constraints, a modification of this algorithm was devised by Brodie *et al.* (2009) to make it suitable for solving the portfolio optimization problem (2.13). For the technical details, we refer the interested reader to the supplementary appendix of that paper.

2.4 Empirical Validation

The sparse portfolio methodology described in the previous Section 2.3 has been validated by an empirical exercise, the results of which are succinctly described here. For a complete description, we refer the reader to the original paper by Brodie *et al.* (2009).

Sparse portfolios were constructed using two benchmark datasets compiled by Fama and French and available from the site http://mba.tuck.dartmouth.edu/pages/faculty/ken.french/data_library.html. They are ensembles of 48 and 100 portfolios and will be referred to as FF48 and FF100, respectively. The out-of-sample performances of the portfolios constructed by solving (2.13) were assessed and compared to the tough benchmark of the Talmudic or equal-weight portfolios for the same period. Using annualized monthly returns from the FF48 and FF100 datasets, the following simulated investment exercise was performed over a period of 30 years between 1976 and 2006. In June of each year, sparse optimal portfolios were constructed for a wide range of values of the regularization parameter τ in order to get different levels of sparsity, namely portfolios containing different numbers K of active positions. To run the regression, historical data from the preceding 5 years (60 months) were used. At the time of each portfolio construction, the target return, ρ, was set to be the average return achieved by the naive, equal-weight portfolio over the same historical period. Once constructed, the portfolios were held until June of the next year, and their monthly out-of-sample returns were observed. The same exercise was repeated each year until June 2005. All the observed monthly returns of the portfolios form a time series from which one can compute the average monthly return $\hat{\rho}$ (over the whole period or a subperiod), the corresponding standard deviation $\hat{\sigma}$, and the Sharpe ratio $S = \hat{\rho}/\hat{\sigma}$. We report some Sharpe ratios obtained when averaging over the whole period 1976–2006. For FF48, the best one was $S = 41$ and was obtained with the no-short portfolio, comprising a number of active assets varying over the years, but typically ranging between 4 and 10. Then, when looking at the performances of sparse portfolios with a given number K of active positions, their Sharpe ratios, lower than for the no-short portfolio, decreased with K, clearly outperforming the equal-weight benchmark (for which $S = 27$) as long as $K \lesssim 25$ but falling below for K larger. For FF100, a different behavior was observed. The Sharpe ratios were maximum and of the order of 40 for a number of active positions K around 30, thus including short positions, whereas $S = 30$ for the no-short portfolio. The sparse portfolios were outperforming the equal-weight benchmark with $S = 28$ as long as $K \lesssim 60$.

In parallel and independently of the paper by Brodie *et al.* (2009), DeMiguel *et al.* (2009b) performed an extensive comparison of the improvement in terms of the Sharpe ratio obtained through various portfolio construction methods, and in particular by imposing constraints on some specific norm of the weight vector, including ℓ_2 and ℓ_1 norms. Subsequent papers confirmed the good performances of the sparse portfolios, also on other and larger datasets and in somewhat different frameworks, such as those by Fan *et al.* (2012), by Gandy and Veraart (2013) and by Henriques and Ortega (2014).

2.5 Variations on the Theme

2.5.1 *Portfolio Rebalancing*

The empirical exercise described in Section 2.4 is not very realistic in representing the behaviour of a single investor since a sparse portfolio would be constructed from scratch each year. Its aim was rather to assess the validity of the investment strategy, as it would be carried out by different investors using the same methodology in different years.

More realistically, an investor already holding a portfolio with weights $\mathbf{w}$ would like to adjust it to increase its performance. This means that one should look for an adjustment $\Delta\mathbf{w}$, so that the new rebalanced portfolio weights are $\mathbf{w} + \Delta\mathbf{w}$. The incurred transaction costs concern only the adjustment and hence can be modelled by the ℓ_1 norm of the vector $\Delta\mathbf{w}$. This means that we must now solve the following optimization problem:

$$\begin{aligned}\Delta\mathbf{w}_{sparse} &= \arg\min_{\Delta\mathbf{w}}[\|\rho\mathbf{1}_T - \boldsymbol{R}(\mathbf{w} + \Delta\mathbf{w})\|_2^2 + \tau\|\Delta\mathbf{w}\|_1]\\ &\text{s. t. } \Delta\mathbf{w}^\top\hat{\boldsymbol{\mu}} = 0\\ &\quad\;\; \Delta\mathbf{w}^\top\mathbf{1}_N = 0\end{aligned}$$

ensuring sparsity in the number of weights to be adjusted and conservation of the total unit capital invested as well as of the target return. The methodology proposed by Brodie *et al.* (2009) can be straightforwardly modified to solve this problem. An empirical exercise on sparse portfolio rebalancing is described by Henriques and Ortega (2014).

2.5.2 Portfolio Replication or Index Tracking

In some circumstances, an investor may want to construct a portfolio that replicates the performances of a given portfolio or of a financial index such as the S&P 500, but is easier to manage, for example because it contains less assets. In such a case, the investor will have at his disposal a time series of index values or global portfolio historical returns, which can be put in a $T \times 1$ column vector $\mathbf{y}$. The time series of historical returns of the assets that he can use to replicate $\mathbf{y}$ will be put in a $T \times N$ matrix $\boldsymbol{R}$, as before. The problem can then be formulated as the minimization of the mean square tracking error augmented by a penalty on the ℓ_1 norm of $\mathbf{w}$, representing the transaction costs and enforcing sparsity:

$$\mathbf{w}_{track} = \arg\min_{\mathbf{w}}[\|\mathbf{y} - \boldsymbol{R}\mathbf{w}\|_2^2 + \tau\|\mathbf{w}\|_1], \quad \text{s. t. } \mathbf{w}^\top \mathbf{1}_N = 1. \tag{2.15}$$

This is a constrained lasso regression that can again be solved by means of the methodology described in Section 2.3. A rebalancing version of this tracking problem could also be implemented.

2.5.3 Other Penalties and Portfolio Norms

A straightforward modification of the previous scheme consists of introducing weights in the ℓ_1 norm used as penalty (i.e. replacing it with):

$$\|\mathbf{w}\|_{1,s} = \sum_{i=1}^{N} s_i |w_i| \tag{2.16}$$

where the positive weights s_i can model either differences in transaction costs or some preferences of the investor. Another extension, considered for example by Daubechies *et al.* (2004) for unconstrained lasso regression, is to use ℓ_p-norm penalties with $1 \le p \le 2$, namely of the type

$$\|\mathbf{w}\|_p^p = \sum_{i=1}^{N} |w_i|^p \tag{2.17}$$

yielding as special cases lasso for $p = 1$ or ridge regression for $p = 2$. The use of values of p less than 1 in (2.17) would reinforce the sparsifying effect of the penalty but would render the optimization problem nonconvex and therefore a lot more cumbersome.

A well-known drawback of variable selection methods relying on an ℓ_1-norm penalty or constraint is the instability in selection in the presence of collinearity among the variables. This means that, in the empirical exercise described here, when recomposing each year a new portfolio, the selection will not be stable over time within a group of potentially correlated assets. The same effect has been noted by De Mol *et al.* (2008) when forecasting macroeconomic variables based on a large ensemble of time series. When the goal is forecasting and not variable selection, such effect is not harmful and would not, for example, affect the out-of-sample returns of a portfolio. When stability in the selection matters, however, a possible remedy to this problem is the so-called elastic net strategy proposed by Zou and Hastie (2005) which consists of adding to the ℓ_1-norm penalty a ℓ_2-norm penalty, the role of which is to enforce democracy in the selection within a group of correlated assets. Since all assets in the group thus tend to be selected, it is clear that, though still sparse, the solution of the scheme using both penalties will in general be less sparse than when using the ℓ_1-norm penalty alone. An application of this strategy to portfolio theory is considered by Li (2014).

Notice that for applying the elastic net strategy as a safeguard against selection instabilities, there is no need to know in advance which are the groups of correlated variables. When the groups are known, one may want to select the complete group composed of variables or assets belonging to some predefined category. A way to achieve this is to use the so-called mixed $\ell_1 - \ell_2$ norm, namely

$$\|\mathbf{w}\|_{1,2} = \sum_j (\sum_l |w_{j,l}|^2)^{1/2} \tag{2.18}$$

where the index j runs over the predefined groups and the index l runs inside each group. Such strategy, called "group lasso" by Yuan and Lin (2006), will sparsify the groups but select all variables within a selected group. For more details about these norms ensuring "structured sparsity" and the related algorithmic aspects, see, for example, the review paper by Bach *et al.* (2012).

2.6 Optimal Forecast Combination

The problem of sparse portfolio construction or replication bears strong similarity with the problem of linearly combining individual forecasts in order to improve reliability and accuracy, as noticed by Conflitti *et al.* (2015). These forecasts can be judgemental (i.e., provided by experts asked in a survey to provide forecasts of some economic variables such as inflation) or else be the output of different quantitative prediction models.

The idea is quite old, dating back to Bates and Granger (1969) and Granger and Ramanathan (1984), and has been extensively discussed in the literature (see, e.g., the review by Clemen 1989 and Timmermann 2006).

The problem can be formulated as follows. We denote by y_{t+h} the variable to be forecast at time t, assuming that the desired forecast horizon is h. We have at hand N forecasters, each delivering at time t a forecast $\hat{y}_{i,t+h}$, using the information about y_t they have at time t. We form with these individual forecasts $\hat{y}_{i,t+h}, i = 1, \cdots, N$, the $N \times 1$-dimensional vector $\mathbf{y}_{t+h}$. These forecasts are then linearly combined using time-independent weights $w_i, i = 1, \cdots, N$, which are assumed to satisfy the contraints $w_i \geq 0$ and $\sum_{i=1}^N w_i = 1$, and which are put into the $N \times 1$ vector $\mathbf{w}$. The aim is to minimize the mean square forecast error $\mathbf{E}[(y_{t+h} - \mathbf{w}^\top \hat{\mathbf{y}}_{t+h})^2]$ achieved by the combination. In empirical applications, the expectation is replaced by the sample mean over some historical period for which both the forecasts and the realization of the real variable are available. Hence the optimal forecast combination problem can be formulated as

$$\mathbf{w}_{opt} = \arg\min_{\mathbf{w}} \left[\sum_{t=1}^{T-h} (y_{t+h} - \mathbf{w}^\top \hat{\mathbf{y}}_{t+h})^2 \right] \tag{2.19}$$

$$\text{s. t. } \mathbf{w} \geq \mathbf{0}$$

$$\mathbf{w}^\top \mathbf{1}_N = 1$$

assuming that the variable y_t is observed for $t = 1, \ldots, T$. The resulting combined forecast for the variable y_t at time $t = T + h$ is then given by $\mathbf{w}_{opt}^\top \, \hat{\mathbf{y}}_{T+h}$.

With the vector of forecasts replacing the vector of returns, the problem is analogous to the problem of portfolio tracking described in Section 2.5, but with an additional no-shorting constraint. Besides, since by combining the two constraints we see that the ℓ_1 norm of the weight vector is fixed to one, problem (2.19) is equivalent to

$$\mathbf{w}_{opt} = \arg\min_{\mathbf{w}} \left[\sum_{t=1}^{T-h} (y_{t+h} - \mathbf{w}^{\top}\hat{\mathbf{y}}_{t+h})^2 + \tau \|\mathbf{w}\|_1 \right] \tag{2.20}$$

$$\text{s. t. } \mathbf{w} \geq \mathbf{0}$$

$$\mathbf{w}^{\top}\mathbf{1}_N = 1$$

for any value of the regularization parameter τ, which means that the weight vector will be sparse. Hence we have to solve a constrained lasso regression, and the modified LARS algorithm proposed by Brodie *et al.* (2009) can again be used to this purpose. Notice, however, that the sparsity level cannot be tuned by adjusting the value of τ. Possible remedies to this drawback would be to give up the nonnegativity constraints on the weights or else to use exact sparse simplex projections as in the paper by Kyrillidis *et al.* (2013).

An empirical exercise using survey data from the Survey of Professional Forecasters (SPF) for the Euro area and concerning the forecast of inflation and of GDP (Gross Domestic Product) growth is described in detail in the paper by Conflitti *et al.* (2015). The findings are that the optimal combinations of more than 50 individual forecasts perform well compared to the equal-weight combinations currently used by the European Central Bank. Nevertheless, the corresponding gains are relatively modest, which shows that the $1/N$ puzzle applies to this situation as well. The paper by Conflitti *et al* (2015) also addresses the problem of optimally combining density forecasts, in which case the least-squares objective function is replaced by a Kullback–Leibler Information Criterion between densities or by a derived "log-score" criterion.

Acknowlegments

I would like to thank my coauthors of the sparse portfolio paper on which most of the material of this chapter is based, namely Joshua Brodie, Ingrid Daubechies, Domenico Giannone, and Ignace Loris. Useful comments by an anonymous referee are also gratefully acknowledged.

This work was supported by the research contracts ARC-AUWB/2010-15/ULB-11 and IAP P7/06 StUDys.

References

Bach, F., Jenatton, R., Mairal, F. and Obozinski, G. (2012). Structured sparsity through convex optimization. *Statistical Science*, 27, 450–468.

Bates, J.M. and Granger, C.W.J. (1969). The combination of forecasts. *Operations Research Quarterly*, 20, 451–468.

Bertero, M. and Boccacci, P. (1998). *Introduction to inverse problems in imaging*. London: Institute of Physics Publishing.

Brodie, J., Daubechies, I., De Mol, C., Giannone, D. and Loris, I. (2009). Sparse and stable Markowitz portfolios. *Proceedings of the National Academy of Science*, 106 (30), 12267–12272.

Campbell, J.Y., Lo, A.W. and MacKinlay, C.A. (1997). *The econometrics of financial markets*. Princeton, NJ: Princeton University Press.

Carrasco, M. and Noumon, N. (2012). Optimal portfolio selection using regularization. https://www.webdepot.umontreal.ca/Usagers/carrascm/MonDepotPublic/carrascm/index.htm

Chen, S., Donoho, D. and Saunders, M. (2001). Atomic decomposition by basis pursuit. *SIAM Review*, 43, 129–159.

Clemen, R.T. (1989). Combining economic forecasts: a review and annotated bibliography. *International Journal of Forecasting*, 5, 559–583.

Conflitti, C., De Mol, C. and Giannone, D. (2015). Optimal combination of survey forecasts. *International Journal of Forecasting*, 31, 1096–1103.

Daubechies, I., Defrise, M. and De Mol, C. (2004). An iterative thresholding algorithm for linear inverse problems with a sparsity constraint. *Communications on Pure and Applied Mathematics*, 57, 1416–1457.

DeMiguel, V., Garlappi, L. and Uppal, R. (2009a). Optimal versus naive diversification: how inefficient is the 1/*N* portfolio strategy? *Review of Financial Studies*, 22, 1915–1953.

DeMiguel, V., Garlappi, L., Nogales, F.J. and Uppal, R. (2009b). A generalized approach to portfolio optimization: improving performance by constraining portfolio norms. *Management Science*, 55, 798–812.

De Mol, C., Giannone, D. and Reichlin, L. (2008). Forecasting using a large number of predictors: is Bayesian shrinkage a valid alternative to principal components? *Journal of Econometrics*, 146, 318–328.

Efron, B., Hastie, T., Johnstone, I. and Tibshirani, R. (2004). Least angle regression. *Annals of Statistics*, 32, 407–499.

Fan, J., Zhang, J. and Yu, K. (2012). Vast portfolio selection with gross-exposure constraints. *Journal of American Statistical Association*, 107, 592–606.

Foucart, S. and Rauhut, H. (2013). *A mathematical introduction to compressive sensing*. Basel: Birkhauser.

Gandy, A. and Veraart, L.A.M. (2013). The effect of estimation in high-dimensional portfolios. *Mathematical Finance*, 23, 531–559.

Granger, C.W.J. and Ramanathan, R. (1984). Improved methods of combining forecasts. *Journal of Forecasting*, 3, 197–204.

Henriques, J. and Ortega, J-P. (2014). Construction, management, and performances of Markowitz sparse portfolios. *Studies in Nonlinear Dynamics and Econometrics*, 18, 383–402.

Jagannathan, R. and Ma, T. (2003). Risk reduction in large portfolios: why imposing the wrong constraints helps. *Journal of Finance*, 58, 1651–1684.

Kyrillidis, A., Becker, S., Cevher, V. and Koch, C. (2013). Sparse projections onto the simplex. Proceedings of the 30th International Conference on Machine Learning (ICML 2013). *JMLR W&CP*, 28, 235–243.

Li, J. (2014). Sparse and stable portfolio selection with parameter uncertainty. *Journal of Business and Economic Statistics*, 33, 381–392.

Markowitz, H. (1952). Portfolio selection. *The Journal of Finance*, 7, 77–91.

Osborne, M.R., Presnell, B. and Turlach, B.A. (2000a). A new approach to variable selection in least squares problems. *IMA Journal of Numerical Analysis*, 20, 389–403.

Osborne, M.R., Presnell, B. and Turlach, B.A. (2000b). On the lasso and its dual. *Journal of Computational and Graphical Statistics*, 9, 319–337.

Park, S. and O'Leary, D.P. (2010). Portfolio selection using Tikhonov filtering to estimate the covariance matrix. *SIAM Journal on Financial Mathematics*, 1, 932–961.

Rosenbaum, M. and Tsybakov, A.B. (2010). Sparse recovery under matrix uncertainty. *Annals of Statistics*, 38, 2620–2651.

Ruppert, D. (2004). *Statistics and finance: an introduction*. Berlin: Springer.

Tibshirani, R. (1996). Regression shrinkage and selection via the lasso. *Journal of the Royal Statistical Society, Series B*, 58, 267–288.

Timmermann, A. (2006). Forecast combination: *Handbook of economic forecasting*, Vol. 1 (ed. G. Elliott, C. Granger and A. Timmermann). Amsterdam: North Holland.

Yuan, M. and Lin, Y. (2006). Model selection and estimation in regression with grouped variables. *Journal of the Royal Statistical Society, Series B*, 68, 49–67.

Zou, H. and Hastie, T. (2005). Regularization and variable selection via the elastic net. *Journal of the Royal Statistical Society, Series B*, 67, 301–320.

3

Mean-Reverting Portfolios

Tradeoffs between Sparsity and Volatility

Marco Cuturi[1] and Alexandre d'Aspremont[2]
[1]*Kyoto University, Japan*
[2]*CNRS - Ecole Normale supérieure, France*

Mean-reverting assets are one of the holy grails of financial markets: if such assets existed, they would provide trivially profitable investment strategies for any investor able to trade them, thanks to the knowledge that such assets oscillate predictably around their long-term mean. The modus operandi of cointegration-based trading strategies (Tsay, 2005, §8) is to create first a portfolio of assets whose aggregate value mean-reverts, and then to exploit that knowledge by selling short or buying that portfolio when its value deviates from its long-term mean. Such portfolios are typically selected using tools from cointegration theory (Engle and Granger, 1987; Johansen, 1991), whose aim is to detect combinations of assets that are stationary and therefore mean-reverting. We argue in this chapter that focusing on stationarity only may not suffice to ensure profitability of cointegration-based strategies. While it might be possible to create synthetically, using a large array of financial assets, a portfolio whose aggregate value is stationary and therefore mean-reverting, trading such a large portfolio incurs in practice important trade or borrow costs. Looking for stationary portfolios formed by many assets may also result in portfolios that have a very small volatility and that require significant leverage to be profitable. We study in this chapter algorithmic approaches that can mitigate these effects by searching for maximally mean-reverting portfolios that are sufficiently sparse and/or volatile.

3.1 Introduction

Mean-reverting assets, namely assets whose price oscillates predictably around a long-term mean, provide investors with an ideal investment opportunity. Because of their tendency

Financial Signal Processing and Machine Learning, First Edition.
Edited by Ali N. Akansu, Sanjeev R. Kulkarni and Dmitry Malioutov.
© 2016 John Wiley & Sons, Ltd. Published 2016 by John Wiley & Sons, Ltd.

to pull back to a given price level, a naive contrarian strategy of buying the asset when its price lies below that mean, or selling short the asset when it lies above that mean, can be profitable. Unsurprisingly, assets that exhibit significant mean reversion are very hard to find in efficient markets. Whenever mean reversion is observed in a single asset, it is almost always impossible to profit from it: the asset may typically have very low volatility, be illiquid, or be hard to short-sell, or its mean reversion may occur at a time scale (months, years) for which the borrow cost of holding or shorting the asset may well exceed any profit expected from such a contrarian strategy.

3.1.1 Synthetic Mean-Reverting Baskets

Since mean-reverting assets rarely appear in liquid markets, investors have focused instead on creating synthetic assets that can mimic the properties of a single mean-reverting asset, and trading such synthetic assets as if they were a single asset. Such a synthetic asset is typically designed by combining long and short positions in various liquid assets to form a *mean-reverting portfolio*, whose aggregate value exhibits significant mean reversion.

Constructing such synthetic portfolios is, however, challenging. Whereas simple descriptive statistics and unit-root test procedures can be used to test whether a single asset is mean-reverting, building mean-reverting portfolios requires finding a proper vector of algebraic weights (long and short positions) that describes a portfolio that has a mean-reverting aggregate value. In that sense, mean-reverting portfolios are made by the investor and cannot be simply chosen among tradable assets. A mean-reverting portfolio is characterized both by the pool of assets the investor has selected (starting with the dimension of the vector) and by the fixed nominal quantities (or weights) of each of these assets in the portfolio, which the investor also needs to set. When only two assets are considered, such baskets are usually known as long-short trading pairs. We consider in this paper baskets that are constituted by more than two assets.

3.1.2 Mean-Reverting Baskets with Sufficient Volatility and Sparsity

A mean-reverting portfolio must exhibit sufficient mean reversion to ensure that a contrarian strategy is profitable. To meet this requirement, investors have relied on cointegration theory (Engle and Granger, 1987; Johansen, 2005; Maddala and Kim, 1998) to estimate linear combinations of assets that exhibit stationarity (and therefore mean reversion) using historical data. We argue in this chapter, as we did in earlier publications (Cuturi and d'Aspremont, 2013; d'Aspremont, 2011), that mean-reverting strategies cannot, however, only rely on this approach to be profitable. Arbitrage opportunities can only exist if they are large enough to be traded without using too much leverage or incurring too many transaction costs. For mean-reverting baskets, this condition translates naturally into a first requirement that the gap between the basket valuation and its long-term mean is large enough on average, namely that the basket price has sufficient variance or volatility. A second desirable property is that mean-reverting portfolios require trading as few assets as possible to minimize costs, namely that the weights vector of that portfolio is sparse. We propose in this work methods that maximize a proxy for mean reversion, and that can take into account at the same time constraints on variance and sparsity.

We propose first in Section 3.2 three proxies for mean reversion. Section 3.3 defines the basket optimization problems corresponding to these quantities. We show in Section 3.4 that each of these problems translate naturally into semidefinite relaxations that produce either exact or approximate solutions using sparse principal component analysis (PCA) techniques. Finally, we present numerical evidence in Section 3.5 that taking into account sparsity and volatility can significantly boost the performance of mean-reverting trading strategies in trading environments where trading costs are not negligible.

3.2 Proxies for Mean Reversion

Isolating stable linear combinations of variables of multivariate time series is a fundamental problem in econometrics. A classical formulation of the problem reads as follows: given a vector valued process $x = (x_t)_t$ taking values in $\mathbb{R}^n$ and indexed by time $t \in \mathbb{N}$, and making no assumptions on the stationarity of each individual component of x, can we estimate one or many directions $y \in \mathbb{R}^n$ such that the univariate process $(y^T x_t)$ is stationary? When such a vector y exists, the process x is said to be cointegrated. The goal of cointegration techniques is to detect and estimate such directions y. Taking for granted that such techniques can efficiently isolate sparse mean-reverting baskets, their financial application can be either straightforward using simple event triggers to buy, sell, or simply hold the basket (Tsay, 2005, §8.6), or more elaborate optimal trading strategies if one assumes that the mean-reverting basket value is a Ohrstein–Ullenbeck process, as discussed in Elie and Espinosa, (2011), Jurek and Yang (2007), and Liu and Timmermann (2010).

3.2.1 Related Work and Problem Setting

Engle and Granger (1987) provided in their seminal work a first approach to compare two nonstationary univariate time series (x_t, y_t), and test for the existence of a term α such that $y_t - \alpha x_t$ becomes stationary. Following this seminal work, several techniques have been proposed to generalize that idea to multivariate time series. As detailed in the survey by Maddala and Kim (1998, §5), cointegration techniques differ in the modeling assumptions they require on the time series themselves. Some are designed to identify only one cointegrated relationship, whereas others are designed to detect many or all of them. Among these references, Johansen (1991) proposed a popular approach that builds upon a vector autoregression (VAR) model, as surveyed in Johansen (2004, 2005). These approaches all discuss issues that are relevant to econometrics, such as detrending and seasonal adjustments. Some of them focus more specifically on testing procedures designed to check whether such cointegrated relationships exist or not, rather than on the robustness of the estimation of that relationship itself. We follow in this work a simpler approach proposed by d'Aspremont (2011), which is to trade off interpretability, testing, and modeling assumptions for a simpler optimization framework that can be tailored to include other aspects than only stationarity. d'Aspremont (2011) did so by adding regularizers to the predictability criterion proposed by Box and Tiao (1977). We follow in this chapter the approach we proposed in Cuturi and d'Aspremont (2013) to design mean reversion proxies that do not rely on any modeling assumption.

Throughout this chapter, we write $\mathbf{S}_n$ for the $n \times n$ cone of positive definite matrices. We consider in the following a multivariate stochastic process $x = (x_t)_{t \in \mathbb{N}}$ taking values in $\mathbb{R}^n$.

We write $\mathcal{A}_k = \mathbf{E}[x_t x_{t+k}^T], k \geq 0$ for the lag-k autocovariance matrix of x_t if it is finite. Using a sample path $\mathbf{x}$ of (x_t), where $\mathbf{x} = (\mathbf{x}_1, \cdots, \mathbf{x}_T)$ and each $\mathbf{x}_t \in \mathbb{R}^n$, we write A_k for the *empirical* counterpart of $\mathcal{A}_k$ computed from $\mathbf{x}$,

$$A_k \stackrel{\text{def}}{=} \frac{1}{T-k-1} \sum_{t=1}^{T-k} \tilde{\mathbf{x}}_t \tilde{\mathbf{x}}_{t+k}^T, \tilde{\mathbf{x}}_t \stackrel{\text{def}}{=} \mathbf{x}_t - \frac{1}{T} \sum_{t=1}^{T} \mathbf{x}_t. \tag{3.1}$$

Given $y \in \mathbb{R}^n$, we now define three measures that can all be interpreted as proxies for the mean reversion of $y^T x_t$. *Predictability* –defined for stationary processes by Box and Tiao (1977) and generalized for nonstationary processes by Bewley *et al.* (1994) – measures how close to noise the series is. The *portmanteau* statistic of Ljung and Box (1978) is used to test whether a process is white noise. Finally, the *crossing statistic* (Ylvisaker, 1965) measures the probability that a process crosses its mean per unit of time. In all three cases, low values for these criteria imply a fast mean reversion.

3.2.2 *Predictability*

We briefly recall the canonical decomposition derived in Box and Tiao (1977). Suppose that x_t follows the recursion:

$$x_t = \hat{x}_{t-1} + \varepsilon_t, \tag{3.2}$$

where $\hat{x}_{t-1}$ is a predictor of x_t built upon past values of the process recorded up to $t-1$, and ε_t is a vector of independent and identically distributed (i.i.d.) Gaussian noise with zero mean and covariance $\Sigma \in \mathbf{S}_n$ independent of all variables $(x_r)_{r<t}$. The canonical analysis in Box and Tiao (1977) starts as follows.

3.2.2.1 Univariate case

Suppose $n = 1$ and thus $\Sigma \in \mathbb{R}_+$; Equation (3.2) leads thus to

$$\mathbf{E}[x_t^2] = \mathbf{E}[\hat{x}_{t-1}^2] + \mathbf{E}[\varepsilon_t^2], \text{ thus } 1 = \frac{\hat{\sigma}^2}{\sigma^2} + \frac{\Sigma}{\sigma^2},$$

by introducing the variances σ^2 and $\hat{\sigma}^2$ of x_t and $\hat{x}_t$, respectively. Box and Tiao measure the *predictability* of x_t by the ratio

$$\lambda \stackrel{\text{def}}{=} \frac{\hat{\sigma}^2}{\sigma^2}.$$

The intuition behind this variance ratio is simple: when it is small, the variance of the noise dominates that of $\hat{x}_{t-1}$ and x_t is dominated by the noise term; when it is large, $\hat{x}_{t-1}$ dominates the noise and x_t can be accurately predicted on average.

3.2.2.2 Multivariate case

Suppose $n > 1$, and consider now the univariate process $(y^T x_t)_t$ with weights $y \in \mathbb{R}^n$. Using (3.2), we know that $y^T x_t = y^T \hat{x}_{t-1} + y^T \varepsilon_t$, and we can measure its predicability as

$$\lambda(y) \stackrel{\text{def}}{=} \frac{y^T \hat{\mathcal{A}}_0 y}{y^T \mathcal{A}_0 y}, \tag{3.3}$$

where $\hat{\mathcal{A}}_0$ and $\mathcal{A}_0$ are the covariance matrices of x_t and $\hat{x}_{t-1}$, respectively. Minimizing predictability $\lambda(y)$ is then equivalent to finding the minimum generalized eigenvalue λ solving

$$\det(\lambda \mathcal{A}_0 - \hat{\mathcal{A}}_0) = 0. \tag{3.4}$$

Assuming that $\mathcal{A}_0$ is positive definite, the basket with minimum predictability will be given by $y = \mathcal{A}_0^{-1/2} y_0$, where y_0 is the eigenvector corresponding to the smallest eigenvalue of the matrix $\mathcal{A}_0^{-1/2} \hat{\mathcal{A}}_0 \mathcal{A}_0^{-1/2}$.

3.2.2.3 Estimation of $\lambda(y)$

All of the quantities used to define λ above need to be estimated from sample paths. $\mathcal{A}_0$ can be estimated by A_0 following Equation (3.1). All other quantities depend on the predictor $\hat{x}_{t-1}$. Box and Tiao assume that x_t follows a vector autoregressive model of order p –VAR(p), in short –and therefore $\hat{x}_{t-1}$ takes the form,

$$\hat{x}_{t-1} = \sum_{k=1}^{p} \mathcal{H}_k x_{t-k},$$

where the p matrices $(\mathcal{H}_k)$ contain each $n \times n$ autoregressive coefficient. Estimating $\mathcal{H}_k$ from the sample path $\mathbf{x}$, Box and Tiao solve for the optimal basket by inserting these estimates in the generalized eigenvalue problem displayed in Equation (3.4). If one assumes that $p = 1$ (the case $p > 1$ can be trivially reformulated as a VAR(1) model with adequate reparameterization), then

$$\hat{\mathcal{A}}_0 = \mathcal{H}_1 \mathcal{A}_0 \mathcal{H}_1^T \text{ and } \mathcal{A}_1 = \mathcal{A}_0 \mathcal{H}_1,$$

and thus the Yule–Walker estimator (Lütkepohl, 2005, §3.3) of $\mathcal{H}_1$ would be $H_1 = A_0^{-1} A_1$. Minimizing predictability boils down to solving in that case

$$\min_y \hat{\lambda}(y),\ \hat{\lambda}(y) \stackrel{\text{def}}{=} \frac{y^T (H_1 A_0 H_1^T) y}{y^T A_0 y} = \frac{y^T (A_1 A_0^{-1} A_1^T) y}{y^T A_0 y},$$

which is equivalent to computing the smallest eigenvector of the matrix $A_0^{-1/2} A_1 A_0^{-1} A_1^T A_0^{-1/2}$ if the covariance matrix A_0 is invertible.

The machinery of Box and Tiao to quantify mean reversion requires defining a model to form $\hat{x}_{t-1}$, the conditional expectation of x_t given previous observations. We consider now two criteria that do without such modeling assumptions.

3.2.3 Portmanteau Criterion

Recall that the *portmanteau* statistic of order p (Ljung and Box 1978) of a centered univariate stationary process x (with $n = 1$) is given by

$$\text{por}_p(x) = \frac{1}{p} \sum_{i=1}^{p} \left(\frac{\mathbf{E}[x_t x_{t+i}]}{\mathbf{E}[x_t^2]} \right)^2$$

where $\mathbf{E}[x_t x_{t+i}]/\mathbf{E}[x_t^2]$ is the ith-order autocorrelation of x_t. The portmanteau statistic of a white noise process is by definition 0 for any p. Given a multivariate ($n > 1$) process x, we write

$$\phi_p(y) = \mathrm{por}_p(y^T x) = \frac{1}{p}\sum_{i=1}^{p}\left(\frac{y^T \mathcal{A}_i y}{y^T \mathcal{A}_0 y}\right)^2,$$

for a coefficient vector $y \in \mathbb{R}^n$. By construction, $\phi_p(y) = \phi_p(ty)$ for any $t \neq 0$, and in what follows, we will impose $\|y\|_2 = 1$. The quantities $\phi_p(y)$ are computed using the following estimates (Hamilton, 1994, p. 110):

$$\hat{\phi}_p(y) = \frac{1}{p}\sum_{i=1}^{p}\left(\frac{y^T A_i y}{y^T A_0 y}\right)^2. \tag{3.5}$$

3.2.4 *Crossing Statistics*

Kedem and Yakowitz (1994, §4.1) define the *zero crossing rate* of a univariate ($n = 1$) process x (its expected number of crosses around 0 per unit of time) as

$$\gamma(x) = \mathbf{E}\left[\frac{\sum_{t=2}^{T}\mathbf{1}_{\{x_t x_{t-1} \leq 0\}}}{T-1}\right]. \tag{3.6}$$

A result known as the cosine formula states that if x_t is an autoregressive process of order 1 (AR(1)), namely if $|a| < 1$, ε_t is i.i.d. standard Gaussian noise and $x_t = a x_{t-1} + \varepsilon_t$, then (Kedem and Yakowitz, 1994, §4.2.2):

$$\gamma(x) = \frac{\arccos(a)}{\pi}.$$

Hence, for AR(1) processes, minimizing the first-order autocorrelation a also directly maximizes the crossing rate of the process x. For $n > 1$, since the first-order autocorrelation of $y^T x_t$ is equal to $y^T \mathcal{A}_1 y$, we propose to minimize $y^T \mathcal{A}_1 y$ and ensure that all other absolute autocorrelations $|y^T \mathcal{A}_k y|$, $k > 1$ are small.

3.3 Optimal Baskets

Given a centered multivariate process $\mathbf{x}$, we form its covariance matrix A_0 and its p autocovariances $(A_1, \cdots, A_p)$. Because $y^T A y = y^T (A + A^T) y/2$, we symmetrize all autocovariance matrices A_i. We investigate in this section the problem of estimating baskets that have maximal mean reversion (as measured by the proxies proposed in Section 3.2), while being at the same time sufficiently volatile and supported by as few assets as possible. The latter will be achieved by selecting portfolios y that have a small "0-norm," namely that the number of nonzero components in y,

$$\|y\|_0 \overset{\text{def}}{=} \#\{1 \leq i \leq d \,|\, y_i \neq 0\},$$

is small. The former will be achieved by selecting portfolios whose aggregated value exhibits a variance over time that exceeds a given threshold $\nu > 0$. Note that for the variance of $(y^T x_t)$ to exceed a level ν, the largest eigenvalue of A_0 must necessarily be larger than ν, which we always assume in what follows. Combining these two constraints, we propose three different mathematical programs that reflect these tradeoffs.

3.3.1 Minimizing Predictability

Minimizing Box–Tiao's predictability $\hat{\lambda}$ defined in Section 3.2.2, while ensuring that both the variance of the resulting process exceeds ν and the vector of loadings is sparse with a 0-norm equal to k, means solving the following program:

$$\begin{array}{ll} \text{minimize} & y^T M y \\ \text{subject to} & y^T A_0 y \geq \nu, \\ & \|y\|_2 = 1, \\ & \|y\|_0 = k, \end{array} \tag{P1}$$

in the variable $y \in \mathbb{R}^n$ with $M \overset{\text{def}}{=} A_1 A_0^{-1} A_1^T$, where $M, A_0 \in \mathbf{S}_n$. Without the normalization constraint $\|y\|_2 = 1$ and the sparsity constraint $\|y\|_0 = k$, problem (P1) is equivalent to a generalized eigenvalue problem in the pair (M, A_0). That problem quickly becomes unstable when A_0 is ill-conditioned or M is singular. Adding the normalization constraint $\|y\|_2 = 1$ solves these numerical problems.

3.3.2 Minimizing the Portmanteau Statistic

Using a similar formulation, we can also minimize the order p portmanteau statistic defined in Section 3.2.3 while ensuring a minimal variance level ν by solving:

$$\begin{array}{ll} \text{minimize} & \sum_{i=1}^{p} (y^T A_i y)^2 \\ \text{subject to} & y^T A_0 y \geq \nu, \\ & \|y\|_2 = 1, \\ & \|y\|_0 = k, \end{array} \tag{P2}$$

in the variable $y \in \mathbb{R}^n$, for some parameter $\nu > 0$. Problem (P2) has a natural interpretation: the objective function directly minimizes the portmanteau statistic, while the constraints normalize the norm of the basket weights to one, impose a variance larger than ν, and impose a sparsity constraint on y.

3.3.3 Minimizing the Crossing Statistic

Following the results in Section 3.2.4, maximizing the crossing rate while keeping the rest of the autocorrelogram low,

$$\begin{array}{ll} \text{minimize} & y^T A_1 y + \mu \sum_{k=2}^{p} (y^T A_k y)^2 \\ \text{subject to} & y^T A_0 y \geq \nu, \\ & \|y\|_2 = 1, \\ & \|y\|_0 = k, \end{array} \tag{P3}$$

in the variable $y \in \mathbb{R}^n$, for some parameters $\mu, \nu > 0$. This will produce processes that are close to being AR(1) while having a high crossing rate.

3.4 Semidefinite Relaxations and Sparse Components

Problems (P1), (P2), and (P3) are not convex and can be in practice extremely difficult to solve, since they involve a sparse selection of variables. We detail in this section convex relaxations to these problems that can be used to derive relevant suboptimal solutions.

3.4.1 A Semidefinite Programming Approach to Basket Estimation

We propose to relax problems (P1), (P2), and (P3) into semidefinite programs (SDPs) (Vandenberghe and Boyd, 1996). We show that these SDPs can handle naturally sparsity and volatility constraints while still aiming at mean reversion. In some restricted cases, one can show that these relaxations are tight, in the sense that they solve exactly the programs described above. In such cases, the true solution $y^\star$ of some of the programs above can be recovered using their corresponding SDP solution $Y^\star$.

However, in most of the cases we will be interested in, such a correspondence is not guaranteed, and these SDP relaxations can only serve as a guide to propose solutions to these hard nonconvex problems when considered with respect to vector y. To do so, the optimal solution $Y^\star$ needs to be *deflated* from a large rank $d \times d$ matrix to a rank one matrix yy^T, where y can be considered a good candidate for basket weights. A typical approach to deflate a positive definite matrix into a vector is to consider its eigenvector with the leading eigenvalue. Having sparsity constraints in mind, we propose to apply a heuristic grounded on sparse PCA (d'Aspremont *et al.*, 2007; Zou *et al.*, 2006). Instead of considering the lead eigenvector, we recover the leading *sparse* eigenvector of $Y^\star$ (with a 0-norm constrained to be equal to k). Several efficient algorithmic approaches have been proposed to solve approximately that problem; we use the SpaSM (sparse statistiscal modeling) toolbox (Sjöstrand *et al.*, 2012) in our experiments.

3.4.2 Predictability

We can form a convex relaxation of the predictability optimization problem (P1) over the variable $y \in \mathbb{R}^n$,

$$\begin{aligned} \text{minimize} \;\; & y^T M y \\ \text{subject to} \;\; & y^T A_0 y \geq v \\ & \|y\|_2 = 1, \\ & \|y\|_0 = k, \end{aligned}$$

by using the lifting argument of Lovász and Schrijver (1991), (i.e., writing $Y = yy^T$) to solve now the problem using a semidefinite variable Y, and by introducing a sparsity-inducing regularizer on Y that considers the L_1 norm of Y,

$$\|Y\|_1 \overset{\text{def}}{=} \sum_{ij} |Y_{ij}|,$$

so that Problem (P1) becomes (here $\rho > 0$),

$$\begin{aligned} \text{minimize} \;\; & \mathbf{Tr}(MY) + \rho \|Y\|_1 \\ \text{subject to} \;\; & \mathbf{Tr}(A_0 Y) \geq v \\ & \mathbf{Tr}(Y) = 1, \mathbf{Rank}(Y) = 1, Y \succeq 0. \end{aligned}$$

We relax this last problem further by dropping the rank constraint, to get

$$\begin{array}{ll}\text{minimize} & \mathbf{Tr}(MY) + \rho\|Y\|_1 \\ \text{subject to} & \mathbf{Tr}(A_0 Y) \geq \nu \\ & \mathbf{Tr}(Y) = 1, Y \succeq 0\end{array} \tag{SDP1}$$

which is a convex semidefinite program in $Y \in \mathbf{S}_n$.

3.4.3 *Portmanteau*

Using the same lifting argument and writing $Y = yy^T$, we can relax problem (P2) by solving

$$\begin{array}{ll}\text{minimize} & \sum_{i=1}^{p} \mathbf{Tr}(A_i Y)^2 + \rho\|Y\|_1 \\ \text{subject to} & \mathbf{Tr}(BY) \geq \nu \\ & \mathbf{Tr}(Y) = 1, Y \succeq 0,\end{array} \tag{SDP2}$$

a semidefinite program in $Y \in \mathbf{S}_n$.

3.4.4 *Crossing Stats*

As above, we can write a semidefinite relaxation for problem (P3):

$$\begin{array}{ll}\text{minimize} & \mathbf{Tr}(A_1 Y) + \mu \sum_{i=2}^{p} \mathbf{Tr}(A_i Y)^2 + \rho\|Y\|_1 \\ \text{subject to} & \mathbf{Tr}(BY) \geq \nu \\ & \mathbf{Tr}(Y) = 1, Y \succeq 0.\end{array} \tag{SDP3}$$

3.4.4.1 Tightness of the SDP Relaxation in the Absence of Sparsity Constraints

Note that for the crossing stats criterion (with $p = 1$ and no quadratic term in Y), the original problem P3 and its relaxation SDP3 are equivalent, taking for granted that no sparsity constraint is considered in the original problems and μ is set to 0 in the relaxations. These relaxations boil down to an SDP that only has a linear objective, a linear constraint, and a constraint on the trace of Y. In that case, Brickman (1961) showed that the range of two quadratic forms over the unit sphere is a convex set when the ambient dimension $n \geq 3$, which means in particular that for any two square matrices A, B of dimension n,

$$\begin{aligned}&\{(y^T Ay, y^T By) : y \in \mathbb{R}^n, \|y\|_2 = 1\} \\ &= \{(\mathbf{Tr}(AY), \mathbf{Tr}(BY)) : Y \in \mathbf{S}_n, \mathbf{Tr}Y = 1, Y \succeq 0\}.\end{aligned}$$

We refer the reader to Barvinok 2002 (§II.13) for a more complete discussion of this result. As remarked in Cuturi and d'Aspremont (2013), the same equivalence holds for P1 and SDP1. This means that, in the case where $\rho, \mu = 0$ and the 0-norm of y is *not* constrained, for any solution $Y^\star$ of the relaxation (SDP1) there exists a vector $y^\star$ that satisfies $\|y\|_2^2 = \mathbf{Tr}(Y^\star) = 1$, $y^{\star T} A_0 y^\star = \mathbf{Tr}(BY^\star)$, and $y^{\star T} M y^\star = \mathbf{Tr}(MY^\star)$, which means that $y^\star$ is an optimal solution of the original problem (P1). Boyd and Vandenberghe (2004, App. B) show how to explicitly

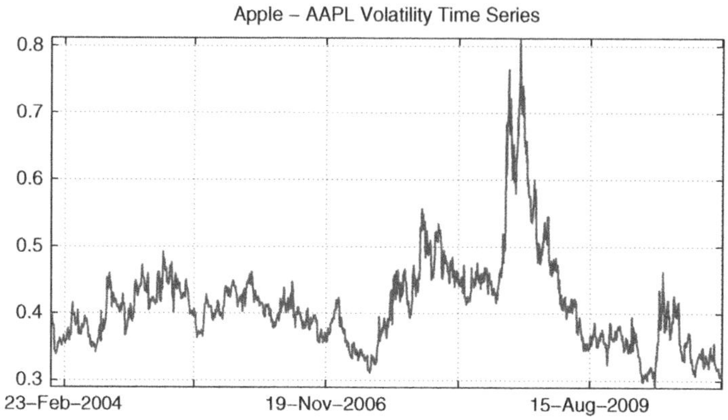

Figure 3.1 Option implied volatility for Apple between January 4, 2004, and December 30, 2010.

extract such a solution $y^{\star}$ from a matrix $Y^{\star}$ solving (SDP1). This result is, however, mostly anecdotical in the context of this chapter, in which we look for sparse and volatile baskets: using these two regularizers breaks the tightness result between the original problems in $\mathbb{R}^d$ and their SDP counterparts.

3.5 Numerical Experiments

In this section, we evaluate the ability of our techniques to extract mean-reverting baskets with sufficient variance and small 0-norm from a universe of tradable assets. We measure performance by applying to these baskets a trading strategy designed specifically for mean-reverting processes. We show that, under realistic trading costs assumptions, selecting sparse and volatile mean-reverting baskets translates into lower incurred costs and thus improves the performance of trading strategies.

3.5.1 Historical Data

We consider daily time series of option implied volatilities for 210 stocks from January 4, 2004, to December 30, 2010. A key advantage of using option implied volatility data is that these numbers vary in a somewhat limited range. Volatility also tends to exhibit regime switching, and hence can be considered piecewise stationary, which helps in extracting structural relationships. We plot a sample time series from this dataset in Figure 3.1 that corresponds to the implicit volatility of Apple's stock. In what follows, we mean by *asset* the implied volatility of any of these stocks whose value can be efficiently replicated using option portfolios.

3.5.2 Mean-reverting Basket Estimators

We compare the three basket selection techniques detailed here, *predictability, portmanteau,* and *crossing statistic*, implemented with varying targets for both sparsity and volatility, with two cointegration estimators that build upon PCA (Maddala and Kim, 1998, §5.5.4).. By the label PCA, we mean in what follows the eigenvector with the smallest eigenvalue of the covariance matrix A_0 of the process (Stock and Watson, 1988). By sPCA, we mean the sparse eigenvector of A_0 with 0-norm k that has the smallest eigenvalue, which can be simply estimated by computing the leading sparse eigenvector of $\lambda I - A_0$ where λ is bigger than the leading eigenvalue of A_0. This sparse principal component of the covariance matrix A_0 should not be confused with our utilization of sparse PCA in Section 3.4.1 as a way to recover a vector solution from the solution of a positive semidefinite problem. Note also that techniques based on principal components do not take explicitly variance levels into account when estimating the weights of a co-integrated relationship.

3.5.3 Jurek and Yang (2007) Trading Strategy

While option implied volatility is not directly tradable, it can be synthesized using baskets of call options, and we assimilate it to a tradable asset with (significant) transaction costs in what follows. For baskets of volatilities isolated by the techniques listed above, we apply the Jurek and Yang (2007) strategy for log utilities to the basket process recording out of sample performance. Jurek and Yang proposed to trade a stationary autoregressive process $(x_t)_t$ of order 1 and mean μ governed by the equation $x_{t+1} = \rho x_t + \sigma \varepsilon_t$, where $|\rho| < 1$, by taking a position N_t in the asset x_t, which is proportional to

$$N_t = \frac{\rho(\mu - x_t)}{\sigma^2} W_t \tag{3.7}$$

In effect, the strategy advocates taking a long (resp. short) position in the asset whenever it is below (resp. above) its long-term mean, and adjust the position size to account for the volatility of x_t and its mean reversion speed ρ. Given basket weights y, we apply standard AR estimation procedures on the in-sample portion of $y^T \mathbf{x}$ to recover estimates for $\hat{\rho}$ and $\hat{\sigma}$ and plug them directly in Equation (3.7). This approach is illustrated for two baskets in Figure 3.2.

3.5.4 Transaction Costs

We assume that fixed transaction costs are negligible, but that transaction costs per contract unit are incurred at each trading date. We vary the size of these costs across experiments to show the robustness of the approaches tested here to trading costs fluctuations. We let the transaction cost per contract unit vary between 0.03 and 0.17 cents by increments of 0.02 cents. Since the average value of a contract over our dataset is about 40 cents, this is akin to considering trading costs ranging from about 7 to about 40 base points (BPs), that is, 0.07 to 0.4%.

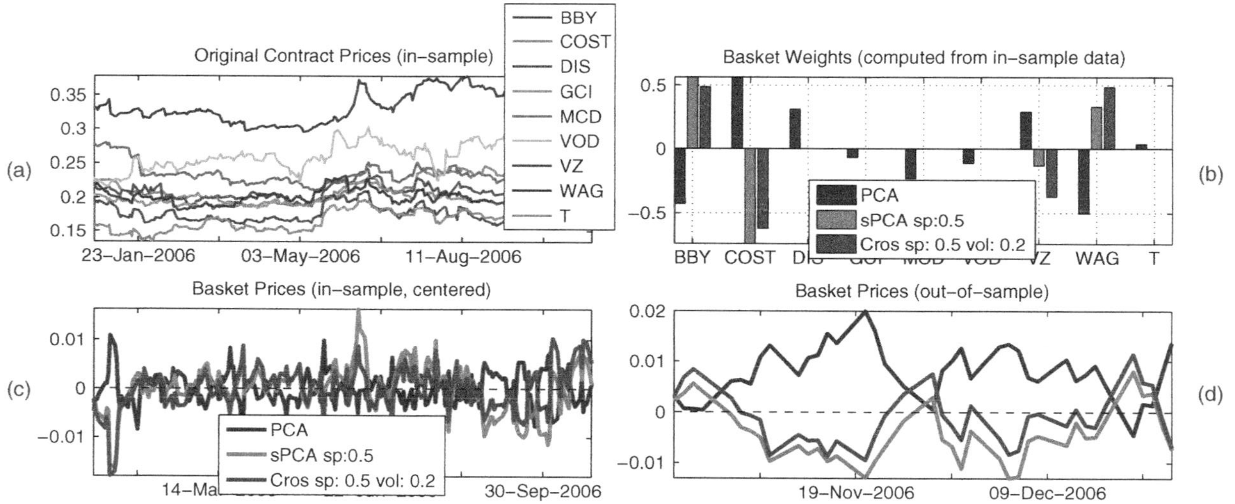

Figure 3.2 Three sample trading experiments, using the PCA, sparse PCA, and crossing statistics estimators. (a) Pool of 9 volatility time series selected using our fast PCA selection procedure. (b) Basket weights estimated with in-sample data using the eigenvector of the covariance matrix with the smallest eigenvalue, the smallest eigenvector with a sparsity constraint of $k = \lfloor 0.5 \times 9 \rfloor = 4$, and the crossing statistics estimator with a volatility threshold of $\nu = 0.2$, (i.e., a constraint on the basket's variance to be larger than $0.2\times$ the median variance of all 8 assets). (c) Using these 3 procedures, the time series of the resulting basket price in the in-sample part (c) and out-of-sample parts (d) are displayed. (e) Using the Jurek and Yang (2007) trading strategy results in varying positions (expressed as units of baskets) during the out-sample testing phase. (f) Transaction costs that result from trading the assets to achieve such positions accumulate over time. (g) Taking both trading gains and transaction costs into account, the net wealth of the investor for each strategy can be computed (the Sharpe ratio over the test period is displayed in the legend). Note how both sparsity and volatility constraints translate into portfolios composed of fewer assets, but with a higher variance.

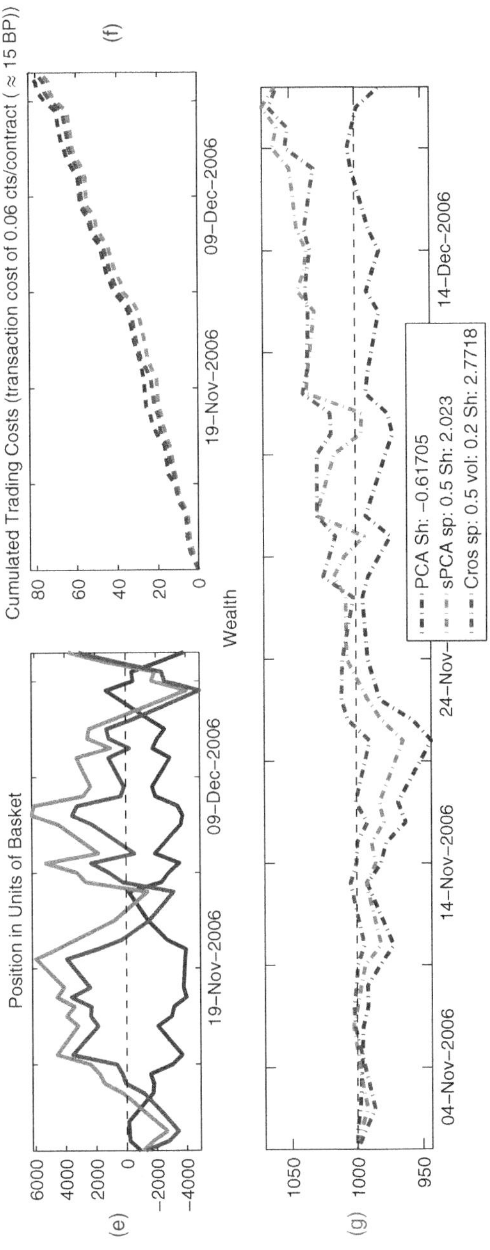

Figure 3.2 *(continued)*

3.5.5 Experimental Setup

We consider 20 sliding windows of one year (255 trading days) taken in the history, and consider each of these windows independently. Each window is split between 85% of days to estimate and 15% of days to test-trade our models, resulting in 38 test-trading days. We do not recompute the weights of the baskets during the test phase. The 210 stock volatilities (assets) we consider are grouped into 13 subgroups, depending on the economic sector of their stock. This results in 13 sector pools whose size varies between 3 assets and 43 assets. We look for mean-reverting baskets in each of these 13 sector pools.

Because all combinations of stocks in each of the 13 sector pools may not necessarily be mean-reverting, we select smaller candidate pools of n assets through a greedy backward-forward minimization scheme, where $8 \leq n \leq 12$. To do so, we start with an exhaustive search of all pools of size 3 within the sector pool, and proceed by adding or removing an asset using the PCA estimator (the smallest eigenvalue of the covariance matrix of a set of assets). We use the PCA estimator in that backward-forward search because it is the fastest to compute. We score each pool using that PCA statistic, the smaller meaning the better. We generate up to 200 candidate pools per each of the 13 sector pools. Out of all these candidate pools, we keep the best 50 in each window, and then use our cointegration estimation approaches separately on these candidates. One such pool was, for instance, composed of the stocks {`BBY`, `COST`, `DIS`, `GCI`, `MCD`, `VOD`, `VZ`, `WAG`, `T`} observed during the year 2006. Figure 3.2 provides a closeup on that universe of stocks, and shows the results of three trading experiments using PCA, sparse PCA, or the Crossing Stats estimator to build trading strategies.

3.5.6 Results

3.5.6.1 Robustness of Sharpe Ratios to Costs

In Figure 3.3, we plot the average of the Sharpe ratio over the 922 baskets estimated in our experimental set versus transaction costs. We consider different PCA settings as well as our three estimators using, in all three cases, the variance bound ν to be 0.3 times the median of all variances of assets available in a given asset pool, and the 0-norm to be equal to 0.3 times the size of the universe (itself between 8 and 12). We observe that Sharpe ratios decrease the fastest for the naive PCA-based method, this decrease being somewhat mitigated when adding a constraint on the 0-norm of the basket weights obtained with sparse PCA. Our methods require, in addition to sparsity, enough volatily to secure sufficient gains. These empirical observations agree with the intuition of this chapter: simple cointegration techniques can produce synthetic baskets with high mean reversion, large support, and low variance. Trading a portfolio with low variance that is supported by multiple assets translates in practice into high trading costs, which can damage the overall performance of the strategy. Both sparse PCA and our techniques manage instead to achieve a tradeoff between desirable mean reversion properties and, at the same time, control for sufficient variance and small basket size to allow for lower overall transaction costs.

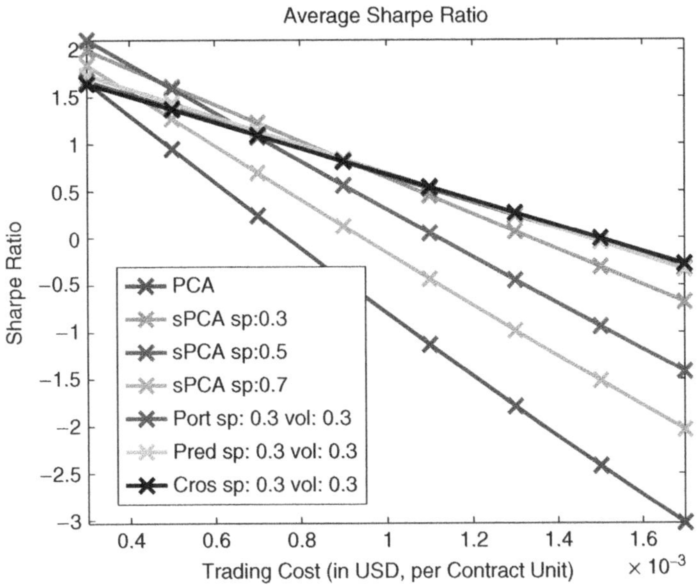

Figure 3.3 Average Sharpe ratio for the Jurek and Yang (2007) trading strategy captured over about 922 trading episodes, using different basket estimation approaches. These 922 trading episodes were obtained by considering 7 disjoint time-windows in our market sample, each of a length of about one year. Each time-window was divided into 85% in-sample data to estimate baskets, and 15% outsample to test strategies. On each time-window, the set of 210 tradable assets during that period was clustered using sectorial information, and each cluster screened (in the in-sample part of the time-window) to look for the most promising baskets of size between 8 and 12 in terms of mean reversion, by choosing greedily subsets of stocks that exhibited the smallest minimal eigenvalues in their covariance matrices. For each trading episode, the same universe of stocks was fed to different mean-reversion algorithms. Because volatility time-series are bounded and quite stationary, we consider the PCA approach, which uses the eigenvector with the smallest eigenvalue of the covariance matrix of the time-series to define a cointegrated relationship. Besides standard PCA, we have also consider sparse PCA eigenvectors with minimal eigenvalue, with the size k of the support of the eigenvector (the size of the resulting basket) constrained to be 30%, 50% or 70% of the total number of considered assets. We consider also the portmanteau, predictability and crossing stats estimation techniques with variance thresholds of $\nu = 0.2$ and a support whose size k (the number of assets effectively traded) is targeted to be about 30% of the size of the considered universe (itself between 8 and 12). As can be seen in the figure, the sharpe ratios of all trading approaches decrease with an increase in transaction costs. One expects sparse baskets to perform better under the assumption that costs are high, and this is indeed observed here. Because the relationship between sharpe ratios and transaction costs can be efficiently summarized as being a linear one, we propose in the plots displayed in Figure 3.4 a way to summarize the lines above with two numbers each: their intercept (Sharpe level in the quasi-absence of costs) and slope (degradation of Sharpe as costs increase). This visualization is useful to observe how sparsity (basket size) and volatility thresholds influence the robustness to costs of the strategies we propose. This visualization allows us to observe how performance is influenced by these parameter settings.

3.5.6.2 Tradeoffs between Mean Reversion, Sparsity, and Volatility

In the plots of Figure 3.4, this analysis is further detailed by considering various settings for v (volatility threshold) and k. To improve the legibility of these results, we summarize, following the observation in Figure 3.3 that the relationship between Sharpes and transactions costs seems almost linear, each of these curves by two numbers: an intercept level (Sharpe ratio when costs are low) and a slope (degradation of Sharpe as costs increase). Using these two

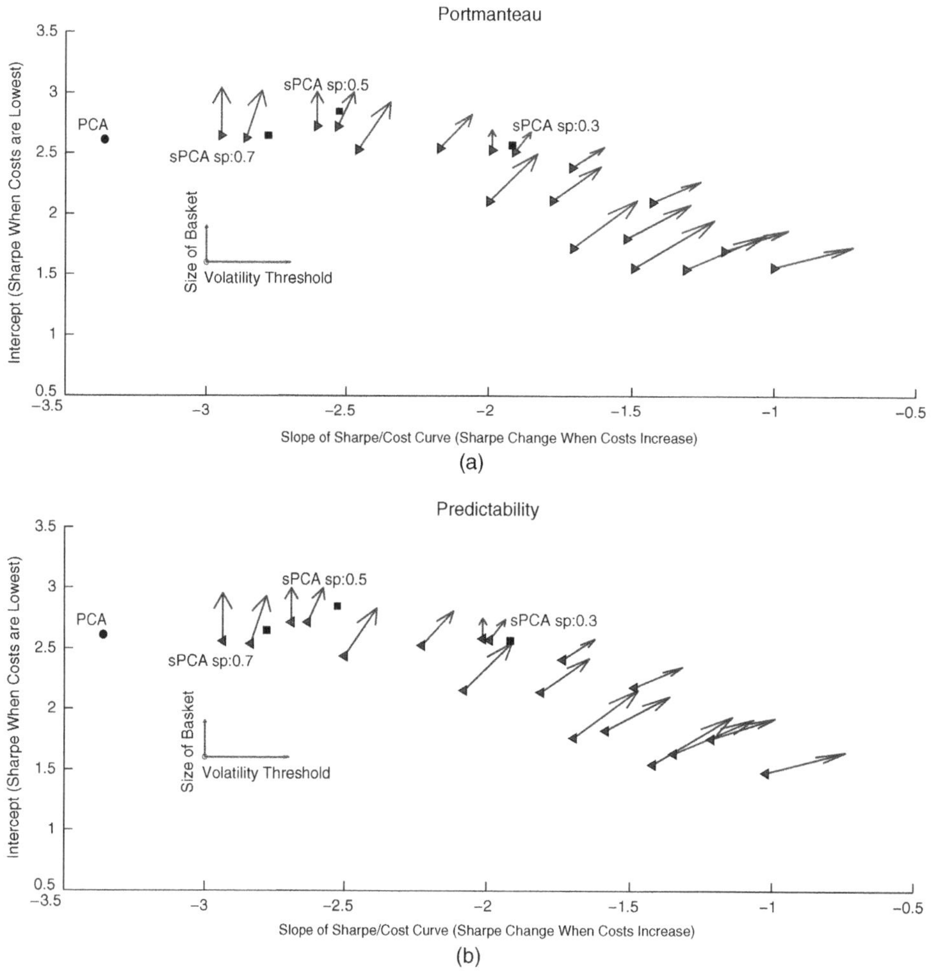

Figure 3.4 Relationships between Sharpe in a low cost setting (intercept) in the x-axis and robustness of Sharpe to costs (slope of Sharpe/costs curve) of a different estimators implemented with varying volatility levels v and sparsity levels k parameterized as a multiple of the universe size. Each colored square in the figures above corresponds to the performance of a given estimator (Portmanteau in subfigure (a), Predictability in subfigure (b) and Crossing Statistics in subfigure (c)) using different parameters for $v \in \{0, 0.1, 0.2, 0.3, 0.4, 0.5\}$ and $u \in \{0.3, 0.5, 0.7\}$. The parameters used for each experiment are displayed using an arrow whose vertical length is proportional to v and horizontal length is proportional to u.

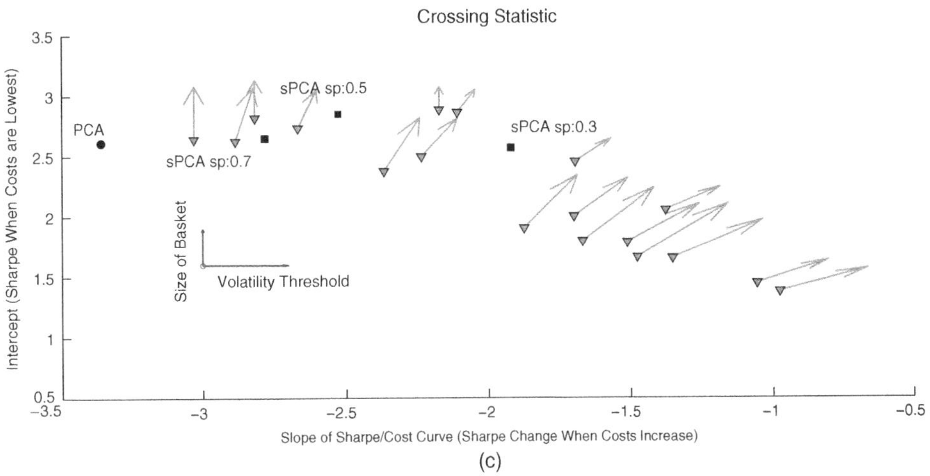

Figure 3.4 (*continued*)

numbers, we locate all considered strategies in the intercept–slope plane. We first show the spectral techniques, PCA and sPCA, with different levels of sparsity, meaning that k is set to $\lfloor u \times d \rfloor$, where $u \in \{0.3, 0.5, 0.7\}$ and d is the size of the original basket. Each of the three estimators we propose is studied in a separate plot. For each, we present various results characterized by two numbers: a volatility threshold $v \in \{0, 0.1, 0.2, 0.3, 0.4, 0.5\}$ and a sparsity level $u \in \{0.3, 0.5, 0.7\}$. To avoid cumbersome labels, we attach an arrow to each point: the arrow's length in the vertical direction is equal to u and characterizes the size of the basket, and the horizontal length is equal to v and characterizes the volatility level. As can be seen in these three plots, an interesting interplay between these two factors allows for a continuum of strategies that trade mean reversion (and thus Sharpe levels) for robustness to cost level.

3.6 Conclusion

We have described three different criteria to quantify the amount of mean reversion in a time series. For each of these criteria, we have detailed a tractable algorithm to isolate a vector of weights that has optimal mean reversion, while constraining both the variance (or signal strength) of the resulting univariate series to be above a certain level and its 0-norm to be at a certain level. We show that these bounds on variance and support size, together with our new criteria for mean reversion, can significantly improve the performance of mean reversion statistical arbitrage strategies and provide useful controls to adjust mean-reverting strategies to varying trading conditions, notably liquidity risk and cost environment.

References

Barvinok, A. (2002). *A course in convexity*. Providence, RI: American Mathematical Society.

Bewley, R., Orden, D., Yang, M. and Fisher, L. (1994). Comparison of Box-Tiao and Johansen canonical estimators of cointegrating vectors in VEC (1). Models. *Journal of Econometrics*, 64, 3–27.

Box, G. and Tiao, G. (1977). A canonical analysis of multiple time series. *Biometrika*, 64(2), 355–365.

Boyd, S. and Vandenberghe, L. (2004). *Convex optimization*. Cambridge: Cambridge University Press.

Brickman, L. (1961). On the field of values of a matrix. *Proceedings of the American Mathematical Society*, 12, 61–66.

Cuturi, M. and d'Aspremont, A. (2013). Mean reversion with a variance threshold. In *Proceedings of the International Conference in Machine Learning 2013*, June, Atlanta, GA.

d'Aspremont, A. (2011). Identifying small mean reverting portfolios. *Quantitative Finance*, 11(3), 351–364.

d'Aspremont, A., El Ghaoui, L., Jordan, M.I. and Lanckriet, G.R. (2007). A direct formulation for sparse PCA using semidefinite programming. *SIAM Review*, 49(3), 434–448.

Elie, R., and Espinosa, G.E. (2011). Optimal stopping of a mean reverting diffusion: minimizing the relative distance to the maximum. hal-00573429.

Engle, R.F. and Granger, C.W.J. (1987). Co-integration and error correction: Representation, estimation, and testing. *Econometrica*, 55(2), 251–276.

Hamilton, J. (1994). *Time series analysis*, Vol. 2. Cambridge: Cambridge University Press.

Johansen, S. (1991). Estimation and hypothesis testing of cointegration vectors in Gaussian vector autoregressive models. *Econometrica*, 59(6), 1551–1580.

Johansen, S. (2004). Cointegration: overview and development. In *Handbook of financial time series* (ed. Andersen, T.G., Davis, R.A., Kreiß, J.P. and Mikosch, T.V.). Berlin: Springer.

Johansen, S. (2005). *Cointegration: a survey*. Palgrave Handbook of Econometrics. London: Palgrave.

Jurek, J.W. and Yang, H. (2007). Dynamic portfolio selection in arbitrage. EFA 2006 Meetings Paper. SSRN eLibrary, http://dx.doi.org/10.2139/ssrn.882536

Kedem, B. and Yakowitz, S. (1994). *Time series analysis by higher order crossings*. Piscataway, NJ: IEEE Press.

Liu, J. and Timmermann, A. (2010). *Optimal arbitrage strategies*. Technical report. San Diego: University of California, San Diego.

Ljung, G. and Box, G. (1978). On a measure of lack of fit in time series models. *Biometrika*, 65(2), 297–303.

Lovász, L. and Schrijver, A. (1991). Cones of matrices and set-functions and 0-1 optimization. *SIAM Journal on Optimization*, 1(2), 166–190.

Lütkepohl, H. (2005). *New introduction to multiple time series analysis*. Berlin: Springer.

Maddala, G. and Kim, I. (1998). *Unit roots, cointegration, and structural change*. Cambridge: Cambridge University Press.

Sjöstrand, K., Clemmensen, L.H., Larsen, R. and Ersbøll, B. (2012). SpaSM: A MATLAB toolbox for sparse statistical modeling. *Journal of Statistical Software*, accepted.

Stock, J. and Watson, M. (1988). Testing for common trends. *Journal of the American Statistical Association*, 83, 1097–1107.

Tsay, R.S. (2005). *Analysis of financial time series*. Weinheim: Wiley-Interscience.

Vandenberghe, L. and Boyd, S. (1996). Semidefinite programming. *SIAM Review*, 38(1), 49–95.

Ylvisaker, N.D. (1965). The expected number of zeros of a stationary Gaussian process. *The Annals of Mathematical Statistics*, 36(3), 1043–1046.

Zou, H., Hastie, T. and Tibshirani, R. (2006). Sparse principal component analysis. *Journal of Computational and Graphical Statistics*, 15(2), 265–286.

4

Temporal Causal Modeling

Prabhanjan Kambadur[1], Aurélie C. Lozano[2] and Ronny Luss[2]
[1]*Bloomberg LP, USA*
[2]*IBM T.J. Watson Research Center, USA*

4.1 Introduction

Discovering causal relationships in multivariate time series data has many important applications in finance. Consider portfolio management, where one of the key tasks is to quantify the risk associated with different portfolios of assets. Traditionally, correlations amongst assets have been used to manage risk in portfolios. Knowledge of causal structures amongst assets can help improve portfolio management as knowing causality – rather than just correlation – can allow portfolio managers to mitigate risks directly. For example, suppose that an index fund "A" is found to be one of the causal drivers of another index fund "B." Then, the variance of B can be reduced by offsetting the variation due to the causal effects of A. In contrast simply knowing that "A" is correlated with "B" provides no guidance on how to *act* on index "B," as this does not mean that the two indexes are connected by a cause-and-effect relationship; hedging solely based on correlation does not protect against the possibility that correlation is driven by an unknown effect. Moreover, causal structures may be more stable across market regimes as they have more chance to capture effective economic relationships.

In order to mitigate risks effectively, we need several enhancements to mere causality detection. First, we need to be able to reason about the "strength" of the causal relationship between two assets using statistical measures such as p-values. Attaching well-founded strengths to causal relationships allows us to focus on the important relationships and serves as a guard against false discovery of causal relationships. Second, we need to be able to infer causality in the presence of heteroscedasticity. Typically, causal relationships are modeled by regressing to the conditional mean; this does not always give us a complete understanding of the conditional distributions of the responses based on their causalities. Finally, as the causality amongst assets might be seasonal, we need to be able to automatically identify

Financial Signal Processing and Machine Learning, First Edition.
Edited by Ali N. Akansu, Sanjeev R. Kulkarni and Dmitry Malioutov.
© 2016 John Wiley & Sons, Ltd. Published 2016 by John Wiley & Sons, Ltd.

regime changes. If we can successfully discover a temporally accurate causal structure that encodes causal strengths, we would be able to enhance the accuracy of tasks such as explaining the effect of political or financial events on different markets and understanding the microstructure of a financial network.

In this chapter, we discuss *temporal causal modeling* (or TCM) (Lozano *et al.*, 2009a,b), an approach that generalizes the notion of Granger causality to multivariate time series by linking the causality inference process to the estimation of sparse vector autoregressive (VAR) models (Section 4.2). Granger causality (Granger, 1980) is an operational definition of causality well known in econometrics, where a source time series is said to "cause" a target time series if it contains additional information for predicting the future values of the target series, beyond the information contained in the past values of the target time series. In essence, TCM combines Granger causality with sparse multivariate regression algorithms, and performs graphical modeling over the lagged temporal variables. We define and use a notion of causal strength modeling (or CSM) for TCM to investigate these potential implications (Section 4.3). We describe how TCM can be extended to the quantile loss function (or Q-TCM) to better model heteroscedactic data (Section 4.4). Finally, we extend TCM to identify regime changes by combining it with a Markov switching modeling framework (Chib, 1998); specifically, we describe a Bayesian Markov switching model for estimating sparse dynamic Bayesian networks (or MS-SDBN) (Section 4.5).

As a concrete case study that highlights the benefits of TCM, consider a financial network of various exchange-traded funds (ETFs) that represent indices tracking a mix of stocks traded on the largest exchanges of various countries. For example, the ticker symbol EWJ represents an ETF that tracks the MSCI Japan Index. We consider a dataset from a family of ETFs called iShares that contains ETF time series of 15 countries as well as an index tracking oil and gas prices and an index tracking the spot price of gold; iShares are managed by Blackrock and the data are publicly available from `finance.yahoo.com`. The causal CSM graphs formed during four different 750-day periods that cover 2005–2008 are shown in Figure 4.1; for interpretability, we use a lag spanning the previous 5 days for vector autoregression. Each feature is a monthly return computed over the previous 22 business days and the lag of 5 days is the monthly return ending on each of those 5 days. Each graph moves the window of data over 50 business days in order to view how time affects the causal networks. Each arc appearing in the CSM causal graphs represents a causal relationship with causal strength greater than a predefined threshold in [0, 1]. Causal strength, as we define it, measures the likelihood that the causal relationship between two nodes is statistically significant. For example, in Figure 4.1a, the causal strength of the relationship directed from South Korea to the United States measures the likelihood that including the South Korea data increases the performance of a United States model. This likelihood is further defined in Section 4.3 (note that this definition of causal strength differs from heuristic notions of causal strength, such as measuring coefficient magnitudes in a linear model). In particular, we are interested in the dependencies of the United States – represented by an ETF that tracks the S&P 500 – during the financial crisis of 2007 and 2008. The panels in Figure 4.1 show an interesting dependence of US-listed equities on Asian-listed equities, which focuses mostly on Japan. To analyze this further, we perform TCM to discover the United States' dependencies over several time periods beginning in 2005 and running through 2015. Table 4.1, which shows the results of our analysis, depicts the causal strength values for the three strongest relationships for each time period.

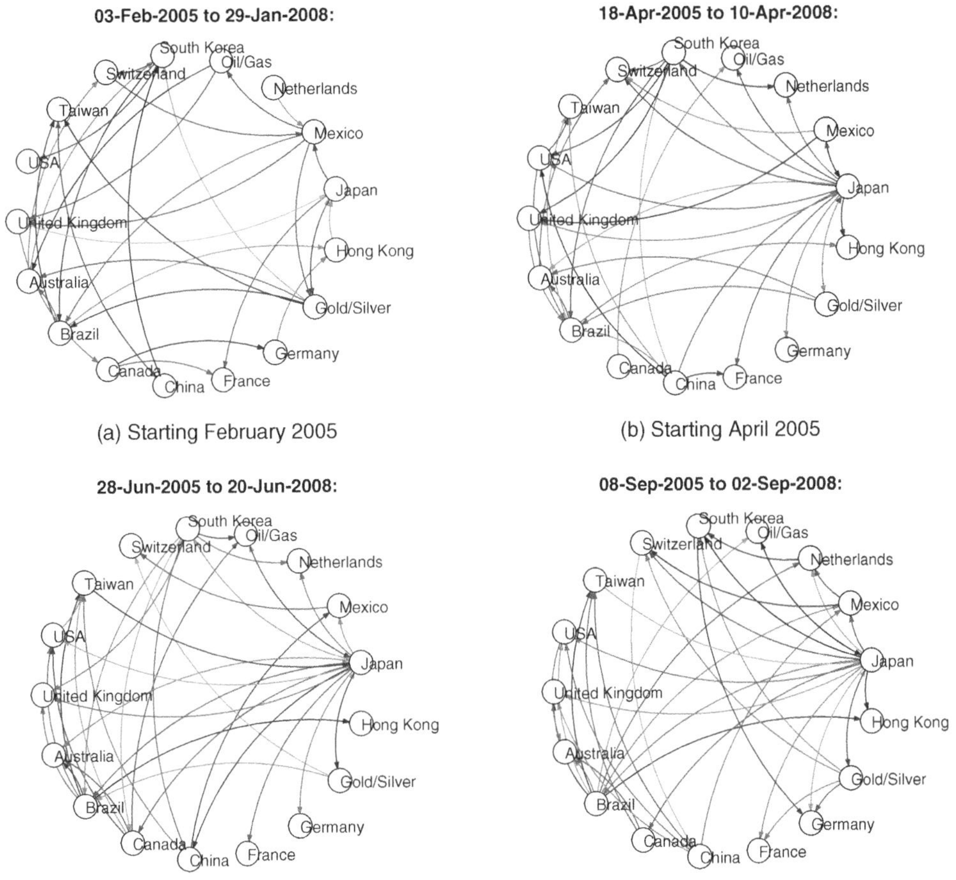

(a) Starting February 2005

(b) Starting April 2005

(c) Starting June 2005

(d) Starting September 2005

Figure 4.1 Causal CSM graphs of ETFs from iShares formed during four different 750-day periods in 2007–2008. Each graph moves the window of data over 50 business days in order to discover the effect of time on the causal networks. The lag used for VAR spans the 5 days (i.e., uses five features) preceding the target day. Each feature is a monthly return computed over the previous 22 business days.

From Table 4.1, we see that causal relationships change very quickly with time (periods are shifted by 50 business days, resulting in rapid changes to the causal networks). For example, we see that Asian dependencies almost always play a major role for the United States, but the specific Asian countries of interest change. In the early ongoing period of the financial crisis, Japan, South Korea, and China all played major roles, but after the brunt of the crisis occurred (2009 and onwards), South Korea and Germany had the biggest influences. In 2012, Japan and China superseded South Korea as the major dependencies, and 2 years later, Hong Kong played the bigger role. We also note from this analysis that, in 2012, Switzerland overtook Germany as the main European factor for US-listed equities.

Table 4.1 Results of TCM modeling on an ETF that tracks the S&P 500 between 2005 and 2015 that depicts the causal strength values for the three strongest relationships are given for each time period.

Time period	Factor 1	Factor 2	Factor 3	Factor 4
03-Feb-2005 to 29-Jan-2008	South Korea (0.9994)	Mexico (0.9264)	Australia (0.9216)	Gold & Silver (0.0136)
18-Apr-2005 to 10-Apr-2008	Japan (1)	South Korea (1)	China (0.968)	Gold & Silver (0.0162)
28-Jun-2005 to 20-Jun-2008	Brazil (1)	Japan (0.9998)	Canada (0.9994)	Gold & Silver (0.0248)
08-Sep-2005 to 02-Sep-2008	Brazil (1)	Japan (0.9998)	China (0.982)	Gold & Silver (0.0486)
17-Nov-2005 to 11-Nov-2008	China (0.9996)	Switzerland (0.993)	South Korea (0.987)	Gold & Silver (0.3646)
01-Feb-2006 to 26-Jan-2009	South Korea (0.987)	Netherlands (0.973)	Oil & Gas (0.8532)	Gold & Silver (0.5068)
13-Apr-2006 to 07-Apr-2009	Netherlands (0.9998)	South Korea (0.9958)	Oil & Gas (0.8332)	Gold & Silver (0.5622)
26-Jun-2006 to 18-Jun-2009	South Korea (0.998)	Netherlands (0.9878)	Oil & Gas (0.8874)	Gold & Silver (0.6324)
06-Sep-2006 to 28-Aug-2009	South Korea (0.979)	Germany (0.9704)	Oil & Gas (0.8208)	Gold & Silver (0.6126)
15-Nov-2006 to 09-Nov-2009	South Korea (0.981)	Germany (0.9734)	Oil & Gas (0.8078)	Gold & Silver (0.676)
31-Jan-2007 to 22-Jan-2010	South Korea (0.986)	Germany (0.9774)	Oil & Gas (0.7702)	Gold & Silver (0.6364)
13-Apr-2007 to 06-Apr-2010	South Korea (0.9962)	Germany (0.9792)	Gold & Silver (0.7118)	Oil & Gas (0.625)
25-Jun-2007 to 16-Jun-2010	South Korea (0.9922)	Germany (0.934)	Oil & Gas (0.762)	Gold & Silver (0.666)
05-Sep-2007 to 26-Aug-2010	South Korea (0.9828)	Germany (0.9524)	Oil & Gas (0.8022)	Gold & Silver (0.7132)
14-Nov-2007 to 05-Nov-2010	South Korea (0.9736)	Germany (0.9424)	Oil & Gas (0.7744)	Gold & Silver (0.745)
29-Jan-2008 to 19-Jan-2011	Germany (0.9868)	South Korea (0.9718)	Oil & Gas (0.898)	Gold & Silver (0.779)
10-Apr-2008 to 31-Mar-2011	Germany (0.953)	South Korea (0.9308)	Oil & Gas (0.9082)	Gold & Silver (0.7752)
20-Jun-2008 to 13-Jun-2011	Germany (0.9674)	Oil & Gas (0.9208)	South Korea (0.909)	Gold & Silver (0.8036)

02-Sep-2008 to 23-Aug-2011	Germany (0.9924)	Oil & Gas (0.9758)	South Korea (0.9178)	Gold & Silver (0.6388)
11-Nov-2008 to 02-Nov-2011	Brazil (0.9382)	Switzerland (0.8766)	South Korea (0.7974)	Gold & Silver (0.4654)
26-Jan-2009 to 17-Jan-2012	Japan (0.9996)	Switzerland (0.797)	South Korea (0.6104)	Gold & Silver (0.3424)
07-Apr-2009 to 28-Mar-2012	Japan (1)	China (0.99)	Switzerland (0.6638)	Gold & Silver (0.3712)
18-Jun-2009 to 08-Jun-2012	Japan (1)	China (0.9902)	Switzerland (0.749)	Gold & Silver (0.7332)
28-Aug-2009 to 20-Aug-2012	Japan (1)	China (0.9662)	Switzerland (0.8934)	Gold & Silver (0.6304)
09-Nov-2009 to 01-Nov-2012	Japan (1)	Switzerland (0.9404)	China (0.9386)	Gold & Silver (0.5754)
22-Jan-2010 to 15-Jan-2013	Japan (1)	Switzerland (0.9576)	China (0.8048)	Gold & Silver (0.4698)
06-Apr-2010 to 28-Mar-2013	Japan (0.9998)	Switzerland (0.9552)	China (0.8902)	Gold & Silver (0.244)
16-Jun-2010 to 10-Jun-2013	Switzerland (0.9996)	Japan (0.9996)	China (0.7656)	Gold & Silver (0.3372)
26-Aug-2010 to 20-Aug-2013	Switzerland (0.9966)	Brazil (0.9872)	China (0.621)	Gold & Silver (0.4168)
05-Nov-2010 to 30-Oct-2013	Switzerland (1)	Brazil (0.9964)	Hong Kong (0.9558)	Gold & Silver (0.185)
19-Jan-2011 to 13-Jan-2014	Switzerland (1)	Hong Kong (0.976)	Brazil (0.9572)	Gold & Silver (0.1658)
31-Mar-2011 to 26-Mar-2014	Switzerland (1)	Brazil (0.996)	Hong Kong (0.946)	Gold & Silver (0.2392)
13-Jun-2011 to 06-Jun-2014	Switzerland (1)	Hong Kong (0.9946)	Brazil (0.9668)	Gold & Silver (0.185)
23-Aug-2011 to 18-Aug-2014	Hong Kong (0.9604)	Taiwan (0.9598)	Switzerland (0.9564)	Gold & Silver (0.3786)
02-Nov-2011 to 28-Oct-2014	Switzerland (1)	Brazil (0.907)	Gold & Silver (0.787)	Taiwan (0.7262)

4.2 TCM

In this section, we exposit the basic methodology of TCM. In Section 4.1, we introduce Granger causality, which forms the underpinning for TCM. In Section 4.2, we expand the notion of Granger causality to grouping, which allows us to determine causality of an entire time series on other time series. Finally, in Section 4.3, we present experiments on synthetic datasets with known ground truths for the basic TCM methods.

4.2.1 Granger Causality and Temporal Causal Modeling

Granger causality (Granger, 1980), which was introduced by the Nobel prize–winning economist Clive Granger, has proven useful as an operational notion of causality in time-series analysis in the area of econometrics. Granger causality is based on the simple intuition that a cause should precede its effect; in particular, if a "source" time-series causally affects another "target" time-series, then the past values of the source should be helpful in predicting the future values of the target, beyond what can be predicted based only on the target's own past values. That is, a time-series **x** "Granger causes" another time series **y**, if the accuracy of regressing for **y** in terms of past values of **y** and **x** is (statistically) significantly better than that of regressing just with past values of **y**. Let $\{x_t\}_{t=1}^{T}$ denote the time-series variables for x and $\{y_t\}_{t=1}^{T}$ the same for y; then, to determine Granger causality, we first perform the following two regressions:

$$y_t \approx \sum_{l=1}^{L} a_l \cdot y_{t-l} + \sum_{l=1}^{L} b_l \cdot x_{t-l} \tag{4.1}$$

$$y_t \approx \sum_{l=1}^{L} a_l \cdot y_{t-l} \tag{4.2}$$

where L is the maximum "lag" allowed in past observations. To determine whether or not (4.1) is more accurate than (4.2) with a statistically significantly advantage, we perform an F-test or another suitable statistical test. We shall use the term *feature* to mean a time-series (e.g., **x**) and use temporal variables or lagged variables to refer to the individual values (e.g., x_t).

The notion of Granger causality, as introduced above, was defined for a pair of time-series; however, we typically want to determine causal relationships amongst several time-series. Naturally, we use graphical modeling over time-series data to determine conditional dependencies between the temporal variables, and obtain insight and constraints on the causal relationship between the time-series. One technique for graphical modeling is to use regression algorithms with variable selection to determine the causal relationships of each variable; for example, lasso (Tibshirani, 1996), which minimizes the sum of squared errors loss plus a sparsity-inducing ℓ_1 norm penalty on the regression coefficients. That is, we can consider the variable selection process in regression for y_t in terms of $y_{t-1}, x_{t-1}^1, x_{t-1}^2$ and so on, as an application of the Granger test on time-series y against the time-series $x^1, x^2, \ldots, x^p$.[1] When a pairwise Granger test is extended to facilitate multiple causal time-series, we can say that x^1 Granger causes y, if x_{t-l}^1 is selected for any time lag $l = \{1, 2, \ldots, L\}$ in the above variable selection. If

[1] Superscripts represent features; for example, x^p is the pth feature or the pth time-series.

such regression-based variable selection coincides with the conditional dependence between the variables, the above operational definition can be interpreted as the key building block of the temporal causal model.

4.2.2 Grouped Temporal Causal Modeling Method

In TCM, we are typically interested in knowing whether an entire time-series $x_{t-1}, x_{t-2}, \ldots, x_{t-L}$ provides information to help predict another time-series y_t; it is of little or no consequence if an individual lag l, x_{t-l} provides additional information for predicting y_t. From a modeling perspective, the relevant variable selection question is not whether an individual lagged variable is to be included in regression, but whether the lagged variables for a given time-series as a *group* are to be included. This would allow us to make statements of the form "**x** Granger causes **y**." Therefore, a more faithful implementation of TCM methods should take into account the group structure imposed by the time-series into the modeling approach and fitting criteria that are used in the variable selection process. This is the motivation for us to turn to the recently developed methodology, group lasso (Yuan and Lin, 2006), which performs variable selection with respect to model-fitting criteria that penalize intragroup and intergroup variable inclusion differently. This argument leads to the generic procedure of the grouped graphical Granger modeling method that is shown in Figure 4.2. We now describe both regularized and greedy regression methods that can serve as **REG** in Figure 4.2. Under regularized methods, we describe with both nongrouped and grouped variable selection techniques: lasso, adaptive lasso, and group lasso. Of these three, we prefer group lasso for the subprocedure **REG** in Figure 4.2 as it performs regression with group variable selection. Lasso and adaptive lasso are not grouped methods and will be used for comparison purposes in the simulations of Section 4.2.3.

1. **Input**
   - Time-series data $\{x_t\}_{t=1,\ldots,T}$ where each $\mathbf{x}_t$ is a p-dimensional vector.
   - A regression method with group variable selection, **REG**.
2. **Initialization**
   Initialize the adjacency matrix for the p features, that is, $G = \langle V, E \rangle$, where V is the set of p features (e.g., by all 0's).
3. **Selection**
   For each feature $\mathbf{y} \in V$, run **REG** on regressing for y_t in terms of the past lagged variables, $x_{t-L}, \cdots, x_{t-1}$, for all the features $\mathbf{x} \in V$ (including $\mathbf{y}$). That is, regress $(y_T, y_{T-1}, \cdots, y_{1+L})^T$ in terms of

$$\begin{pmatrix} x^1_{T-1} & \cdots & x^1_{T-L} & \cdots & x^p_{T-1} & \cdots & x^p_{T-L} \\ x^1_{T-2} & \cdots & x^1_{T-1-L} & \cdots & x^p_{T-2} & \cdots & x^p_{T-1-L} \\ \vdots & \vdots & \vdots & \vdots & \vdots\,\vdots & \vdots & \vdots \\ x^1_{L} & \cdots & x^1_{1} & \cdots & x^p_{L} & \cdots & x^p_{1} \end{pmatrix}$$

   where $V = \{\mathbf{x}^j, j = 1, \ldots, p\}$. For each feature $\mathbf{x}^j \in V$, place an edge $\mathbf{x}^j \rightarrow \mathbf{y}$ into E, if and only if $\mathbf{x}^j$ was selected as a group by the grouped variable selection method **REG**.

Figure 4.2 Generic TCM algorithm.

Regularized Least-Squares Methods

Let $\mathbf{y} = (y_1, \ldots, y_n)^T \in \mathbb{R}^n$ be a response vector and let $\mathbf{X} = [\mathbf{x}^1, \mathbf{x}^2, \ldots, \mathbf{x}^p] \in \mathbb{R}^{n\times p}$ be the predictor matrix, where $\mathbf{x}^j = (x_1^j, \ldots, x_n^j)^T, j = 1, \ldots p$, are the covariates. Typically the pairs $(\mathbf{X}_i, \mathbf{y}_i)$ are assumed to be independently identically distributed (i.i.d.) but most results can be generalized to stationary processes given a reasonable decay rate of dependencies such as conditions on the mixing rates. As we are interested in selecting the most important predictors in a high-dimensional setting, the ordinary-least-squares (OLS) estimate is not satisfactory; instead, procedures performing coefficient shrinkage and variable selection are desirable. A popular method for variable selection is the lasso (Tibshirani, 1996), which is defined as:

$$\hat{\theta}_{\text{lasso}}(\lambda) = \arg\ \min_{\theta}(\|\mathbf{y} - \mathbf{X}\theta\|^2 + \lambda\|\theta\|_1),$$

where λ is a penalty parameter. Here the ℓ_1 norm penalty $\|\theta\|_1$ automatically introduces variable selection, that is $\hat{\theta}_j(\lambda) = 0$ for some $j's$, leading to improved accuracy and interpretability. The lasso procedure – with lag $L = 1$ – has been used for causality analysis in Fujita *et al.* (2007). Unfortunately, lasso tends to overselect the variables and to address this issue, Zou (2006) proposed the adaptive lasso, a two-stage procedure solving:

$$\hat{\theta}_{\text{adapt}}(\lambda) = \arg\ \min_{\theta}\left(\|\mathbf{y} - \mathbf{X}\theta\|^2 + \lambda\sum_{j=1}^{p}\frac{|\theta_j|}{|\hat{\theta}_{\text{init},j}|}\right),$$

where $\hat{\theta}_{\text{init}}$ is an initial root-n consistent estimator such as that obtained by OLS or Ridge Regression. Notice that if $\hat{\theta}_{\text{init},j} = 0$ then $\forall\lambda > 0$, $\hat{\theta}_{\text{adapt}}(\lambda) = 0$. In addition if the penalization parameter λ is chosen appropriately, adaptive lasso is consistent for variable selection, and enjoys the "Oracle Property", which (broadly) signifies that the procedure performs as well as if the true subset of relevant variables were known. Our final regression method – group lasso (Yuan and Lin, 2006; Zhao *et al.*, 2006) – shines in situations where natural groupings exist between variables, and variables belonging to the same group should be either selected or eliminated as a whole. Given J groups of variables that partition the set of predictors, the group lasso estimate of Yuan and Lin (2006) solves:

$$\hat{\theta}_{\text{group}}(\lambda) = \arg\ \min_{\theta}\|\mathbf{y} - \mathbf{X}\theta\|^2 + \lambda\sum_{j=1}^{J}\|\theta_{\mathcal{G}_j}\|_2,$$

where $\theta_{\mathcal{G}_j} = \{\theta_k; k \in \mathcal{G}_j\}$ and $\mathcal{G}_j$ denotes the set of group indices. Notice that the penalty term $\lambda\sum_{j=1}^{J}\|\theta_{\mathcal{G}_j}\|_2$ in the above equation corresponds to the sparsity-inducing ℓ_1 norm applied to the J groups of variables, where the ℓ_2 norm is used as the intragroup penalty. In TCM, groups are of equal length as they correspond to the maximum lag that we wish to consider, so the objective does not need to account for unequal group size. By electing to use the ℓ_2 norm as the intragroup penalty, group lasso encourages the coefficients for variables within a given group to be similar in amplitude (as opposed to using the ℓ_1 norm, for example). Note that Granger causality always includes previous values of $\mathbf{y}$ in the model for $\mathbf{y}$; we omit this in the above equation as we assume that the effect of the previous values of $\mathbf{y}$ has "removed" from $\mathbf{y}$. This is done as there is no means to force $\mathbf{y}$ into the model for $\mathbf{y}$ using group-lasso.

Greedy Methods

In lieu of regularized least-squares methods, we could use greedy methods such as the orthogonal patching pursuit algorithm (Lozano *et al.*, 2009d) (or OMP) and its variant for group variable selection, group OMP (Lozano *et al.*, 2009d). These procedures are iterative and pick the best feature (or feature group) in each iteration, with respect to reduction of the residual error, and then re-estimate the coefficients, $\theta^{(k)}$, via OLS on the restricted sets of selected features (or feature groups). The group OMP procedure is described in Figure 4.3; the classical OMP version can be recovered from Figure 4.3 by considering groups of individual features. Note that – to strictly satisfy the definition of Granger causality – we can forcibly select $\mathbf{y}$ as one of the selected predictors of $\mathbf{y}$ by initializing $\theta^{(0)} = \mathbf{X}_{G_y}$ in Figure 4.3.

4.2.3 Synthetic Experiments

We conducted systematic experimentation using synthetic data in order to test the performance of group lasso and group OMP against that of the nongroup variants (lasso and adaptive lasso) for TCM. We present our findings in this section.

Data Synthesis

As models for data generation, we employed the vector autoregression (VAR) models (Enders, 2003). Specifically, let $\mathbf{x}_t$ denote the vector of all feature values at time t, then a VAR model is defined as $\mathbf{x}_t = \Theta_{t-1} \cdot \mathbf{x}_{t-1} + \ldots + \Theta_{t-T} \cdot \mathbf{x}_{t-T}$, where $\mathbf{\Theta}$s are coefficient matrices over the features. We randomly generate an adjacency matrix over the features that determines the structure

1. **Input**
   - The data matrix $\mathbf{X} = [\mathbf{f}_1, \ldots, \mathbf{f}_p] \in \mathbb{R}^{n \times p}$,
   - Group structure $G_1, \ldots, G_J$,
   - The response $\mathbf{y} \in \mathbb{R}^n$.
   - Precision $\epsilon > 0$ for the stopping criterion.
2. **Output**
   - The selected groups $\mathcal{G}^{(k)}$.
   - The regression coefficients $\theta^{(k)}$.
3. **Initialization**
   - $\mathcal{G}^{(0)} = \emptyset, \theta^{(0)} = 0$.
4. **Selection**

   For $k = 1, 2, \ldots$
   1. Let $r^{(k-1)} = \mathbf{X}\theta^{(k-1)} - \mathbf{y}$.
   2. Let $j^{(k)} = \arg\min_j \|\mathbf{r}^{(k-1)} - \mathbf{X}_{G_j}\mathbf{X}^+_{G_j}\mathbf{r}^{(k-1)}\|_2$. That is, $j^{(k)}$ is the group that minimizes the residual for the target $\mathbf{r}^{(k-1)}$. $\mathbf{X}^+_{G_j} = (\mathbf{X}^\top_{G_j}\mathbf{X}_{G_j})^{-1}\mathbf{X}_{G_j}$.
   3. **If** $(\|\mathbf{r}^{(k-1)} - \mathbf{X}_{G_j}\mathbf{X}^+_{G_j}\mathbf{r}^{(k-1)}\|_2 \leq \epsilon)$ **break**,
   4. Set $\mathcal{G}^{(k)} = \mathcal{G}^{(k-1)} \cup G_{j(k)}$; $\theta^{(k)} = \mathbf{X}^+_{\mathcal{G}^{(k)}}\mathbf{y}$.

   End

Figure 4.3 Method *group OMP*.

of the true VAR model, and then randomly assign the coefficients – $\mathbf{\Theta}$ – to each edge in the graph. We use the model thus generated on a random initial vector $\mathbf{x}_1$ to generate time-series data $\mathbf{X} = \{\mathbf{x}_t\}_{t=1,\ldots,T}$ of a specified length T. Following Arnold *et al.* (2007), during data generation, we made use of the following parameters: (1) *affinity*, which is the probability that each edge is included in the graph, was set at 0.2; and (2) *sample size per feature per lag*, which is the total data size per feature per maximum lag allowed, was set at 10. We sampled the coefficients of the VAR model according to a normal distribution with mean 0 and standard deviation 0.25. The noise standard deviation was set at 0.1, and so was the standard deviation of the initial distribution.

Evaluation

For all the variable selection subprocedures, the penalty parameter λ is tuned so as to minimize the *BIC* criterion (as recommended in Zou *et al.* (2006)), with degrees of freedom estimated as in Zou *et al.* (2006) for lasso and adaptive lasso, and as in Yuan and Lin (2006) for group lasso. Following Arnold *et al.* (2007), we evaluate the performance of all methods using the F_1 measure, viewing the causal modeling problem as that of predicting the inclusion of the edges in the true graph, or the corresponding adjacency matrix. Briefly, given precision P and recall R, the F_1-measure is defined as $F_1 = \frac{2PR}{(P+R)}$, and hence strikes a balance in the trade-off between the two measures.

Results

Table 4.2 summarizes the results of our experiments, which reports the average F_1 values over 18 runs along with the standard error. These results clearly indicate that there is a significant gap in performance between group lasso and the nongroup counterparts (lasso and adaptive lasso). Figure 4.4 shows some typical output graphs along with the true graph. In this particular instance, it is rather striking how the nongroup methods tend to overselect, whereas the grouped method manages to obtain a perfect graph.

Our experiments demonstrate the advantage of using the proposed TCM method (using group lasso and group OMP) over the standard (nongrouped) methods (based on lasso or adaptive lasso). Note that the nongrouped method based on lasso can be considered as an extension of the algorithm proposed in Fujita *et al.* (2007) to lags greater than one time unit.

Remark 4.1 *In the remainder of this chapter, we present various extensions of TCM, where we alternatively employ group OMP, group lasso, and Bayesian variants to serve as* ***REG*** *in Figure 4.2. We do so in order to expose the reader to the variety of group variable selection approaches. In general, the TCM extensions presented here can be extended to use any of the group OMP, group lasso, or Bayesian variants interchangeably.*

Table 4.2 The accuracy (F_1) and standard error in identifying the correct model of the two nongrouped TCM methods, compared to those of the grouped TCM methods on synthetic data

Method	lasso	adalasso	grplasso	grpOMP
Accuracy (F_1)	0.62 ± 0.09	0.65 ± 0.09	0.92 ± 0.19	0.92 ± 0.08

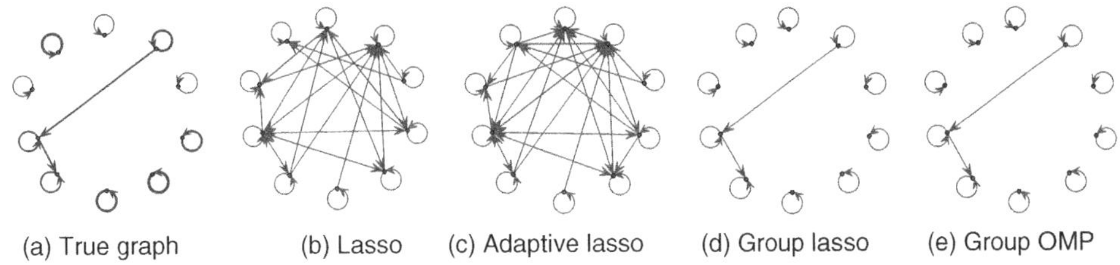

Figure 4.4 Output causal structures on one synthetic dataset by the various methods. In this example, the group-based method exactly reconstructs the correct graph, while the nongroup ones fail badly.

4.3 Causal Strength Modeling

In Section 4.2, we discussed modeling of causal relationships using Granger causality. In particular, in Figure 4.3, we discussed group OMP, a greedy mechanism to compute the causal graph. As causal relationships are defined by multiple lags of each time series, the next logical statistic to consider is the strength of a group, where a group of variables is defined by the different lags of the same time series. In the iShares example discussed in Section 4.1, the causal graphs made use of a notion called causal strength modeling (or CSM). To recap, the causal graphs in Figure 4.1 maintain only those relationships where the causal strength is greater than a given threshold. In this section, we define the notion of causal strength for each causal relationship discovered when *applying the group OMP method to TCM*.[2] The focus here is to determine the likelihood that a selected group of features is a true predictor for the model.

Method

Group OMP (see Figure 4.3) is a greedy feature selection procedure that produces, at each iteration, a new linear model that is fit using least-squares regression. Causal strength is a concept that tells information about the significance of a causal relationship. There are various ways to describe this relationship; a simple heuristic, for example, would be to measure coefficient magnitudes. In this framework, we measure causal strength as the likelihood that the coefficients for a group G in the linear model are nonzero. This likelihood is estimated by testing the probability of the null hypothesis $H_0 : \theta_G = 0$ for all coefficient indices in group G. In general, the distribution of the coefficients of a group of variables can be shown to be Gaussian (assuming Gaussian noise), and the null hypothesis can be tested using an F-test. However, as group OMP selects group G in a greedy fashion, the distribution of the coefficients of a group of variables is only *conditionally* Gaussian and the F-test offers conservative estimates for p-values, which leads to incorrectly declaring certain relationships as causal. To estimate the p-values accurately, we use Monte Carlo simulation. Loftus and Taylor (2014) analyze the conditional Gaussian distribution of the coefficients that group OMP (which they refer to as forward stepwise model selection) discovers at each iteration and discuss how to compute what they term the truncated χ test statistic. This statistic has a uniform distribution and leads to unbiased p-values for the corresponding hypothesis tests.[3]

[2] Although causal strengths can be estimated for group lasso, we have not tried it in practice and therefore omit discussion.

[3] While this procedure is significantly less computationally intensive, the Monte Carlo simulations are sufficient for our purposes. However, we have not had the opportunity to try out this method.

The proxy for testing whether the coefficients of a group added at iteration k are zero tests the statistic $\max_{j \in \bar{J}_{sel}} \|\mathbf{X}^*_{G_j} \hat{\mathbf{r}}^{(k-1)}\|_2$ as a proxy for $\|\boldsymbol{\theta}^{(k)}_{G_j}\|_2$, where $\bar{J}_{sel}$ is the set of remaining groups that can be selected and includes the group selected at iteration k, and $\hat{\mathbf{r}}^{(k-1)}$ is the normalized residual $\mathbf{X}\boldsymbol{\theta}^{(k-1)} - \mathbf{y}$. We want to test whether the normalized residual $\hat{\mathbf{r}}^{(k-1)}$ is noise or contains information. At each iteration, the probability that the null hypothesis holds is estimated by Monte Carlo simulation. First, random vectors are generated in order to create an empirical distribution of $\max_{j \in \bar{J}_{sel}} \|\mathbf{X}^*_{G_j} \mathbf{z}\|_2$ where $\mathbf{z}$ has a standard normal distribution. This represents the distribution of the above statistic that would be observed if the normalized residual were Gaussian. Then, the probability that the null hypothesis should be rejected can be computed from the quantiles of this empirical distribution. Refer to Loftus and Taylor (2014) for further discussion of this Monte Carlo estimation as well as their novel test statistic for testing the null hypothesis. Note that our values of causal strength are defined as $1 - p$ where p is the p-value corresponding to the null hypothesis.

4.4 Quantile TCM (Q-TCM)

The causal models discussed in Section 4.2 consider the causal relationship of the conditional mean of the responses given predictors; yet in many relevant applications, interest lies in the causal relationships for certain quantiles. These relationships may differ from those for the conditional mean or might be more or less pronounced. Therefore, it is critical to develop a complete understanding of the conditional distributions of the responses based on their predictors. In addition, real-world time-series often deviate from the Gaussian distribution, while quantile estimation obtained via the distribution function of the conditional mean regression models is very sensitive to these distributional assumptions. However, quantile regression does not rely on a specific distributional assumption on the data and provides more robust quantile estimation, and is thus more applicable to real-world data. In view of the above desiderata, we present in this section a quantile TCM approach, which extends the traditional VAR model to estimate quantiles, and promotes sparsity by penalizing the VAR coefficients (Aravkin *et al.*, 2014). In particular, we extend the group OMP algorithm in Figure 4.3 by replacing the ordinary-least-squares solution for $\hat{\boldsymbol{\theta}}^{(k)}$ at each iteration k with the solution to quantile regression. Section 4.4.1 details the new algorithm for quantile TCM, and Section 4.4.2 uses quantile TCM to perform feature selection on the example from Section 4.1 with additional outliers.

4.4.1 *Modifying Group OMP for Quantile Loss*

Overview

Figure 4.3 details the group OMP algorithm for ℓ_2 norm loss function. We modify two steps of this algorithm in order to apply it to Q-TCM. First, we generalize the group selection step (4.4.1) for the quantile loss function. Second, the refitting step (4.4.3) is modified to learn the new linear model using a quantile loss function. We begin with a discussion of quantile regression. Let $\mathbf{y}$ and $\mathbf{X}$ denote the response vector and data matrix, respectively. Quantile regression assumes that the τ-th quantile is given by

$$F^{-1}_{\mathbf{y}|\mathbf{X}}(\tau) = \mathbf{X}\bar{\boldsymbol{\theta}}_\tau \tag{4.3}$$

where $\bar{\theta}_\tau \in \mathbb{R}^p$ is the coefficient vector that we want to estimate in p dimensions and $F_{\mathbf{y}|\mathbf{X}}$ is the cumulative distribution function for a multivariate random variable with the same distribution as $\mathbf{y}|\mathbf{X}$. Let $\mathbf{r} = \mathbf{y} - \mathbf{X}\boldsymbol{\theta}$ be the vector of residuals. Quantile regression is traditionally solved using the following "check-function":

$$c_\tau(\mathbf{r}) = \sum_i (-\tau + 1\{r_i \geq 0\})r_i,$$

where the operations are taken element-wise; note that setting $\tau = 0.5$ yields the least absolute deviation (LAD) loss. Denote the quantile regression solution for a linear model as:

$$\hat{\boldsymbol{\theta}}_{QR,X}(\mathcal{G}, \mathbf{y}) = \arg\ \min_{\boldsymbol{\theta}} c_\tau(\mathbf{y} - \mathbf{X}_\mathcal{G}\boldsymbol{\theta}).$$

Quantile regression can be solved efficiently using various methods; for example, the regression problem can be rewritten as a linear program. In the case of Q-TCM, quantile regression is used to fit a linear model with the currently selected groups at each iteration. Thus, each iteration solves a larger regression problem, but in practice the total number of groups selected in most applications tends to be small. Our implementation uses an interior point (IP) method described in Aravkin *et al.* (2013) to solve the quantile regression problem. In short, the IP method rewrites the check function in a variational form,

$$c_\tau(\mathbf{r}) = \max_{\mathbf{u}} \{\langle \mathbf{u}, \mathbf{r}\rangle\ :\ \mathbf{u} \in [-\tau, 1-\tau]^n\},$$

and applies the Newton method to solve the optimality conditions of the resulting min-max optimization problem. Structure of the problem is crucial to an efficient implementation. We now discuss how to modify Algorithm 4.3 for Q-TCM. The group selection step is generalized to select the group that maximizes the projection onto the direction of steepest descent (i.e., gradient) with respect to the loss function. In the case of Q-TCM, the selection step becomes:

$$j^{(k)} = \arg\ \max_j \|\mathbf{X}^*_{G_j}((1-\tau)(\mathbf{r}^{(k-1)})_+ + \tau(\mathbf{r}^{(k-1)})_-)\|_2,$$

where $\mathbf{r}^{(k-1)} = \mathbf{y} - \mathbf{X}\boldsymbol{\theta}^{(k-1)}$ and $(1-\tau)(\mathbf{r}^{(k-1)})_+ + \tau(\mathbf{r}^{(k-1)})_-$ is a subgradient of the quantile loss evaluated at $\mathbf{r}^{(k-1)}$. The second change replaces the refitting step with $\boldsymbol{\theta}^{(k)} = \hat{\boldsymbol{\theta}}_{QR,\mathbf{X}}(\mathcal{G}, y)$. The fully modified version of group OMP tailored for Q-TCM is given in Figure 4.5.

4.4.2 Experiments

We analyze the iShares dataset used in Section 4.1, after introducing a few outliers, and illustrate how using a quantile loss function can discover the same causal relationships in the presence of outliers as an l_2 loss function can discover without the outliers (but cannot discover with them). United States The focus is on learning a model for the United States for the period April 18, 2005, through April 10, 2008 (the second row of Table 4.1). In Section 4.1, we showed that a TCM model for the United States is (statistically) significantly improved by including time-series data pertaining to Japan, South Korea, and China. This is a period where Asian-traded companies have a major influence on US-traded companies. *We introduce three outliers into the original data.* Figure 4.6 displays the original time-series, during the period of interest, with the noisy dates represented by red circles. Note that each data point is a monthly

1. **Input**
   - Data $\mathbf{X} = [\mathbf{f}_1, \ldots, \mathbf{f}_p] \in \mathbb{R}^{n \times p}$
   - Group structure $G_1, \ldots, G_J$, such that $\mathbf{X}^*_{G_j}\mathbf{X}_{G_j} = \mathbf{I}_{d_j}$.
   - The response $\mathbf{y} \in \mathbb{R}^n$
   - Precision $\epsilon > 0$ for the stopping criterion.
2. **Output**
   - The selected groups $\mathcal{G}^{(k)}$.
   - The regression coefficients $\boldsymbol{\theta}^{(k)}$.
3. **Initialization**
   $\mathcal{G}^{(0)} = \emptyset$, $\boldsymbol{\theta}^{(0)} = 0$.
4. **Selection**
   For $k = 1, 2, \ldots$
   - Let $j^{(k)} = \arg\max_j \|\mathbf{X}^*_{G_j}((1-\tau)(\mathbf{r}^{(k-1)})_+ + \tau(\mathbf{r}^{(k-1)})_-)\|_2$, where $\mathbf{r}^{(k-1)} = \mathbf{y} - \mathbf{X}\boldsymbol{\theta}^{(k-1)}$.
   - **If** $(\|\mathbf{X}^*_{G_{j(k)}}(\mathbf{X}\boldsymbol{\theta}^{(k-1)} - \mathbf{y})\|_2 \le \epsilon)$ **break**,
   - Set $\mathcal{G}^{(k)} = \mathcal{G}^{(k-1)} \cup G_{j^{(k)}}$. Let $\boldsymbol{\theta}^{(k)} = \hat{\boldsymbol{\theta}}_{QR,X}(\mathcal{G}, \mathbf{y})$.

   End

Figure 4.5 Method *Quantile group OMP.*

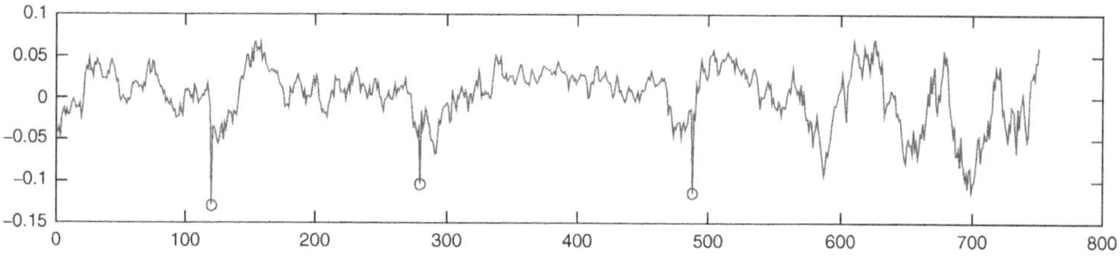

Figure 4.6 Log-returns for ticker IVV (which tracks S&P 500) from April 18, 2005, through April 10, 2008. Outliers introduced on 10/26/2005, 12/14/2007, and 01/16/2008 are represented by red circles.

return and that consecutive points are consecutive days, so that the period for computing the return overlaps on all but one day (and hence there are long periods of correlation as opposed to daily returns that would appear as white noise). While these outliers have been introduced, they are on the same magnitude with the largest magnitude dates of the original data. Outlier detection algorithms could likely detect at least two of the outliers simply because they occur suddenly. The outlier on 01/16/2008 occurs during an event with an already highly negative return.

A look at the data shows possible outliers and that Q-TCM should be considered. The grid of quantiles considered here is a default grid that can be used when no other information is known. Given extra information about the noise, a particular set of quantiles could be considered. We consider the following quantiles: $\tau \in \{0.10, 0.25, 0.35, 0.50, 0.65, 0.75, 0.90\}$. Models here are limited to selecting a total of five regressor time-series, including the target IVV (which tracks S&P 500), so only four additional time-series can be included. The selected regressors for each of the models after running Q-TCM and TCM are given in Table 4.3.

Table 4.3 Time-series selected for IVV, which tracks S&P 500, using Q-TCM and TCM on noisy data. The correct features are South Korea, Japan, and China, which are discovered by Q-TCM

Q-TCM $\tau = 0.10$	Q-TCM $\tau = 0.25$	Q-TCM $\tau = 0.35$	Q-TCM $\tau = 0.50$	Q-TCM $\tau = 0.65$	Q-TCM $\tau = 0.75$	Q-TCM $\tau = 0.90$	TCM
Gld & Slvr	Gld & Slvr	Gld & Slvr	Gld & Slvr	Gld & Slvr	Gld & Slvr	Gld & Slvr	Gld & Slvr
Japan	Oil & Gas	Japan	South Korea	South Korea	South Korea	South Korea	Taiwan
Brazil	France	Mexico	Japan	Japan	Japan	Japan	Canada
South Korea	Mexico	South Korea	China	China	China	China	China

Table 4.4 MSE on test period for Q-TCM and TCM models for IVV on noisy data

Q-TCM $\tau = 0.10$	Q-TCM $\tau = 0.25$	Q-TCM $\tau = 0.35$	Q-TCM $\tau = 0.50$	Q-TCM $\tau = 0.65$	Q-TCM $\tau = 0.75$	Q-TCM $\tau = 0.90$	TCM
0.1229e-3	0.1246e-3	0.1226e-3	0.1228e-3	0.1226e-3	0.1225e-3	0.1226e-3	0.3956e-3

In order to select the best model, we use data for the three months following the model generation period to learn which model fits best. The test period is 04/11/2008 through 07/11/2008. The Q-TCM and TCM models learned above are used to predict returns in the test period, and the mean squared error (MSE) is computed for the predictions on the test data for each model. The losses for the seven Q-TCM models and TCM are given in Table 4.4. TCM clearly has the worst loss on the test data and is thus not a good model to use. The best model, by a very small margin, is the Q-TCM model for the 0.90 quantile, so that model is selected. As we have seen here, this model selects Japan, South Korea, and China as significant factors, which is consistent with the uncorrupted data (see Section 4.1). Note that models with each of the quantiles $\tau \in \{0.50, 0.65, 0.75, 0.90\}$ select the same model as TCM selected on the uncorrupted data (which, while not a ground truth, is an estimate). In this example, the squared loss grossly penalized the outliers, but various versions of an absolute value penalty (i.e., quantile penalty at different quantiles) would have sufficed. With additional noise, Q-TCM was required here to learn a sufficient model, whereas if TCM was used here with the default squared loss, then Taiwan, Canada, and China would have been selected instead of the model selected on the uncorrupted data.

4.5 TCM with Regime Change Identification

In Section 4.2, we described TCM to accurately model causal relationships. However, we did not attempt to capture time-dependent variations in causal relationships; in Section 4.1, we saw an example of iShares, where the causal relationships vary with time. In this section, we extend TCM to incorporate such temporal information and identity regime changes (Jiang *et al.*, 2012). Formally, the main goal of this section is to describe a computationally efficient methodology to model the time-varying dependency structure underlying multivariate time-series data, with a particular focus on regime change identification. For this purpose, we marry the TCM framework with the Markov switching modeling framework; specifically, we describe a Bayesian Markov switching model for estimating TCMs (MS-TCM).

The key idea is to introduce a latent state variable that captures the regime from which observations at each time period are drawn (Chib, 1998). We allow this latent variable to return to any previous state, which closely resembles reality and allows one to overcome sample scarcity by borrowing strength across samples that are not adjacent in time. Each regime is governed by a TCM model. For group variable selection with TCM, we extend a hierarchical Bayesian framework that adds flexibility to the original group lasso (Yuan and Lin, 2006; Zhao *et al.*, 2006). Briefly, a hierarchical prior is specified for the regression coefficients, which results in *maximum a posteriori* (MAP) estimation with sparsity-inducing regularization; this can be seen as an iteratively reweighted adaptive group lasso estimator. Here, adaptivity refers to the fact that the penalty amount may differ across groups of regression coefficients, similar to adaptive lasso (Zou, 2006). Moreover, the penalty parameter λ is iteratively updated, therefore alleviating the need for parameter tuning (as opposed to non-Bayesian approaches). An additional benefit of such a quasi-Bayesian approach is its computational efficiency, which allows for graceful accommodation of high-dimensional datasets.

By combining a Markov-switching framework with Bayesian group lasso, MS-TCM provides a natural and integrated modeling framework to both capture regime changes and estimate TCM. The rest of this section is laid out as follows. In Section 4.5.1, we present the combined Markov switching model for TCM. In Section 4.5.2, we present algorithms to efficiently solve the combined model. In Sections 4.5.3 and 4.5.4, we present experiments that demonstrate the power of TCM with regime change identification.

4.5.1 Model

Markov-switching Model for TCM

To extend the TCM model in Section 4.2, we propose a Markov switching VAR model as follows: introduce a latent state variable S_t, $S_t \in \{1, 2, \ldots, K\}$ for each time point, where K is the total number of possible states and S_t stands for the state at time t. Given the state variables $\{S_t\}$, we can model the observed data $y_{j,t}$ (j^{th} time series at time t) using the VAR model,

$$y_{j,t} = \sum_{i=1}^{p} \sum_{l=1}^{L} \theta_{ijS_t,l} y_{i,t-l} + \epsilon_{j,t}; \quad \epsilon_{j,t} \sim N(0, \sigma_{S_t}^2). \tag{4.4}$$

As before, p is the number of features, L is the maximum lag, and θ_{ijS_tl} is the coefficient of the lth lagged variable of the ith time series for the model of the jth response variable when the regime is given by S_t. Note that the state variables $S_t \in \{1, \ldots, K\}$ are defined jointly on all responses; that is, they do not depend on j. This introduces a tight coupling amongst models for the different time-series, which is a departure from Sections 4.2 and 4.4 where a model for each response could be estimated independently. Note that without introducing such coupling amongst the different responses, we would not be able to define and identify regimes and associated change points that are common across all responses. The states S_t are modeled as a Markov chain using $\mathbf{P} \in \mathbb{R}^{K \times K}$ as the transition probability matrix, where P_{ij} is the transition probability from state i to j:

$$P_{ij} = \mathbb{P}(S_t = i | S_{t-1} = j), \forall i, j \in \{1, \ldots, K\}.$$

We do not impose any restrictions on the structure of $\mathbf{P}$, which allows for the random process to go back to a previous state or forward to a new state (unlike in Chib (1998). From (4.4), if two time points are from the same state, they will have the same set of autoregression coefficients; that is, $\theta_{ijS_{t1}l} = \theta_{ijS_{t2}l} = \theta_{ijkl}$ iff $S_{t1} = S_{t2} = k$. For simplicity, we denote the regression coefficients at state k by θ_{ijkl} in the rest of this chapter.

Temporal Causal Modeling via the Bayesian Group Lasso

To map the VAR model coefficients into the dependency structure of the TCMs, we make use of a group lasso technique for variable group selection. For a given (i, j, k), we define $\boldsymbol{\theta}_{ijk} = [\theta_{ijk1}, \ldots, \theta_{ijkL}]$ as a coefficient group. We adapt the Bayesian hierarchical framework for group variable selection in Lee *et al.* (2010) as follows:

$$\begin{aligned}
\boldsymbol{\theta}_{ijk}|\sigma^2_{ijk} &\sim N(0, \sigma^2_{ijk}),\\
\sigma^2_{ijk}|\tau_{ijk} &\sim \mathrm{G}\left(\frac{L+1}{2}, 2\tau^2_{ijk}\right),\\
\tau_{ijk}|a_{ijk}, b_{ijk} &\sim \mathrm{IG}(a_{ijk}, b_{ijk}),
\end{aligned} \tag{4.5}$$

where $\mathrm{G}(a, b)$ represents a gamma distribution with density function $f(x) = x^{a-1}b^{-a}\Gamma(a)^{-1}$ exp $(-x/b)$, and $\mathrm{IG}(a, b)$ represents an inverse gamma distribution whose density function is $f(x) = \frac{b^a}{\Gamma(a)}x^{-a-1}\exp\,(-b/x)$. This hierarchical formulation implies an adaptive version of the group lasso algorithm and allows for automatic update of the smoothing parameters. As suggested in Lee *et al.* (2010), $\boldsymbol{\theta}_{ijk}$ can be estimated by the following MAP estimate:

$$\mathrm{argmax}_{\boldsymbol{\theta}_{ijk}} \log\, \mathcal{L}(\boldsymbol{\theta}_{ijk}|a_{ijk}, b_{ijk}) + \log\, \mathbb{P}(\boldsymbol{\theta}_{ijk}|a_{ijk}, b_{ijk})\,.$$

Integrating out σ^2_{ijk}, τ_{ijk} in (4.5), the marginal density for $\boldsymbol{\theta}_{ijk}$ can be written as

$$\mathbb{P}(\boldsymbol{\theta}_{ijk}|a_{ijk}, b_{ijk}) = \frac{(2b_{ijk})^{-L}\pi^{-(L-1)/2}\Gamma(L + a_{ijk})}{\Gamma((L+1)/2)\Gamma(a_{ijk})}\left(\frac{\|\boldsymbol{\theta}_{ijk}\|_2}{b_{ijk}} + 1\right)^{-a_{ijk}-L},$$

where $\|\boldsymbol{\theta}_{ijk}\|_2 = \sqrt{\sum_{l=1}^{L}\theta^2_{ijkl}}$ is the ℓ_2 norm of $\boldsymbol{\theta}_{ijk}$. We note that the marginal distribution includes the ℓ_2 norm of $\boldsymbol{\theta}_{ijk}$, which is directly related to the penalty term in the group lasso. However, the marginal likelihood resulting from the hierarchical group lasso prior is not concave, which means that search for the global mode by direct maximization is not feasible. An alternative approach proposed in Lee *et al.* (2010) is to find local modes of the posterior using the expectation–maximization (EM) algorithm (McLachlan and Krishnan, 2008) with τ_{ijk} being treated as latent variables. This leads to the following iteratively reweighted minimization algorithm,

$$\boldsymbol{\theta}^{(m+1)}_{ijk} = \mathrm{argmax}_{\boldsymbol{\theta}_{ijk}} \log\, \mathbb{P}(\mathbf{Y}|\boldsymbol{\theta}_{ijk}) - w^{(m)}_{ijk}\|\boldsymbol{\theta}_{ijk}\|_2$$

where $w^{(m)}_{ijk} = \frac{a_{ijk}+L}{\|\theta^{(m)}_{ijk}\|_2 + b_{ijk}}$. For the parameters in the transition probability matrix, we assign the Dirichlet distribution as their prior distributions: for a given state k, the transition probabilities to all possible states $\mathbf{P}_k\cdot = [P_{k1}, \ldots, P_{kK}]$ take the form $\mathbf{P}_k\cdot \propto \prod_{k'=1}^{K} P_{kk'}^{\alpha_{k'}-1}$, where $\alpha_{k'}$ are the hyperparameters in the Dirichlet distribution. A popular choice is $\alpha_{k'} = 1$, corresponding to a noninformative prior on the $P_{kk'}$s. The Dirichlet distribution – the conjugate prior for the multinomial distribution – is a popular prior for the probabilities of discrete random variables due to its computational efficiency. Note that the noninformative prior on the transition probability does not imply that the states are equally likely at a transition; the transition probabilities will also be updated according to the data likelihood. Finally, to complete the Bayesian hierarchical model, we assign the following noninformative prior to the variances, $q(\sigma_k^2) \propto \frac{1}{\sigma_k^2}$.[4]

Choosing the Number of States

An important parameter in the previous sections was K, the number of states in the Markov switching network. To determine this, we utilize the Bayesian information criterion (BIC) (Schwarz, 1978), which is defined as $-2\log\ \mathcal{L}(\hat{\psi}_K) + d_K \log\ (N)$, where $\log\ \mathcal{L}(\hat{\psi}_K)$ is the log likelihood of the observed data under $\hat{\psi}_K$, d_K is the number of parameters, and N is the number of observations. The first term in BIC measures the goodness of fit for a Markov model with K states, while the second term is an increasing function of the number of parameters d_K, which penalizes the model complexity. Following (Yuan and Lin, 2006), the complexity of our model with an underlying group sparse structure can be written as:

$$d_K = \sum_{k=1}^{K} \sum_{i=1}^{p} \sum_{j=1}^{p} I(\|\theta_{ijk}\| > 0) + \sum_{k=1}^{K} \sum_{i=1}^{p} \sum_{j=1}^{p} \frac{\|\theta_{ijk}\|}{\|\theta_{ijk}^{LS}\|}(L-1),$$

where θ_{ijk}^{LS} are the parameters estimated by ordinary-least-squares estimates. We thus estimate $\hat{K}$, the total number of states, by the value that has minimum BIC value.

4.5.2 Algorithm

Let $\Theta = \{\theta_{ijk}\}_{i,j=1,\ldots,d;k=1,\ldots,K}$, be the tensor of parameters, $\sigma^2 = \{\sigma_k^2\}_{k=1,\ldots,K}$, and recall that $\mathbf{P}$ is the transition matrix and $\mathbf{Y}$ is the observed data. The unknown parameters in our model are $\psi = (\Theta, \sigma^2, \mathbf{P})$, which we estimate using MAP estimates obtained by maximizing the posterior distribution $q(\psi|\mathbf{Y})$. In this section, we develop an efficient algorithm to find the MAP estimates using an EM approach (McLachlan and Krishnan, 2008) where the state variables S_t are treated as missing data. When the goal is to find MAP estimates, the EM algorithm converges to the local modes of the posterior distribution by iteratively alternating between an expectation (E) step and a maximization (M) step, as follows. In the E-step, we compute the expectation of the joint posterior distribution of latent variables and unknown parameters, conditional on the observed data, denoted by $Q(\psi; \psi^{(m)})$. Let $\mathbf{y}_t$ be the p-dimensional vector of observations at time t, and $\mathbf{Y}_{t_1:t_2}$ is the collection of the measurements from time t_1 to t_2. For simplicity, we set $\mathbf{D}_0 = \{\mathbf{y}_1, \ldots, \mathbf{y}_L\}$ to be the initial information consisting of the first L observations and then relabel $\mathbf{y}_t \Rightarrow \mathbf{y}_{t-L}$ for $t > L$. Then we have,

$$\begin{aligned} Q(\psi; \psi^{(m)}) &= \mathbb{E}_{\mathbf{S},\tau|\mathbf{Y},\mathbf{D}_0,\psi^{(m)}}[\log\ \mathbb{P}(\psi, S, \tau|\mathbf{Y}_{1:T}, \mathbf{D}_0)] \\ &= \sum_{t=1}^{T} \sum_{k=1}^{K} \mathcal{L}_{tk} \log\ f(\mathbf{y}_t|\mathbf{Y}_{t-1:t-L}, S_t = k, \theta, \sigma^2) \end{aligned}$$

[4] We use $q(\cdot)$ instead of the commonly used $p(\cdot)$ to denote probability distributions as we use p to refer to the number of features/time-series and P_{ij} to refer to the individual probabilities in $\mathbf{P}$, the probability transition matrix.

$$+\sum_{t=2}^{T}\sum_{k=1}^{K}\sum_{k'=1}^{K} H_{t,k'k}\log\, \mathbb{P}(S_t = k|S_{t-1} = k', \mathbf{P})$$
$$+\sum_{k=1}^{K}\mathcal{L}_{1k}\log\, \pi_k - \sum_{i=1}^{p}\sum_{j=1}^{p}\sum_{k=1}^{K} w_{ijk}\|\boldsymbol{\theta}_{ijk}\|_2$$
$$-\sum_{k=1}^{K}\log\, \sigma_k^2 + \sum_{k=1}^{K}\sum_{k'=1}^{K}(\alpha_k - 1)\log\, P_{k'k} + \text{constant},$$

where $f(\cdot)$ denotes the probability density function, $\mathcal{L}_{tk} = \mathbb{P}(S_t = k|\mathbf{Y}_{1:T}, \mathbf{D}_0, \boldsymbol{\psi}^{(m)})$ and $H_{t,k'k} = \mathbb{P}(S_{t-1} = k', S_t = k|\mathbf{Y}_{1:T}, \mathbf{D}_0, \boldsymbol{\psi}^{(m)})$ is the posterior probability of all hidden state variables, and $\pi_k = \mathbb{P}(S_1 = k|\mathbf{Y}_{1:T}, \mathbf{D}_0)$ is the probability of the initial state being k. In the E-step, the posterior probability $\mathcal{L}_{tk}$ and $H_{t,k'k}$ can be calculated using the three-step backward-and-forward algorithm (Baum *et al.*, 1970) as follows:

1. Compute the forward probability $\alpha_k^{(m+1)}(t) = \mathbb{P}(\mathbf{Y}_{1:t}, S_t = k|\mathbf{D}_0, \boldsymbol{\psi}^{(m)})$ by going forward iteratively in time

$$\alpha_k^{(m+1)}(1) = \mathbb{P}(\mathbf{y}_1, S_1 = k|\mathbf{D}_0, \boldsymbol{\psi}^{(m)})$$
$$= \pi_k^{(m)}\mathbb{P}(\mathbf{y}_1|S_1 = k, \mathbf{D}_0, \boldsymbol{\psi}^{(m)})$$
$$\alpha_k^{(m+1)}(t) = \mathbb{P}(\mathbf{Y}_{1:t}, S_t = k|\mathbf{D}_0, \boldsymbol{\psi}^{(m)})$$
$$= \sum_{k'=1}^{K} f(\mathbf{y}_t|\mathbf{Y}_{t-L:t-1}, S_t = k, \mathbf{D}_0, \boldsymbol{\psi}^{(m)}) \times P_{k'k}^{(m)}\alpha_{k'}^{(m+1)}(t-1)$$

2. Compute the backward probability $\beta_k^{(m+1)}(t) = f(\mathbf{Y}_{t+1:T}|\ \mathbf{Y}_{1:t}, \mathbf{D}_0, S_t = k, \boldsymbol{\psi}^{(m)})$ by going backward iteratively in time

$$\beta_k^{(m+1)}(T) = 1$$
$$\beta_k^{(m+1)}(t) = f(\mathbf{Y}_{t+1:T}|\mathbf{Y}_{1:t}, \mathbf{D}_0, S_t = k, \boldsymbol{\psi}^{(m)})$$
$$= \sum_{k'=1}^{K} f(\mathbf{y}_{t+1}|\mathbf{Y}_{t:t-L+1}, \mathbf{D}_0, S_{t+1} = k', \boldsymbol{\psi}^{(m)}) \times \beta_{k'}(t+1)P_{kk'}^{(m)}$$

3. Compute the posterior probability

$$\mathcal{L}_{tk}^{(m+1)} = \mathbb{P}(S_t = k|\mathbf{Y}_{1:T}, \mathbf{D}_0, \boldsymbol{\psi}^{(m)})$$
$$= \frac{\alpha^{(m+1)}(t)\beta^{(m+1)}(t)}{f(\mathbf{Y}_{1:T}|\mathbf{D}_0, \boldsymbol{\psi}^{(m)})},$$

and

$$H_{t,k'k}^{(m+1)} = \mathbb{P}(S_{t-1} = k', S_t = k|\mathbf{Y}_{1:T}, \mathbf{D}_0, \boldsymbol{\psi}^{(m)})$$
$$= f(\mathbf{y}_t|S_t = k, \mathbf{Y}_{t-1:t-L}, \boldsymbol{\psi}^{(m)}) \times \frac{\alpha_{k'}^{(m+1)}(t-1)p_{k'k}^{(m)}\beta_k^{(m+1)}(t)}{f(\mathbf{Y}|\mathbf{D}_0, \boldsymbol{\psi}^{(m)})}$$

In the M-step, we update $\boldsymbol{\psi}$ by maximizing $Q(\boldsymbol{\psi}; \boldsymbol{\psi}^{(m)})$, as follows.

1. The VAR coefficients $\boldsymbol{\theta}_{ijk}^{(m+1)}$ are estimated by minimizing

$$\sum_{t=1}^{T} \frac{1}{2} \mathcal{L}_{tk}^{(m+1)} \left(y_{j,t} - \sum_{i=1}^{p} \mathbf{x}_{ti} \boldsymbol{\theta}_{ijk} \right)^2 / \sigma_k^{2,(m)} + \sum_{i=1}^{p} w_{ijk}^{(m+1)} \|\boldsymbol{\theta}_{ijk}\|_2 ,$$

   where $\mathbf{x}_{ti} = (y_{i,t-1}, \ldots, y_{i,t-L})$ and the updated weights are calculated as $w_{ijk}^{(m+1)} = \frac{a_{ijk}+L}{\|\boldsymbol{\theta}_{ijk}^{(m)}\|_2 + b_{ijk}}$. This regularized minimization problem can be transformed into a standard group lasso formulation (Yuan and Lin, 2006) by appropriately rescaling $y_{j,t}$ and $\mathbf{x}_{ti}$. The resulting group lasso problem can be solved efficiently by using the optimization procedure proposed by Meier *et al.* (2008).
2. The variance $\sigma_{0,k}^{2,(m+1)}$ of each of the Markov states is updated as

$$\sigma_k^{2,(m+1)} = \sum_{j=1}^{p} \sum_{t=1}^{T} \frac{\mathcal{L}_{tk}^{(m+1)}}{pT_k^{(m+1)} + 2} \left(y_{j,t} - \sum_{i=1}^{p} \mathbf{x}_{ti} \boldsymbol{\theta}_{ijk}^{(m+1)} \right)^2$$

   where $T_k^{(m+1)} = \sum_{t=1}^{T} \mathcal{L}_{tk}^{(m+1)}$.
3. The transition probability $P_{k'k}^{(m+1)}$ is updated as

$$P_{k'k}^{(m+1)} = \frac{\sum_{t=1}^{T} H_{t,k'k}^{(m+1)} + \alpha_k - 1}{\sum_{t=1}^{T} \mathcal{L}_{tk'} + \sum_{k=1}^{K} (\alpha_k - 1)}$$

4. The initial probability $\pi_k^{(m+1)} = \mathcal{L}_{1k}^{(m+1)}$.

To summarize, the proposed EM algorithm is computationally efficient; the $Q(\boldsymbol{\psi}, \boldsymbol{\psi}^{(m)})$ can be derived in closed form in the E-step, the maximization in the M-step can be transformed into a standard group lasso formulation (Yuan and Lin, 2006), and the maximization can be carried out very efficiently by using the optimization procedure in Meier *et al.* (2008). The algorithm iterates between the E-step and M-step until it converges.

4.5.3 *Synthetic Experiments*

Data Synthesis

We investigate the performance of MS-TCM on synthetic data with respect to identifying the switching states and the resulting TCM networks. We considered $K = 2$ and $K = 5$ states while generating the synthetic data. The synthetic data are generated according to the following steps.

1. *Generate the state assignment sequence.* Instead of assuming that the true generative process for the states is a Markov switching process, we randomly sampled $T/60$ change points and randomly assign states to each block. We relax the Markov assumption to test if our model still enables us to identify the underlying true process under a more general and realistic condition.

2. *Generate a random directed graph (the true network) with a specified edge probability.* We generated a set of $p \times p$ adjacency matrices $\mathbf{A}_1, \ldots, \mathbf{A}_K$, where the entry $a_{ijk} = 1$ indicates that at Markov state k, $\mathbf{y}_i$ influences $\mathbf{y}_j$, and $a_{ijk} = 0$ otherwise. The value of each entry was chosen by sampling from a binomial distribution, where the probability that an entry equals to one was set to 0.2.
3. *Create a sparse Markov switching VAR model that corresponds to the Markov states and the networks generated in the previous two steps.* For each Markov state, the VAR coefficients $a_{ijkl}, l = 1, \ldots, L$ were sampled according to a normal distribution with mean 0 and SD 0.20 if $a_{ijkl} = 1$, and set to be 0 otherwise. The noise SD was set at 0.01 for all states.
4. *Simulate data from the above switching VAR model.* The sample size per feature per lag per state was set at 10. We considered $p = 12$ and maximum lags $L = 1$ and $L = 3$.

Evaluation

To evaluate the accuracy of the switching states estimation, we use the Rand index (Rand, 1971), where we treat the switching modeling problem as clustering T multivariate data vector $\mathbf{y}_t = (y_{1,t}, \cdots, y_{p,t})$ into K Markov states. The Rand index is often used to measure the clustering error and is defined as the fraction of all misclustered pairs of data vectors $(\mathbf{y}_t, \mathbf{y}_s)$. Letting C^* and $\hat{C}$ denote the true and estimated clustering maps, respectively, the Rand index is defined by $\mathcal{R} = \frac{\sum_{t<s} I(\hat{C}(\mathbf{y}_t, \mathbf{y}_s) \neq C^*(\mathbf{y}_t, \mathbf{y}_s))}{\binom{T}{2}}$. To evaluate the accuracy of the DBN estimation, we use the F_1 score: larger values of the Rand index and F_1 score indicate higher accuracy.

We compare our method (MS-TCM) with two comparison methods: Fused-DBN and TV-DBN. Fused-DBN is a change point detection method extending the method of Kolar *et al.* (2009) to the VAR setting, which estimates change points via fused lasso penalized regression and subsequently estimates sparse static networks for each segment separately. TV-DBN (Song *et al.*, 2009) is used, although the simulation setting is not favorable to it, only to illustrate that in real-world settings where networks are not smoothly and constantly varying over time, approaches like TV-DBN usually fall short.

Results

The results of our experiments are summarized in Table 4.5, where we show respectively the average F_1 score for MS-TCM, Fused-DBN, and TV-DBN, and the average Rand index for our method only (since to our knowledge our method is the only method allowing for regime identification in DBNs) over 100 runs, along with the standard errors. Overall, MS-TCM has very good accuracy under a varying number of Markov states. Rand indices do not change much from small- to large-number states, which suggests similar accuracy in change point detection. The F_1 scores, smaller under $L = 1$, suggest that the dependency structure may be more accurately recovered when the relationships involve multiple time lags, illustrating the value of our Bayesian group lasso subprocedure. This approach achieves significantly better F_1 accuracy than the other methods for all the cases considered. Computationally, our algorithm is more efficient when compared to Fused-DBN as Fused-DBN involves applying randomized lasso to a transformed data matrix of dimensions $T \times Tp$; this implies that the number of features effectively considered is T times higher than the actual number of features. This can be a significant impediment even in low-dimensional settings; for example, for 10 features and 1000 time points, one has to work with 10,000 features after the transformation. In addition, Fused-DBN is unable to leverage the grouping structure corresponding to dependencies with $L > 1$.

Table 4.5 Accuracy of comparison methods in identifying the correct Bayesian networks measured by the average Rand index and F_1 score on synthetic data with a varying number of Markov states (K) and lags (L). The numbers in the parentheses are standard errors

Type	Method	K = 2 L = 1	K = 2 L = 3	K = 5 L = 1	K = 5 L = 3
Rand index	MS-TCM	0.94 (0.02)	0.96 (0.02)	0.97 (0.005)	0.97 (0.06)
	Fused-DBN	N/A	N/A	N/A	N/A
	TV-DBN	N/A	N/A	N/A	N/A
F_1 score	MS-TCM	0.76 (0.05)	0.91 (0.12)	0.76 (0.03)	0.87 (0.17)
	Fused-DBN	0.48 (0.02)	0.50 (0.03)	0.53 (0.04)	0.52 (0.02)
	TV-DBN	0.36 (0.02)	0.38 (0.02)	0.50 (0.06)	0.40 (0.04)

Another strength of MS-TCM, compared to Fused-DBN and other change point–based methods proposed in the literature, is in the setting of regime change identification. In the existing methods, once the change points have been estimated, the coefficients are estimated individually for each interval. Hence, if some of the intervals between two subsequent change points are small (which may happen in many practical situations), the algorithms may be forced to work with extremely small sample size, thus leading to poor estimates. In contrast, our method considers all states and allows for return to previous states. It is thus able to borrow strength across a wider number of samples that may be far away in time.

We now turn our attention to the results obtained by MS-TCM, using the synthetic data with $K = 2$ and $L = 1$. According to BIC, the number of states K is estimated as 2. In the left panel of Figure 4.7, we show the Markov path estimated by our method and the true path for a particular simulation run, where the transition jumps highlighted in red in the true path (upper panel) are those missed by our method. As shown in the plot, MS-TCM is able to detect the change points with very little delay. We also observe that MS-TCM tends to miss a transition when the process remains in a single state for too short a duration. In practice, however, such transient jumps rarely happen or may be of less interest in real applications. In the right panel of Figure 4.7 we show the corresponding estimated Bayesian networks along with the true networks. In the true networks (left column), we highlighted in red the *false negatives* (edges that exist in the true graphs but are missed in the estimated graphs) and in the estimated networks (right column) we highlighted in green the *false positives* (edges that do not exist in the true graphs but are selected in the method). As we can see from the plot, the estimated Bayesian networks exhibit reasonable agreement with the true networks.

4.5.4 Application: Analyzing Stock Returns

To demonstrate MS-TCM's utility in finance, we apply MS-TCM on monthly stock return data from 60 monthly stock observations from 2004-10-01 to 2009-09-01 for 24 stocks in six

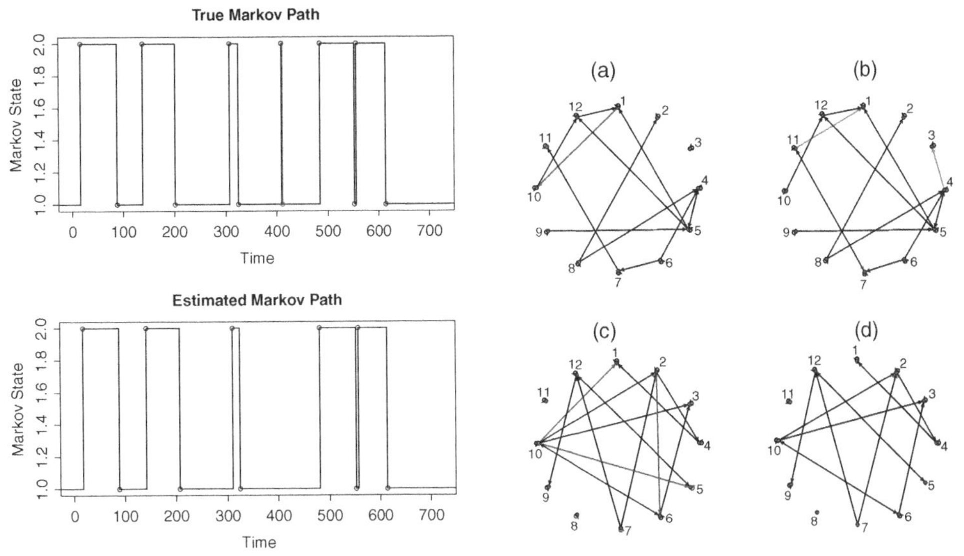

Figure 4.7 (Left) Output switching path on one synthetic dataset with two Markov states. Transition jumps missing in the estimated Markov path are highlighted in red. (Right) The corresponding output networks: (a) true network at state 1; (b) estimated network at state 1; (c) true network at state 2; and (d) estimated network at state 2. Edges coded in red are the false positives, and those in green are the false negatives.

industries. For simplicity, we consider that each state is associated to a VAR model with lag 1. MS-TCM identifies one change point at the 19th time point. The VAR model coefficients for the TCM models concerning Citigroup (C), before and after that change point, are presented in Figure 4.8 along with the model produced when assuming a single static causal model (i.e., running vanilla TCM). By examining the models, one can get some insights on the varying causal relationships. Another benefit of this approach is the potential to improve forecasting accuracy. For instance, if we treat the 60th time point as unobserved, then MS-TCM can forecast this point with relative error of 0.5%, while a static TCM approach has a higher relative error of 26%.[5]

4.6 Conclusions

In this chapter, we presented TCM, a method that builds on Granger causality and generalizes it to multivariate time-series by linking the causality inference process to the estimation of sparse VAR models. We also presented extensions to TCM that allow users to determine causal strengths of causal relationships, use quantile loss functions, and automatically identify and efficiently model regime changes with TCM. TCM and its extensions that were presented in this chapter can be an important tool in financial analysis.

[5] It is possible to leave out and predict more points than 1; we choose to predict just the 60th point for simplicity.

	Model 1	Model 2	Model all
C	-0.2627136	0.21964992	-0.1238139
KEY	0	0	0
WFC	0	0	0
JPM	0	0	0
SO	-1.7279345	-0.1946859	-0.7924643
DUK	-0.7895888	0	-0.2395446
D	0	0	0
HE	1.71362655	0	1.40037993
EIX	0.72517224	0	0.2604309
LUV	1.38827242	0	0.11521195
CAL	0	0	0
AMR	0.39029193	0.11319507	0.3189362
AMGN	-1.3494361	0.02012697	-0.1708315
GILD	0	0	-0.2630309
CELG	0	0	0
GENZ	0	0	0
BIIB	-0.8281948	0	-0.401407
CAT	0	0	0
DE	-0.0239606	0	0
HIT	-0.3431499	0	0
IMO	-0.0428805	0	0
MRO	0	0.00209084	0
HES	0	0	0
YPF	0.34016702	0	0
X.GSPC	0	0	0.41745692

Figure 4.8 Results of modeling monthly stock observations using MS-TCM. MS-TCM uncovered a regime change after the 19th time step; columns `Model 1` and `Model 2` contain the coefficients of the corresponding two TCM models. The column `Model all` gives the coefficients when plain TCM without regime identification is used. The symbols C, KEY, WFC, and JPM are money center banks; SO, DUK, D, HE, and EIX are electrical utilities companies; LUX, CAL, and AMR are major airlines; AMGN, GILD, CELG, GENZ, and BIIB are biotechnology companies; CAT, DE, and HIT are machinery manufacturers; IMO, HES, and YPF are fuel refineries; and X.GPSC is an index.

Further Reading

There are some extensions to TCM that were not covered in this chapter. For example, in order to expose the readers to a variety of approaches, we presented some extensions of TCM using group OMP, and others using group lasso or Bayesian variants. We encourage the readers to experiment with these methods to determine the "best" method for their specific problems. Other extensions that were not covered include extending TCM for modeling spatiotemporal data (Lozano *et al.*, 2009c), modeling multiple related time-series datasets (Liu *et al.*, 2010), and modeling nonlinear VAR models (Sindhwani *et al.*, 2013).

References

Aravkin, A. Y., Burke, J. V. and Pillonetto, G. (2013). Sparse/robust estimation and Kalman smoothing with non-smooth log-concave densities: modeling, computation, and theory. *Journal of Machine Learning Research*, 14, 2689–2728.

Aravkin, A.Y., Kambadur, A., Lozano, A.C., Luss, R. (2014). Orthogonal matching pursuit for sparse quantile regression. IEEE International Conference on Data Mining (ICDM).

Arnold, A., Liu, Y. and Abe, N. (2007). Temporal causal modeling with graphical Granger methods. Proceedings of the Thirteenth ACM SIGKDD International Conference on Knowledge Discovery and Data Mining. http://www.cs.cmu.edu/~aarnold/cald/frp781-arnold.pdf

Baum, L., Petrie, T., Soules, G. and Weiss, N. (1970). A maximization technique occurring in the statistical analysis of probabilistic functions of Markov chains. *The Annals of Mathematical Statistics*, 41(1), 164–171.

Biard, D. and Ducommun, B. (2008). Moderate variations in CDC25B protein levels modulate the response to DNA damaging agents. *Cell Cycle*, 7(14), 2234–2240.

Carpenter, J. and Bithell, J. (2000). Bootstrap confidence intervals: when, which, what? A practical guide for medical statisticians. *Statistics in Medicine*, 19, 1141–1164.

Chib, S. (1998). Estimation and comparison of multiple change-point models. *Journal of Econometrics*, 86, 221–241.

Dahlhaus, R. and Eichler, M. (2003). Causality and graphical models in time series analysis. In *Highly structured stochastic systems* (ed. P. Green, N. Hjort and S. Richardson). Oxford: Oxford University Press.

Davison, A.C. and Hinkley, D. (2006). *Bootstrap methods and their applications*. Cambridge Series in Statistical and Probabilistic Mathematics. Cambridge: Cambridge University Press.

Enders, W. (2003). *Applied econometric time series*, 2nd ed. Hoboken, NJ: John Wiley & Sons.

Fujita, A., *et al.* (2007). Modeling gene expression regulator networks with the sparse vector autoregressive model. *BMC Systems Biology*, 1, 39.

Furstenthal, L., Kaiser, B.K., Swanson, C. and Jackson, P.K.J. (2001). Cyclin E uses Cdc6 as a chromatin-associated receptor required for DNA replication. *Journal of Cell Biology*, 152(6), 1267–1278.

Granger, C. (1980). Testing for causality: a personal viewpoint. *Journal of Economic Dynamics and Control*, 2, 329–352.

Green, P.J. and Silverman, B.W. (1994). *Nonparametric regression and generalized linear models: a roughness penalty approach*. London: Chapman and Hall.

Jiang, H., Lozano, A.C. and Liu, F. (2012). A Bayesian Markov-switching model for sparse dynamic network estimation. Proceedings of 2012 SIAM International Conference on Data Mining. http://citeseerx.ist.psu.edu/viewdoc/download?doi=10.1.1.370.3422&rep=rep1&type=pdf

Kolar, M., Song, L. and Xing, E. P. (2009). Sparsistent learning of varying-coefficient models with structural changes. *Advances in Neural Information Processing Systems*, 1006–1014.

Lee, A., Caron, F., Doucet, A. and Holmes, C. (2010). A hierarchical Bayesian framework for constructing sparsity-inducing priors. http://arxiv.org/abs/1009.1914

Li, X., *et al.* (2006). Discovery of time-delayed gene regulatory networks based on temporal gene expression profiling. *BMC Bioinformatics*, 7, 26.

Liu, Y., Fan, S.W., Ding, J.Y., Zhao, X., Shen, W.L. (2007). Growth inhibition of MG-63 cells by cyclin A2 gene-specific small interfering RNA. *Zhonghua Yi Xue Za Zhi*, 87(9), 627–33 (in Chinese).

Liu, Y., Niculescu-Mizil, A., Lozano, A.C. and Lu, Y. (2010). Learning temporal graphs for relational time-series analysis. Proceedings of the 27th International Conference on Machine Learning (ICML). http://www.niculescu-mizil.org/papers/hMRF-ICML_final.pdf

Loftus, J. R. and Taylor J. E. (2014). A significance test for forward stepwise model selection. arXiv:1405.3920

Lozano, A.C., Abe N., Liu Y. and Rosset, S. (2009a). Grouped graphical Granger modeling for gene expression regulatory network discovery. Proceedings of 17th Annual International Conference on Intelligent System for Molecular Biology (ISMB) and Bioinformatics, 25 (12). http://bioinformatics.oxfordjournals.org/content/25/12/i110.short

Lozano, A.C., Abe N., Liu Y., Rosset, S. (2009b). Grouped graphical Granger modeling methods for temporal causal modeling. In *Proceedings of 15th ACM SIGKDD Conference on Knowledge Discovery and Data Mining 2009*, 577–586.

Lozano, A.C., Li, H., Niculescu-Mizil, A., Liu Y., Perlich C., Hosking, J.R.M. and Abe, N. (2009c). Spatial-temporal causal modeling for climate change attribution. In *Proceedings of 15th ACM SIGKDD Conference on Knowledge Discovery and Data Mining 2009*, 587–596.

Lozano, A.C., Swirszcz, G.M. and Abe, N. (2009d). Group orthogonal matching pursuit for variable selection and prediction. *Proceedings of the Neural Information Processing Systems Conference (NIPS)*. http://papers.nips.cc/paper/3878-grouped-orthogonal-matching-pursuit-for-variable-selection-and-prediction.pdf

McLachlan, G.J. and Krishnan, T. (2008). *The EM algorithm and extensions*. New York: Wiley-Interscience.

Meier, L., van de Geer, S. and Bühlmann, P. (2008). The group lasso for logistic regression. *Journal of the Royal Statistical Society*, 70(1), 53–71.

Meinshausen, N. and Buhlmann, P. (2006). High dimensional graphs and variable selection with the lasso. *Annals of Statistics*, 34(3), 1436–1462.

Meinshausen, N. and Yu, B. (2006). Lasso-type recovery of sparse representations for high-dimensional data. Technical Report, Statistics. Berkeley: University of California, Berkeley.

Mukhopadhyay, N.D. and Chatterjee, S. (2007). Causality and pathway search in microarray time series experiment. *Bioinformatics*, 23(4).

Ong, I.M., Glasner, J.D. and Page, D. (2002) Modelling regulatory pathways in *E. coli* from time series expression profiles. *Bioinformatics*, 18, S241–S248.

Opgen-Rhein, R. and Strimmer, K. (2007). Learning causal networks from systems biology time course data: an effective model selection procedure for the vector autoregressive process. *BMC Bioinformatics*, 8(Suppl. 2), S3.

Rand, W. (1971). Objective criteria for the evaluation of clusterings methods. *Journal of the American Statistical Association*, 66(336), 846–850.

Ray, D. and Kiyokawa, H. (2007). CDC25A levels determine the balance of proliferation and checkpoint response. *Cell Cycle*, 156(24), 3039–3042.

Salon, C., *et al.* (2007). Links E2F-1, Skp2 and cyclin E oncoproteins are upregulated and directly correlated in high-grade neuroendocrine lung tumors. *Oncogene*, 26(48), 6927–6936.

Sambo, F., Di Camillo, B. and Toffolo G. (2008). CNET: an algorithm for reverse engineering of causal gene networks. In: *Bioinformatics methods for biomedical complex systems applications*, 8th Workshop on Network Tools and Applications in Biology NETTAB2008, Varenna, Italy, 134–136.

Schwarz, G. E. (1978). Estimating the dimension of a model. *Annals of Statistics*, 6(2).

Segal, E., *et al.* (2003). Module networks: identifying regulatory modules and their condition-specific regulators from gene expression data. *Nature Genetics*, 34, 166–176.

Sindhwani, V., Minh, H.W., Lozano, A.C. (2013). Scalable matrix-valued kernel learning and high-dimensional nonlinear causal inference. In *Proceedings of the Conference on Uncertainty in Artificial Intelligence (UAI)*. http://arxiv.org/ftp/arxiv/papers/1408/1408.2066.pdf

Song, L., Kolar, M. and Xing, E. P. (2009). Time-varying dynamic Bayesian networks. *Advances in Neural Information Processing Systems*, 1732–1740.

Tibshirani, R. (1996). Regression shrinkage and selection via the lasso. *Journal of the Royal Statistical Society, Series B*, 58(1), 267–288.

Whitfield, M.L., *et al.* (2002). Identification of genes periodically expressed in the human cell cycle and their expression in tumors, *Molecular Biology of the Cell*, 13(6). Dataset available at http://genome-www.stanford.edu/Human-CellCycle/Hela/

Xu, X., Wang, L. and Ding, D. (2004). Learning module networks from genome-wide location and expression data. *FEBS Letters*, 578, 297–304.

Yamaguchi, R., Yoshida, R., Imoto, S., Higuchi, T. and Miyano, S. (2007). Finding module-based gene networks in time-course gene expression data with state space models. *IEEE Signal Processing Magazine*, 24, 37–46.

Yuan, M. and Lin, Y. (2006). Model selection and estimation in regression with grouped variables. *Journal of the Royal Statistical Society, Series B*, 68, 49–67.

Zou, H. (2006). The adaptive lasso and its oracle properties. *Journal of the American Statistical Association*, 101(476), 1418–1429.

Zou, H., Hastie, T. and Tibshirani, R. (2007). On the "degrees of freedom" of the lasso. *Annals of Statistics*, 35(5), 2173–2192.

Zhao, P., Rocha, G. and Yu, B. (2006). Grouped and hierarchical model selection through composite absolute penalties. http://statistics.berkeley.edu/sites/default/files/tech-reports/703.pdf

5

Explicit Kernel and Sparsity of Eigen Subspace for the AR(1) Process

Mustafa U. Torun, Onur Yilmaz and Ali N. Akansu
New Jersey Institute of Technology, USA

5.1 Introduction

Karhunen–Loeve Transform (KLT), also called Eigen decomposition or principal component analysis (PCA), is the optimal orthogonal subspace method (block transform) that maps wide-sense stationary (WSS) stochastic signals with correlations into nonstationary and pairwise uncorrelated transform coefficients. The coefficient with the highest variance corresponds to the most covariability among the observed signals, hence the most meaningful information (Akansu and Haddad, 1992). Therefore, the coefficients with large variances are kept and the ones with low variances corresponding to noise are discarded in noise-filtering and compression applications (Jolliffe, 2002). KLT basis functions are the eigenvectors of the given signal covariance matrix that define the corresponding unique eigen subspace. Therefore, it is a signal-dependent transform as opposed to some other popular transforms like discrete Fourier transform (DFT) and discrete cosine transform (DCT). DFT and DCT have their kernels that are independent of signal statistics. They are called fixed transforms, and their good performance with efficient implementations make them desirable choices for various applications (Akansu and Haddad, 1992). Fast implementation of KLT is of great interest to several disciplines, and there were attempts to derive closed-form kernel expressions for certain classes of stochastic processes. Such kernels for continuous and discrete stochastic processes with exponential autocorrelation function were reported in the literature (Davenport and Root, 1958; Pugachev, 1959a,b; Ray and Driver, 1970; Wilkinson, 1965). We focus on the discrete autoregressive order one, AR(1), and the process and derivation of its

Financial Signal Processing and Machine Learning, First Edition.
Edited by Ali N. Akansu, Sanjeev R. Kulkarni and Dmitry Malioutov.
© 2016 John Wiley & Sons, Ltd. Published 2016 by John Wiley & Sons, Ltd.

explicit eigen kernel, in this chapter. In Section 5.4, we investigate the sparsity of such eigen subspace and present a rate-distortion theory-based sparsing method. Moreover, we highlight the merit of the method for the AR(1) process as well as for the empirical correlation matrix of stock returns in the NASDAQ-100 index.

5.2 Mathematical Definitions

5.2.1 Discrete AR(1) Stochastic Signal Model

Random processes and information sources are often described by a variety of stochastic signal models, including autoregressive (AR), moving average (MA), and autoregressive moving average (ARMA) types. Discrete AR source models, also called all-pole models, have been successfully used in applications including speech processing for decades (Atal and Hanauer, 1971). The first-order AR model, AR(1), is a first approximation to many natural signals and has been widely employed in various disciplines. Its continuous analogue is called the Ornstein–Uhlenbeck (OU) process with popular use in physical sciences and mathematical finance (Doob, 1942; Uhlenbeck and Ornstein, 1930). The AR(1) signal is generated through the first-order regression formula written as (Akansu and Haddad, 1992; Kay, 1988):

$$x(n) = \rho x(n-1) + \xi(n), \tag{5.1}$$

where $\xi(n)$ is a zero-mean white noise sequence, that is,

$$\begin{aligned} E\{\xi(n)\} &= 0, \\ E\{\xi(n)\xi(n+k)\} &= \sigma_\xi^2 \delta_k. \end{aligned} \tag{5.2}$$

$E\{\cdot\}$ is the expectation operator, and δ_k is the Kronecker delta function. The first-order correlation coefficient, ρ, is real in the range of $-1 < \rho < 1$, and the variance of $x(n)$ is given as follows:

$$\sigma_x^2 = \frac{1}{(1-\rho^2)}\sigma_\xi^2. \tag{5.3}$$

The autocorrelation sequence of $x(n)$ is expressed as

$$R_{xx}(k) = E\{x(n)x(n+k)\} = \sigma_x^2 \rho^{|k|};\ k = 0, \pm 1, \pm 2, \ldots. \tag{5.4}$$

The resulting $N \times N$ Toeplitz autocorrelation matrix for the AR(1) process is expressed as

$$\mathbf{R}_x = \sigma_x^2 \begin{bmatrix} 1 & \rho & \rho^2 & \ldots & \rho^{N-1} \\ \rho & 1 & \rho & \ldots & \rho^{N-2} \\ \rho^2 & \rho & 1 & \ldots & \rho^{N-3} \\ \vdots & \vdots & \vdots & \ddots & \vdots \\ \rho^{N-1} & \rho^{N-2} & \rho^{N-3} & \ldots & 1 \end{bmatrix}. \tag{5.5}$$

5.2.2 Orthogonal Subspace

A family of linearly independent N orthonormal real discrete-time sequences (vectors), $\{\phi_k(n)\}$, on the interval $0 \leq n \leq N-1$, satisfies the inner product relationship (Akansu and Haddad, 1992)

$$\sum_{n=0}^{N-1} \phi_k(n)\phi_l(n) = \delta_{k-l} = \begin{cases} 1 & k = l \\ 0 & k \neq l \end{cases}. \tag{5.6}$$

Equivalently, the orthonormality can also be expressed on the unit circle of the complex plane, $z = e^{j\omega}; -\pi \leq \omega \leq \pi$, as follows:

$$\sum_{n=0}^{N-1} \phi_k(n)\phi_l(n) = \frac{1}{2\pi}\int_{-\pi}^{\pi} \Phi_k(e^{j\omega})\Phi_l(e^{j\omega})d\omega = \delta_{k-l}, \tag{5.7}$$

where $\Phi_k(e^{j\omega})$ is the discrete time Fourier transform (DTFT) of $\phi_k(n)$. In matrix form, $\{\phi_k(n)\}$ are the rows of the transform matrix, also called basis functions:

$$\Phi = [\phi_k(n)] \ : \ k, n = 0, 1, \ldots, N-1, \tag{5.8}$$

with the matrix orthonormality property stated as

$$\Phi\Phi^{-1} = \Phi\Phi^{\mathrm{T}} = \mathbf{I}, \tag{5.9}$$

where T indicates a transposed version of a matrix or a vector. A random signal vector

$$\mathbf{x} = \left[x(0) \ x(1) \ \ldots \ x(N-1)\right]^{\mathrm{T}}, \tag{5.10}$$

is mapped onto the orthonormal subspace through a forward transform operator (projection)

$$\boldsymbol{\theta} = \Phi\mathbf{x}, \tag{5.11}$$

where $\boldsymbol{\theta}$ is the transform coefficients vector given as

$$\boldsymbol{\theta} = \left[\theta_0 \ \theta_1 \ \ldots \ \theta_{N-1}\right]^{\mathrm{T}}. \tag{5.12}$$

Similarly, the inverse transform reconstructs the signal vector

$$\mathbf{x} = \Phi^{-1}\boldsymbol{\theta} = \Phi^{\mathrm{T}}\boldsymbol{\theta}. \tag{5.13}$$

Hence, one can derive the correlation matrix of transform coefficients as follows:

$$\mathbf{R}_\theta = E\{\boldsymbol{\theta}\boldsymbol{\theta}^{\mathrm{T}}\} = E\{\Phi\mathbf{x}\mathbf{x}^{\mathrm{T}}\Phi^{\mathrm{T}}\} = \Phi E\{\mathbf{x}\mathbf{x}^{\mathrm{T}}\}\Phi^{\mathrm{T}} = \Phi\mathbf{R}_x\Phi^{\mathrm{T}}. \tag{5.14}$$

Furthermore, total energy represented by the transform coefficients is written as:

$$E\{\boldsymbol{\theta}^{\mathrm{T}}\boldsymbol{\theta}\} = \sum_{k=0}^{N-1} E\{\theta_k^2\} = \sum_{k=0}^{N-1} \sigma_k^2. \tag{5.15}$$

It follows from (5.9) and (5.11) that

$$E\{\boldsymbol{\theta}^{\mathrm{T}}\boldsymbol{\theta}\} = E\{\mathbf{x}^{\mathrm{T}}\Phi^{\mathrm{T}}\Phi\mathbf{x}\} = E\{\mathbf{x}^{\mathrm{T}}\mathbf{x}\} = \sum_{n=0}^{N-1} \sigma_x^2(n) = N\sigma^2, \tag{5.16}$$

where $\sigma_x^2(n)$ is the variance of the nth element of the signal vector given in (5.10) that is equal to σ^2. It follows from (5.15) and (5.16) that

$$\sigma^2 = \frac{1}{N}\sum_{k=0}^{N-1}\sigma_k^2. \tag{5.17}$$

The energy-preserving property of an orthonormal transform is evident in (5.17). It is also noted in (5.11) that the linear transformation of the stationary random vector process $\mathbf{x}$ results in a nonstationary random coefficient vector process $\boldsymbol{\theta}$, that is, $\sigma_k^2 \neq \sigma_l^2$ for $k \neq l$ (Akansu and Haddad, 1992).

5.2.2.1 Karhunen–Loeve Transform (Eigen Subspace)

KLT jointly provides the (i) optimal geometric mean of coefficient variances with a diagonal correlation matrix, $\mathbf{R}_\theta$, in (5.14); and (ii) best possible repacking of signal energy into as few transform coefficients as possible. KLT minimizes the energy of the approximation error due to use of only L basis functions $L \leq N$ in order to approximate covariance subject to the orthonormality constraint given in (5.9). Hence, the cost function is defined as (Akansu and Haddad, 1992):

$$J = \sum_{k=L}^{N-1} J_k = E\{\mathbf{e}_L^{\mathrm{T}}\mathbf{e}_L\} - \sum_{k=L}^{N-1}\lambda_k(\boldsymbol{\phi}_k^{\mathrm{T}}\boldsymbol{\phi}_k - 1), \tag{5.18}$$

where λ_k is the kth Lagrangian multiplier. (5.18) can be rewritten as

$$J = \sum_{k=L}^{N-1} J_k = \sum_{k=L}^{N-1}\boldsymbol{\phi}_k^{\mathrm{T}}\mathbf{R}_x\boldsymbol{\phi}_k - \sum_{k=L}^{N-1}\lambda_k(\boldsymbol{\phi}_k^{\mathrm{T}}\boldsymbol{\phi}_k - 1). \tag{5.19}$$

Taking gradient of one of the components of the error J (i.e., J_k), with respect to $\boldsymbol{\phi}_k$ and setting it to zero as follows (Akansu and Haddad, 1992):

$$\nabla J_k = \frac{\partial J_k}{\partial \boldsymbol{\phi}_k} = 2\mathbf{R}_x\boldsymbol{\phi}_k - 2\lambda_k\boldsymbol{\phi}_k = 0, \tag{5.20}$$

yields

$$\mathbf{R}_x\boldsymbol{\phi}_k = \lambda_k\boldsymbol{\phi}_k, \tag{5.21}$$

which implies that $\boldsymbol{\phi}_k$ is one of the eigenvectors of $\mathbf{R}_x$ and λ_k is the corresponding eigenvalue. It is evident from (5.21) that the basis set for KLT comprises the eigenvectors of the autocorrelation matrix of the input (i.e., $\mathbf{R}_x$), and it needs to be recalculated whenever signal statistics change. It follows from (5.21) that

$$\mathbf{R}_x\mathbf{A}_{KLT}^{\mathrm{T}} = \mathbf{A}_{KLT}^{\mathrm{T}}\boldsymbol{\Lambda},$$

$$\mathbf{R}_x = \mathbf{A}_{KLT}^{\mathrm{T}}\boldsymbol{\Lambda}\mathbf{A}_{KLT} = \sum_{k=0}^{N-1}\lambda_k\boldsymbol{\phi}_k\boldsymbol{\phi}_k^{\mathrm{T}}, \tag{5.22}$$

where $\boldsymbol{\Lambda} = diag(\lambda_k); k = 0, 1, \ldots, N-1$; and the kth column of $\mathbf{A}_{KLT}^{\mathrm{T}}$ matrix is the kth eigenvector $\boldsymbol{\phi}_k$ of $\mathbf{R}_x$ with the corresponding eigenvalue λ_k. It is noted that $\{\lambda_k = \sigma_k^2\}\forall k$, for the given $\mathbf{R}_x$ where σ_k^2 is the variance of the k*th* transform coefficient, θ_k (Akansu and Haddad, 1992).

5.2.2.2 Performance Metrics for Orthogonal Subspaces

In practice, it is desired that variances of the transform coefficients decrease as the coefficient index k increases, and so the signal energy is consolidated into as small a number of transform coefficients as possible (Akansu and Haddad, 1992). In other words, it is desired to minimize the energy of the approximation error defined as:

$$\mathbf{e}_L = \mathbf{x} - \hat{\mathbf{x}}_L = \sum_{k=0}^{N-1} \theta_k \boldsymbol{\phi}_k - \sum_{k=0}^{L-1} \theta_k \boldsymbol{\phi}_k = \sum_{k=L}^{N-1} \theta_k \boldsymbol{\phi}_k, \tag{5.23}$$

where $0 < L \leq N-1$. There are three commonly used metrics to measure the performance of a given orthonormal transform (subspace) (Akansu and Haddad, 1992). The compaction efficiency of a transform, that is, the ratio of the energy in the first L transform coefficients to the total energy, is defined as

$$\eta_E(L) = 1 - \frac{E\{\mathbf{e}_L^{\mathrm{T}} \mathbf{e}_L\}}{E\{\mathbf{e}_0^{\mathrm{T}} \mathbf{e}_0\}} = \frac{\sum_{k=0}^{L-1} \sigma_k^2}{N\sigma_x^2}. \tag{5.24}$$

This is an important metric to assess the efficiency of a transform for the given signal type.

The gain of transform coding (TC) over pulse code modulation (PCM) performance of an $N \times N$ unitary transform for a given input correlation is particularly significant and widely utilized in transform coding applications as defined:

$$G_{TC}^N = \frac{\frac{1}{N}\sum_{k=0}^{N-1} \sigma_k^2}{\left(\prod_{k=0}^{N-1} \sigma_k^2\right)^{1/N}}. \tag{5.25}$$

Similarly, decorrelation efficiency of a transform is defined as

$$\eta_c = 1 - \frac{\sum_{k=0}^{N-1} \sum_{l=1;l\neq k}^{N-1} |\mathbf{R}_\theta(k,l)|}{\sum_{k=0}^{N-1} \sum_{l=1;l\neq k}^{N-1} |\mathbf{R}_x(k,l)|}. \tag{5.26}$$

It is desired to have high compaction efficiency, $\eta_E(L)$; high gain of TC over PCM, G_{TC}^N; and high decorrelation efficiency, η_c, for a given $N \times N$ orthonormal transform. Detailed discussion on the performance metrics for the orthonormal transforms can be found in Akansu and Haddad, 1992.

In Figure 5.1a, $\eta_E(L)$ of KLT and the popular fixed transform DCT (Akansu and Haddad, 1992) are displayed for the AR(1) process with various correlation coefficients ρ and transform size $N = 31$. Similarly, Figure 5.1b depicts G_{TC}^N performances of KLT and DCT as a function of ρ for $N = 31$. This figure justifies the use of fixed transform DCT as a replacement to signal-dependent KLT in applications where signal samples are highly correlated. Moreover, it is noted that the energy-packing performances of both transforms degrade for lower values of correlation coefficient. Hence, subspace methods bring less value for applications where signal correlation is low.

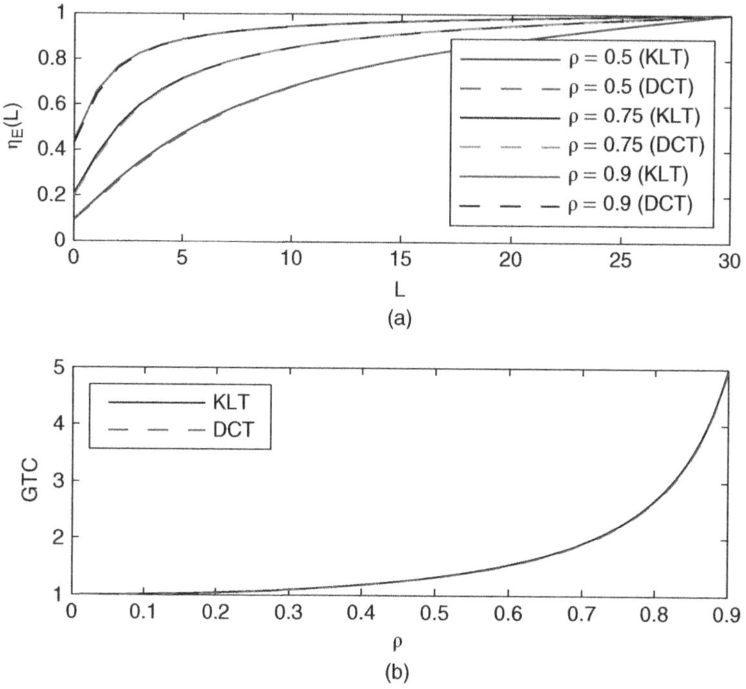

Figure 5.1 (a) $\eta_E(L)$ Performance of KLT and DCT for an AR(1) process with various values of ρ and $N = 31$; (b) G_{TC}^N performance of KLT and DCT as a function of ρ for $N = 31$.

5.3 Derivation of Explicit KLT Kernel for a Discrete AR(1) Process

A KLT matrix $\mathbf{A}_{KLT}$ of size $N \times N$ for an AR(1) process is expressed with the closed-form kernel as (Ray and Driver, 1970):

$$\mathbf{A}_{KLT} = [A(k,n)] = \left(\frac{2}{N+\lambda_k}\right)^{1/2} \sin\left[\omega_k\left(n - \frac{N-1}{2}\right) + \frac{(k+1)\pi}{2}\right], \tag{5.27}$$

where $0 \leq k, n \leq N-1$. Corresponding transform coefficient variances (i.e., the eigenvalues of the autocorrelation matrix given in (5.5), λ_k) are derived to be in the closed form (Ray and Driver, 1970):

$$\sigma_k^2 = \lambda_k = \frac{1-\rho^2}{1 - 2\rho\cos(\omega_k) + \rho^2}, \tag{5.28}$$

where $\{\omega_k\}$ are the positive roots of the following transcendental equation (Ray and Driver, 1970):

$$\tan(N\omega) = -\frac{(1-\rho^2)\sin(\omega)}{\cos(\omega) - 2\rho + \rho^2\cos(\omega)}. \tag{5.29}$$

The derivation steps leading to the equations (5.27), (5.28), and (5.29) will be explained throughout this section. Moreover, there is a need to derive an explicit expression for the roots of the transcendental equation in (5.29) such that the kernel (eigenvectors) and the corresponding variances (eigenvalues) are expressed accordingly.

5.3.1 A Simple Method for Explicit Solution of a Transcendental Equation

A simple method of formulating an explicit solution for the roots of transcendental equations using Cauchy's integral theorem from complex analysis (Strang, 1986) that was introduced by Luck and Stevens (2002) is highlighted in this section. The method determines the roots of a transcendental function by locating the singularities of a reciprocal function. Although derivation steps are detailed in Luck and Stevens (2002), a summary is given here for the completeness of presentation.

Cauchy's theorem states that if a function is analytic in a simple connected region containing the closed curve C, the path integral of the function around the curve C is zero. On the other hand, if a function, $f(z)$, contains a single singularity at z_0 somewhere inside C but analytic elsewhere in the region, then the singularity can be removed by multiplying $f(z)$ with $(z - z_0)$ (i.e., by a pole-zero cancellation). Cauchy's theorem implies that the path integral of the new function $(z - z_0)f(z)$ around C must be zero:

$$\oint_C (z - z_0)f(z)dz = 0. \tag{5.30}$$

Evaluation of the integral given in (5.30) yields a first-order polynomial in z_0 with constant coefficients, and its solution for z_0 provides the location of the singularity as given (Luck and Stevens, 2002)

$$z_0 = \frac{\oint_C zf(z)dz}{\oint_C f(z)dz}. \tag{5.31}$$

This is an explicit expression for the singularity of the function $f(z)$. A root-finding problem is restated as a singularity at the root. It is noted that (5.31) gives the location of the desired root, and it can be evaluated for any closed path by employing either an analytical or a numerical technique. Luck and Stevens (2002) suggested to use a circle in the complex plane with its center h and radius R as the closed curve C, expressed as

$$\begin{aligned} z &= h + Re^{j\theta}, \\ dz &= jRe^{j\theta}d\theta, \end{aligned} \tag{5.32}$$

where $0 \leq \theta \leq 2\pi$; $h \in \mathbb{R}$; and $R \in \mathbb{R}$; Values of h and R do not matter as long as the circle circumscribes the root z_0. Cauchy's argument principle (Brown and Churchill, 2009) or graphical methods may be used to determine the number of roots enclosed by the path C. A function in θ is defined as

$$w(\theta) = f(z)|_{z=h+Re^{j\theta}} = f(h + Re^{j\theta}). \tag{5.33}$$

Then (5.31) becomes (Luck and Stevens, 2002):

$$z_0 = h + R\left[\frac{\int_0^{2\pi} w(\theta)e^{j2\theta}d\theta}{\int_0^{2\pi} w(\theta)e^{j\theta}d\theta}\right]. \tag{5.34}$$

One can easily evaluate (5.34) by employing Fourier analysis since the *nth* Fourier series coefficient for any $x(t)$ is calculated as

$$A_n = \frac{1}{2\pi}\int_0^{2\pi} x(t)e^{jnt}dt. \tag{5.35}$$

It is observed that the term in brackets in (5.34) is equal to the ratio of the second Fourier series coefficient over the first one for the function $w(\theta)$. Fourier series coefficients can be easily calculated numerically by using DFT or by using its fast implementation, Fast Fourier transform (FFT), as is suggested in Luck and Stevens (2002). However, it is observed from (5.34) that one does not need all DFT coefficients to solve the problem since it requires only two Fourier series coefficients. Therefore, it is possible to further improve the computational cost by employing a discrete summation operator to implement (5.34) numerically. Hence, the algorithm would have a computational complexity of $O(N)$ instead of $O(NlogN)$ required for FFT algorithms.

It is also noted that given $f(z)$ is analytic at h, multiplying $f(z)$ by a factor $(z-h) = Re^{j\theta}$ does not change the location of the singularities of $f(z)$. It means that for a given singularity, the term in brackets is also equal to any ratio of the $(m+1)th$ to the mth Fourier series coefficients of $w(\theta)$ for $m \geq 1$ (Luck and Stevens, 2002). The MATLAB™ code given in Torun and Akansu (2013) to calculate the roots of (5.43) shows the simplicity of the method to solve such transcendental equations.

5.3.2 *Continuous Process with Exponential Autocorrelation*

The classic problem of deriving explicit solutions for characteristic values and functions of a continuous random process with an exponential autocorrelation function provides the foundation for the derivation of explicit KLT kernel for a discrete AR(1) process. This problem is discussed in Davenport and Root (1958) and Van Trees (2001). Similar discussions can also be found in Pugachev (1959a,b). After some derivation steps, the characteristic equation of the process is expressed as

$$\phi''(t) + \frac{\alpha(2-\alpha\lambda)}{\lambda}\phi(t) = 0, \tag{5.36}$$

and (5.36) has a solution only in the range of $0 < \lambda < \frac{2}{\alpha}$ and is rewritten as (Davenport and Root, 1958):

$$\phi''(t) + b^2\phi(t) = 0, \tag{5.37}$$

where

$$b^2 = \frac{\alpha(2-\alpha\lambda)}{\lambda}, 0 < b^2 < \infty. \tag{5.38}$$

A general solution to (5.37) is given as

$$\phi(t) = c_1 e^{jbt} + c_2 e^{-jbt}, \tag{5.39}$$

where c_1 and c_2 are arbitrary constants. A solution is possible only when $c_1 = c_2$ or $c_1 = -c_2$. For $c_1 = c_2$, it is shown that one of the unknowns in the general solution is given in (5.39), b, satisfies the equation (Davenport and Root, 1958)

$$b \tan b\frac{T}{2} = \alpha. \tag{5.40}$$

It follows from (5.39) that for every positive b_k that satisfies the transcendental equation (5.40), there is a characteristic function that satisfies the characteristic equation and given as (Davenport and Root, 1958):

$$\phi_k(t) = c_k \cos b_k t, \tag{5.41}$$

where integer $k \geq 0$. Similarly, for $c_1 = -c_2$, b, one can derive the resulting characteristic functions as

$$\phi_k(t) = c_k \sin b_k t. \tag{5.42}$$

The steps required to determine the roots of (5.40) are summarized next. We can rewrite (5.40) for $\alpha = B$ and $T = 2$ as

$$b \tan b = B. \tag{5.43}$$

Positive roots of (5.43), $b_m > 0$, must be calculated in order to determine the even indexed characteristic values and functions. Figure 5.2 displays functions $\tan(b)$ and B/b for various values of B. It is observed from the figure that for the m*th* root, a suitable choice for the closed path C is a circle of radius $R = \pi/4$ centered at $h_m = (m - 3/4)\pi$, as suggested in (Luck and Stevens, 2002). A straightforward way to configure (5.43) to introduce singularities is to use the inverse of (5.43) rearranged as follows (Luck and Stevens, 2002):

$$f(b) = \frac{1}{b \sin(b) - B \cos(b)}. \tag{5.44}$$

Applying (5.34) to (5.44) results in an explicit expression for the m*th* root. This expression can be evaluated by calculating a pair of adjacently indexed DFT coefficients (coefficients of

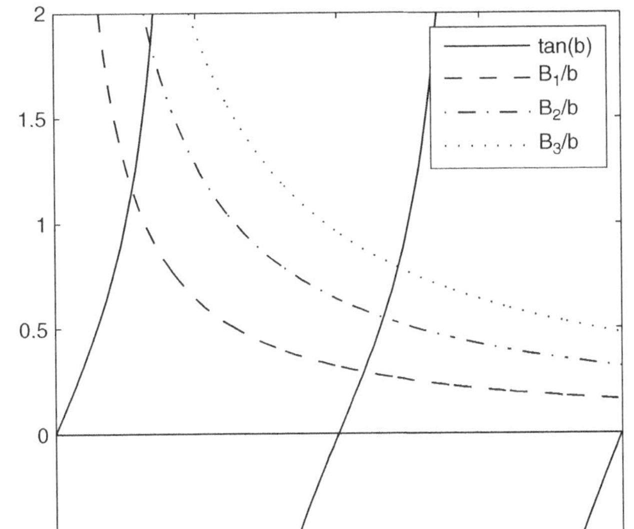

Figure 5.2 Functions $\tan(b)$ and B/b for various values of B where $B_1 = 1$, $B_2 = 2$, and $B_3 = 3$.

two adjacent harmonics), as described in Section 5.3.1, or by using a numerical integration method. Therefore, by setting $b = h + Re^{j\theta}$, $w_m(\theta)$ of (5.33) for this case is defined as:

$$\begin{aligned} w_m(\theta) &= f(h_m + Re^{j\theta}), \\ &= \frac{1}{(h_m + Re^{j\theta})\sin(h_m + Re^{j\theta}) - B\cos(h_m + Re^{j\theta})}, \end{aligned} \tag{5.45}$$

where $0 \leq \theta \leq 2\pi$. Hence, the location of the mth root is explicitly defined as

$$b_m = h_m + R\left[\frac{\int_0^{2\pi} w_m(\theta)e^{j2\theta}d\theta}{\int_0^{2\pi} w_m(\theta)e^{j\theta}d\theta}\right]. \tag{5.46}$$

5.3.3 *Eigenanalysis of a Discrete AR(1) Process*

In this section, derivations of (5.27), (5.28), and (5.29) are given in detail. For a a discrete random signal (process), $x(n)$, a discrete Karhunen–Loeve (K-L) series expansion is given as follows (Akansu and Haddad, 1992):

$$\sum_{m=0}^{N-1} R_{xx}(n,m)\phi_k(m) = \lambda_k\phi_k(n), \tag{5.47}$$

where m and n are the independent discrete variables;

$$R_{xx}(n,m) = E\{x(n)x(m)\}, m, n = 0, 1, \ldots, N-1, \tag{5.48}$$

is the autocorrelation function of the random signal; λ_k is the kth eigenvalue; and $\phi_k(n)$ is the corresponding kth eigenfunction. The autocorrelation function of the stationary discrete AR(1) process is given as (Papoulis, 1991):

$$R_x(n,m) = R_x(n-m) = \rho^{|n-m|}. \tag{5.49}$$

Hence, the discrete K-L series expansion for an AR(1) process, from (5.47) and (5.49), is stated as follows

$$\sum_{m=0}^{N-1} \rho^{|n-m|}\phi_k(m) = \lambda_k\phi_k(n). \tag{5.50}$$

In order to eliminate the magnitude operator, (5.50) can be rewritten in the form

$$\sum_{m=0}^{n} \rho^{n-m}\phi_k(m) + \sum_{m=n+1}^{N-1} \rho^{m-n}\phi_k(m) = \lambda_k\phi_k(n). \tag{5.51}$$

It follows from the continuous case presented in Section 5.3.2 that the general solution for the kth eigenvector is given as (Davenport and Root, 1958; Van Trees, 2001):

$$\phi_k(t) = c_1e^{j\omega_k t} + c_2e^{-j\omega_k t}, \tag{5.52}$$

where c_1 and c_2 are arbitrary constants; t is the independent continuous variable; $-T/2 \leq t \leq T/2$; and $\omega_k = b_k$. The eigenfunction given in (5.52) is shifted by $T/2$ and sampled at

$t_n = nT_s$, $0 \le n \le N-1$, where $T_s = T/(N-1)$. Accordingly, the sampled eigenfunction is written as

$$\phi_k(n) = c_1 e^{j\omega_k(n-\frac{N-1}{2})} + c_2 e^{-j\omega_k(n-\frac{N-1}{2})}. \quad (5.53)$$

The solution to (5.50) exists only when $c_1 = \pm c_2$. In the following discussions, the case for $c_1 = c_2$ is considered given the fact that the sister case for $c_1 = -c_2$ is similar. For $c_1 = c_2$, it follows from (5.53) that

$$\phi_k(n) = c_1 \cos\left[\omega_k\left(n - \frac{N-1}{2}\right)\right]. \quad (5.54)$$

By substituting (5.54) in (5.51) and defining a new independent discrete variable $p = m - (N-1)/2$, (5.50) can be rewritten as follows:

$$\sum_{p=-\frac{N-1}{2}}^{n-\frac{N-1}{2}} \rho^{n-p-\frac{N-1}{2}} \cos(\omega_k p) + \sum_{p=n+1-\frac{N-1}{2}}^{\frac{N-1}{2}} \rho^{p+\frac{N-1}{2}-n} \cos(\omega_k p) \quad (5.55)$$

$$= \lambda_k \cos\left[\omega_k\left(n - \frac{N-1}{2}\right)\right]. \quad (5.56)$$

The first summation on the left in (5.55) is rewritten as

$$\frac{1}{2}\rho^{n-\frac{N-1}{2}}\left[\sum_{p=-\frac{N-1}{2}}^{n-\frac{N-1}{2}} (\rho^{-1}e^{j\omega_k})^p + \sum_{p=-\frac{N-1}{2}}^{n-\frac{N-1}{2}} (\rho^{-1}e^{-j\omega_k})^p\right]. \quad (5.57)$$

Using the fact that

$$\sum_{n=N_1}^{N_2} \beta^n = \frac{\beta^{N_1} - \beta^{N_2+1}}{1-\beta}, \quad (5.58)$$

and following simple derivation steps, it can be shown that (5.57), hence the first summation on the left in (5.55), is equal to

$$\frac{\rho^{n+2}\cos\omega_1 - \rho\cos\omega_2 - \rho^{n+1}\cos\omega_3 + \cos\omega_4}{1 - 2\rho\cos\omega_k + \rho^2}. \quad (5.59)$$

Similarly, the second summation on the left in (5.55) is equal to

$$\frac{\rho^{N-n+1}\cos\omega_1 + \rho\cos\omega_2 - \rho^{N-n}\cos\omega_3 - \rho^2\cos\omega_4}{1 - 2\rho\cos\omega_k + \rho^2}, \quad (5.60)$$

where

$$\begin{aligned} \omega_1 &= \omega_k[(N-1)/2], \\ \omega_2 &= \omega_k[n - (N-1)/2 + 1], \\ \omega_3 &= \omega_k[(N-1)/2 + 1], \\ \omega_4 &= \omega_k[n - (N-1)/2], \end{aligned} \quad (5.61)$$

for both (5.59) and (5.60). It is possible to express λ_k on the right-hand side of (5.55) in terms of ρ and ω_k by taking the discrete K-L expansion given in (5.50) into the frequency domain via DTFT as follows:

$$S_x(e^{j\omega})\Phi_k(e^{j\omega}) = \lambda_k\Phi_k(e^{j\omega}), \tag{5.62}$$

where $S_x(e^{j\omega})$ is the power spectral density (PSD) of the discrete AR(1) process and expressed as

$$S_x(e^{j\omega}) = \mathcal{F}\{\rho^{|n-m|}\} = \frac{1-\rho^2}{1-2\rho\cos\omega+\rho^2}. \tag{5.63}$$

$\mathcal{F}\{\cdot\}$ is the DTFT operator (Akansu and Haddad, 1992). The Fourier transform of the eigenfunction in (5.54) is calculated as

$$\begin{aligned}\Phi_k(e^{j\omega}) &= \mathcal{F}\{\phi_k(n)\},\\ &= c_1 e^{-j\omega_k\frac{N-1}{2}}[\delta(\omega-\omega_k)+\delta(\omega+\omega_k)],\end{aligned} \tag{5.64}$$

where $\delta(\omega-\omega_0)$ is an impulse function of frequency located at ω_0. By substituting (5.63) and (5.64) into (5.62), λ_k is derived as

$$\lambda_k = \frac{1-\rho^2}{1-2\rho\cos\omega_k+\rho^2}. \tag{5.65}$$

Equation (5.65) shows that the eigenvalues are the samples of the PSD given in (5.63). Moreover, (5.28) and (5.65) are identical. By substituting (5.59), (5.60), and (5.65) in (5.51), one can show that

$$\rho = \frac{\cos(\omega_k N/2+\omega_k/2)}{\cos(\omega_k N/2-\omega_k/2)}. \tag{5.66}$$

Using trigonometric identities, the relationship between ω_k and ρ in (5.66) is rewritten as follows

$$\tan\left(\omega_k\frac{N}{2}\right) = \left(\frac{1-\rho}{1+\rho}\right)\cot\left(\frac{\omega_k}{2}\right). \tag{5.67}$$

Similarly, for the case of $c_1 = -c_2$, following the same procedure, the relationship between ω_k and ρ is shown as

$$\tan\left(\omega_k\frac{N}{2}\right) = -\left(\frac{1+\rho}{1-\rho}\right)\tan\left(\frac{\omega_k}{2}\right). \tag{5.68}$$

Finally, from (5.67) and (5.68), it is observed that ω_k are the positive roots of the equation

$$\left[\tan\left(\omega\frac{N}{2}\right)+\frac{1+\rho}{1-\rho}\tan\left(\frac{\omega}{2}\right)\right]\left[\tan\left(\omega\frac{N}{2}\right)-\frac{1-\rho}{1+\rho}\cot\left(\frac{\omega}{2}\right)\right] = 0. \tag{5.69}$$

Using trigonometric identities, (5.69) can be rewritten as

$$\tan(N\omega) = -\frac{(1-\rho^2)\sin(\omega)}{\cos(\omega)-2\rho+\rho^2\cos(\omega)}, \tag{5.70}$$

which is the same transcendental equation expressed in (5.29). The roots $\{\omega_k\}$ of the transcendental tangent equation in (5.70) are required in the KLT kernel expressed in (5.27). There are well-known numerical methods like the secant method (Allen and Isaacson, 1997) to approximate roots of the equation given in (5.70) in order to solve it implicitly rather than explicitly. A method to find explicit solutions to the roots of transcendental equations, including (5.70), is revisited next. That method leads to an explicit definition of KLT kernel given in (5.27) for an AR(1) process.

5.3.4 *Fast Derivation of KLT Kernel for an AR(1) Process*

In this section, the fast derivation method of KLT kernel for a discrete AR(1) process is explained. Moreover, a step-by-step implementation of the technique is presented.

5.3.4.1 Discrete AR(1) Process

In order to define an explicit expression for the discrete KLT kernel of (5.27), first, one must find $N/2$ positive roots of the following two transcendental equations:

$$\tan\left(\omega\frac{N}{2}\right) = \frac{1}{\gamma}\cot\left(\frac{\omega}{2}\right), \tag{5.71}$$

$$\tan\left(\omega\frac{N}{2}\right) = -\gamma\tan\left(\frac{\omega}{2}\right), \tag{5.72}$$

as discussed in Section 5.3.3. In both equations, N is the transform size and

$$\gamma = (1+\rho)/(1-\rho), \tag{5.73}$$

where ρ is the first-order correlation coefficient of the AR(1) process. Roots of (5.71) and (5.72) correspond to the even and odd indexed eigenvalues and eigenvectors, respectively. Figure 5.3 displays functions $\tan(\omega N/2)$ and $-\gamma\tan(\omega/2)$ for $N = 8$ and various values of ρ. It is apparent from the figure that for the mth root of (5.72), a suitable choice for the closed path C in (5.31) is a circle of radius

$$R_m = \begin{cases} \pi/2N & m \leq 2 \\ \pi/N & m > 2 \end{cases}, \tag{5.74}$$

centered at $h_m = (m - 1/4)(2\pi/N)$, where $1 \leq m \leq N/2$. It is worth noting that for positively correlated signals, $0 < \rho < 1$, the ratio given in (5.73) is always greater than 1, $\gamma > 1$. However, for negatively correlated signals, $-1 < \rho < 0$, the ratio is between 0 and 1 (i.e., $0 < \gamma < 1$). Therefore, for $\rho < 0$, the last two roots must be smaller than the rest as

$$R_m = \begin{cases} \pi/N & m < N/2 - 1 \\ \pi/2N & m \geq N/2 - 1 \end{cases}. \tag{5.75}$$

Similar to the continuous case, (5.72) is reconfigured and the poles of the following inverse function are calculated

$$g(\omega) = \frac{1}{\tan(\omega N/2) + \gamma\tan(\omega/2)}. \tag{5.76}$$

By setting $\omega = h + Re^{j\theta}$, the function $w(\theta)$ of (5.33) for this case is defined as

$$\begin{aligned} w_m(\theta) &= g(h_m + R_m e^{j\theta}), \\ &= \frac{1}{\tan\left[\left(h_m + R_m e^{j\theta}\right)\frac{N}{2}\right] + \gamma\tan\left[\left(h_m + R_m e^{j\theta}\right)\frac{1}{2}\right]}, \end{aligned} \tag{5.77}$$

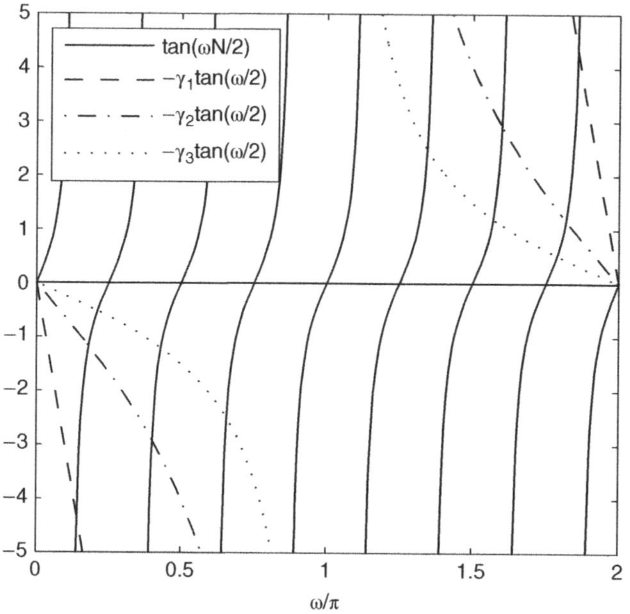

Figure 5.3 Functions $\tan(\omega N/2)$ and $-\gamma\tan(\omega/2)$ for the AR(1) process with $N = 8$ and various values of ρ, where $\rho_1 = 0.9$, $\rho_2 = 0.6$, $\rho_3 = 0.2$, and $\gamma_i = (1+\rho_i)/(1-\rho_i)$, $i = 1, 2, 3$.

where $0 \le \theta \le 2\pi$. Hence, the m*th* root is located at

$$\omega_m = h_m + R_m \left[\frac{\int_0^{2\pi} w_m(\theta) e^{j2\theta} d\theta}{\int_0^{2\pi} w_m(\theta) e^{j\theta} d\theta} \right]. \tag{5.78}$$

The procedure is the same as finding the roots of (5.71) with the exceptions that (5.77) must be modified as follows:

$$w_m(\theta) = \frac{1}{\tan\left[\left(h_m + R_m e^{j\theta}\right)\frac{N}{2}\right] - \frac{1}{\gamma}\cot\left[\left(h_m + R_m e^{j\theta}\right)\frac{1}{2}\right]}, \tag{5.79}$$

and a suitable choice for the closed path C is a circle of radius $R_m = \pi/N$ centered at

$$h_m = \begin{cases} (m - 1/2)\ (2\pi/N) & m \le 2 \\ (m-1)\ (2\pi/N) & m > 2 \end{cases}, \tag{5.80}$$

which can be determined by generating a plot similar to the ones in Figures 5.2 and 5.3.

Finally, the steps to derive an explicit KLT kernel of dimension N for an arbitrary discrete dataset by employing an AR(1) approximation are summarized as follows

1. Estimate the first-order correlation coefficient of an AR(1) model for the given dataset $\{x(n)\}$ as

$$\rho = \frac{R_{xx}(1)}{R_{xx}(0)} = \frac{E\{x(n)x(n+1)\}}{E\{x(n)x(n)\}}, \tag{5.81}$$

where n is the index of random variables (or discrete-time) and $-1 < \rho < 1$.

2. Calculate the positive roots $\{\omega_k\}$ of the polynomial given in (5.29) by substituting (5.77) and (5.79) into (5.78) for odd and even values of k, respectively, and use the following indexing:

$$m = \begin{cases} k/2+1 & k \text{ even} \\ (k+1)/2 & k \text{ odd} \end{cases}. \tag{5.82}$$

3. Plug in the values of ρ and $\{\omega_k\}$ in (5.28) and (5.27) to calculate the eigenvalues λ_k and eigenvectors, respectively.

MATLAB™ code for steps 2 and 3 with DFT (FFT) used in solving (5.78) are provided in (Torun and Akansu, 2013).

The computational cost of deriving KLT matrix for an arbitrary signal source has two distinct components: the calculation of the first-order correlation coefficient ρ for the given signal set, and the calculation of the roots $\{\omega_k\}$ of (5.29) that are plugged in (5.27) to generate the resulting transform matrix $\mathbf{A}_{KLT}$. The roots $\{\omega_k\}$ of the transcendental tangent equation (5.29), calculated by using (5.78), as a function of ρ and for $N = 8$ are displayed in Figure 5.4.

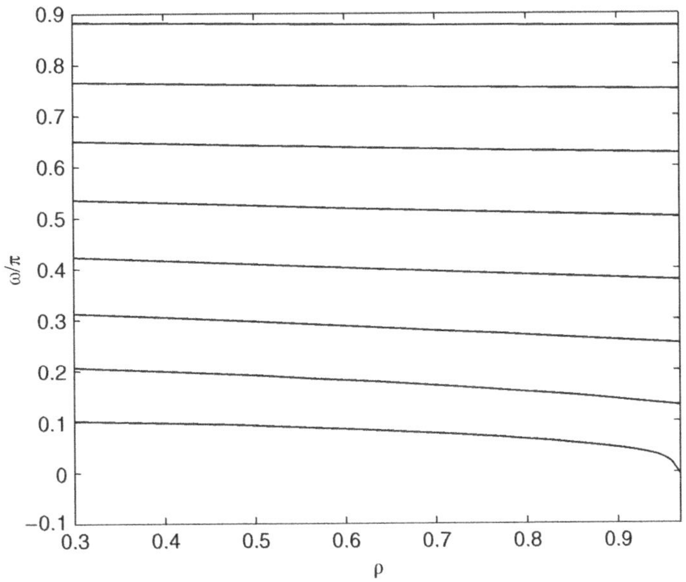

Figure 5.4 The roots of the transcendental tangent equation 5.29, $\{\omega_k\}$, as a function of ρ for $N = 8$.

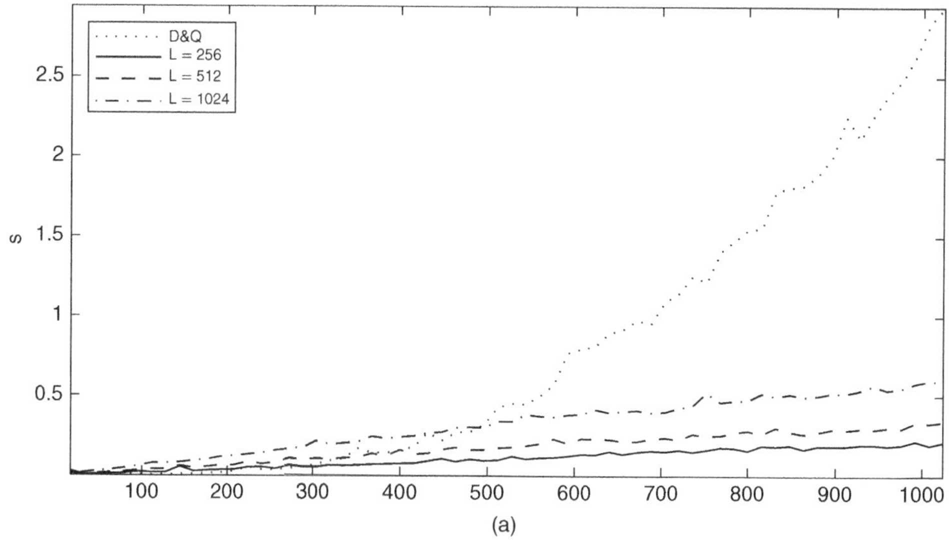

Figure 5.5 Computation time, in seconds, to calculate $\mathbf{A}_{KLT,DQ}$ and $\mathbf{A}_{KLT,E}$ for an AR(1) process with $\rho = 0.95$, and different values of N ($16 \leq N \leq 1024$) and $L = 256, 512, 1024$ (Torun and Akansu, 2013).

The computational cost of generating KLT kernel for the given statistics by using the method presented in this section ($\mathbf{A}_{KLT,E}$) is compared with a widely used numerical algorithm called divide and conquer (D&Q) (Golub and Loan, 1996).

Computation times (in seconds) to generate $\mathbf{A}_{KLT,DQ}$ and $\mathbf{A}_{KLT,E}$ (DFT sizes of $L = 256, 512, 1024$) for the case of $\rho = 0.95$ and $16 \leq N \leq 1024$ are displayed in Figure 5.5. Both computations are performed by using one thread on a single processor. The machine used for the simulations has an Intel® Core™ i5-520M CPU and 8 GB of RAM. It is displayed in the figure that the explicit KLT kernel derivation method significantly outperforms the D&Q algorithm for larger values of N. Furthermore, it has a so-called *embarrassingly parallel* nature. Hence, it can be easily computed on multiple threads and processors for any k. Therefore, by implementing it on a parallel device such as a GPU and FPGA, its speed can be significantly improved.

5.4 Sparsity of Eigen Subspace

The sparse representation of signals has been of great research interest in signal processing and other disciplines. KLT has been one of the most popular mathematical tools employed in multivariate data processing and dimension reduction. The application-specific interpretation of eigenvectors with eigenvalues and their vector components is often a critical task (Cadima and Jolliffe, 1995; d'Aspremont *et al.*, 2007; Trendafilov *et al.*, 2003; Zou *et al.*, 2006). Moreover, small but nonzero loadings (elements) of each principal component (PC) (or eigenvector) bring implementation cost that is hard to justify in applications such as generation and maintenance (rebalancing) of eigenportfolios with a large number of financial instruments (d'Aspremont *et al.*, 2007; Torun *et al.*, 2011). This and other data-intensive applications that utilize loading coefficients have motivated researchers to study the sparsity of PCs in eigenanalysis of measured covariance matrices. Furthermore, unevenness of signal energy distributed among PCs in eigen subspace is reflected in eigenvalues (coefficient variances) that lead to dimension reduction. The latter is the very foundation of transform coding successfully used in visual signal processing and data compression (Akansu and Haddad, 1992; Clarke, 1985; Jayant and Noll, 1984). Therefore, both dimension reduction and sparsity of basis functions (vectors) are significant attributes of orthogonal transforms widely utilized in many fields. This recent development has paved the way for the development of several popular methods for sparse eigen subspace. A rate-distortion-based framework to sparse subspaces is presented in this section. The challenge is to maximize explained variance (eigenvalue) by a minimum number of PCs, also called energy compaction (dimension reduction), while replacing the less significant samples (loading coefficients) of basis functions with zero to achieve the desired level of sparsity (cardinality reduction) in representation.

5.4.1 Overview of Sparsity Methods

We provide an overview of the recent literature on sparsity in this subsection for readers with more interest on the subject. Regularization methods have been used to make an ill-conditioned matrix invertible or to prevent overfitting (Bertero and Boccacci, 1998; Engl *et al.*, 1996). More recently, regularization methods also have been utilized for sparsity. ℓ_0 regularizer leads to a sparse solution. On the other hand, it makes the optimization problem nonconvex. Eigenfiltering is another popular method employed for regularization (Bertero and Boccacci, 1998; Engl *et al.*, 1996). ℓ_1 regularizer, so-called lasso, is widely used as an approximation (convex relaxation) of ℓ_0 regularizer (Tibshirani, 1996; Trendafilov *et al.*, 2003). Another ℓ_1-based method was proposed in (Brodie *et al.*, 2009) for sparse portfolios. SCoTLASS (Trendafilov *et al.*, 2003) and SPCA (Zou *et al.*, 2006) utilize the ℓ_1 and ℓ_2 regularizers for sparse approximation to PCs, respectively.

The sparse PCA is modeled in (Trendafilov *et al.*, 2003; Zou *et al.*, 2006) as an explained variance maximization problem where the number of nonzero elements in the PCs is considered as a basis design constraint. These methods suffer from potentially being stuck in local minima due to the nonconvex nature of the optimization. A convex relaxation method called SDP relaxations for sparse PCA (DSPCA) using semidefinite programming (SDP) was proposed to deal with a simpler optimization (d'Aspremont *et al.*, 2007). Empirical performance results for certain cases indicate that DSPCA may generate sparse PCs that preserve slightly more explained variance than SCoTLASS (Trendafilov *et al.*, 2003) and SPCA (Zou *et al.*, 2006) for the same sparsity level. A nonnegative variant of the sparse PCA problem, which forces the elements of each PC to be nonnegative, is introduced in Zass and Shashua (2006). Nonnegative sparse PCA (NSPCA) offers competitive performance to SCoTLASS, SPCA, and DSPCA in terms of explained variance for a given sparsity. However, signs of the PC elements bear specific information for the applications of interests such as eigenportfolios. Thus, NSPCA is not applicable for all types of applications. Another lasso-based approach, so-called sparse PCA via regularized SVD (sPCA–rSVD), is proposed in Shen and Huang (2008). Simulation results for certain cases show that sPCA–rSVD provides competitive results to SPCA. A variation of sPCA–rSVD, so-called sparse principal components (SPCs), that utilizes the penalized matrix decomposition (PMD) is proposed in Witten *et al.* (2009). PMD that computes the rank K approximation of a given matrix is proposed in Witten *et al.* (2009). It utilizes the lasso penalty for sparsity. Unfortunately, none of these methods result in guaranteed sparsity regardless of their prohibitive computational cost. Moreover, the lack of mathematical framework to measure distortion, or explained variance loss, for a desired sparsity level makes sparse PCA methods of this kind quite ad hoc and difficult to use. On the other hand, the simple thresholding technique is easy to implement (Cadima and Jolliffe, 1995). It performs better than SCoTLASS and slightly worse than SPCA (Zou *et al.*, 2006). Although simple thresholding is easy to implement, it may cause unexpected distortion levels as variance loss. Soft thresholding (ST) is another technique that is utilized for sparse representation in Zou *et al.*, (2006). Certain experiments show that ST offers slightly better performance than simple thresholding (Zou *et al.*, 2006). Therefore, threshold selection plays a central role in sparsity performance.

In this section, we present in detail a subspace sparsing framework based on the rate-distortion theory (Akansu and Haddad, 1992; Berger, 2003; Lloyd, 1982; Max, 1960). It may be considered as an extension of the simple or soft thresholding method to combine a sparse representation problem with an optimal quantization method used in the source coding field (Akansu and Haddad, 1992; Berger, 2003; Cadima and Jolliffe, 1995; Clarke, 1985; Jayant and Noll, 1984). The method employs a varying-size midtread (zero-zone) probability density function (pdf)-optimized (Lloyd–Max) quantizer designed for a component histogram of each eigenvector (or the entire eigenmatrix) to achieve the desired level of distortion (sparsity) in the subspace with reduced cardinality (Hajnal, 1983; Lloyd, 1982; Max, 1960; Sriperumbudur *et al.*, 2009). Herein, we focus specifically on eigen subspace of a discrete AR(1) process with closed–form expressions for its eigenvectors and eigenvalues as derived earlier in the chapter. It is known that the AR(1) process approximates well many real-world signals (Akansu and Haddad, 1992). We also sparse eigenportfolios of the NASDAQ-100 index by using this method. Note that the method to sparse a subspace through quantization of its basis functions is a marked departure from the traditional transform coding where transform coefficients, in the subspace, are quantized for dimension reduction (Akansu and Haddad, 1992; Clarke, 1985; Jayant and Noll, 1984). Therefore, we investigate the trade-off between subspace orthogonality and sparsity from a rate-distortion perspective. Then, we provide a comparative performance of the presented method along with the various methods reported in the literature, such as ST (Zou *et al.*, 2006), SPCA (Zou *et al.*, 2006), DSPCA (d'Aspremont *et al.*, 2007), and SPC (Witten *et al.*, 2009), with respect to the metrics of nonsparsity (NS) and variance loss (VL).

5.4.2 pdf-Optimized Midtread Quantizer

Quantizers may be categorized as midrise and midtread (Jayant and Noll, 1984). A midtread quantizer is preferred for applications requiring entropy reduction and noise filtering (or sparsity) simultaneously (Gonzales and Akansu, 1997). In this section, we utilize a midtread type to quantize basis function (vector) samples (components) of a transform (subspace) to achieve sparse representation in the signal domain.

A celebrated design method to calculate optimum intervals (bins) and representation (quanta) values of a quantizer for the given input signal pdf, a so-called pdf-optimized quantizer, was independently proposed by Max and Lloyd (Lloyd, 1982; Max, 1960). It assumes a random information source X with zero-mean and a known pdf function $p(x)$. Then, it minimizes quantization error in the mean squared error (mse) sense and also makes sure that all bins of a quantizer have the same level of error. The quantization error of an L-bin pdf-optimized quantizer is expressed as follows:

$$\sigma_q^2 = \sum_{k=1}^{L} \int_{x_k}^{x_{k+1}} (x - y_k)^2 p(x)dx, \tag{5.83}$$

where quantizer bin intervals, x_k, and quanta values, y_k, are calculated iteratively. The necessary conditions for an mse-based pdf-optimized quantizer are given as (Lloyd, 1982; Max, 1960):

$$\begin{aligned}\frac{\partial \sigma_q^2}{\partial x_k} &= 0;\ k = 2, 3, \ldots, L,\\ \frac{\partial \sigma_q^2}{\partial y_k} &= 0;\ k = 1, 2, 3, \ldots, L,\end{aligned} \tag{5.84}$$

leading to the optimal unequal intervals and resulting quanta values as

$$x_{k,opt} = \frac{1}{2}(y_{k,opt} + y_{k-1,opt});\ k = 2, 3, \ldots, L, \tag{5.85}$$

$$y_{k,opt} = \frac{\int_{x_k}^{x_{k+1,opt}} xp(x)dx}{\int_{x_k}^{x_{k+1,opt}} p(x)dx};\ k = 1, 2, \ldots, L, \tag{5.86}$$

where $x_{1,opt} = -\infty$ and $x_{L+1,opt} = \infty$. A sufficient condition to avoid local optimum in (5.84) is the log-concavity of the pdf function $p(x)$. The log-concave property holds for uniform, Gaussian, and Laplacian pdf types (Jayant and Noll, 1984). The representation point (quantum) of a bin in such a quantizer is its centroid that minimizes the quantization noise for the interval. We are interested in pdf-optimized quantizers with an adjustable zero-zone, odd L, or midtread quantizer, to sparse (quantize) eigenvectors of an eigen subspace. One can adjust zero-zone(s) of the quantizer(s) to achieve the desired level of sparsity in a transform matrix with the trade-off of resulting imperfectness in orthogonality and explained variance. We will present design examples by using the presented technique to sparse subspaces in Section 5.4.3.

The discrepancy between input and output of a quantizer is measured by the signal-to-quantization-noise ratio (SQNR) (Berger, 2003)

$$SQNR(dB) = 10\log_{10}\left(\frac{\sigma_x^2}{\sigma_q^2}\right), \tag{5.87}$$

where σ_x^2 is the variance of an input with zero-mean and known pdf type, and is expressed as

$$\sigma_x^2 = \int_{-\infty}^{\infty} x^2 p(x)dx. \tag{5.88}$$

The first-order entropy (rate) of the output for an L-level quantizer with such an input is calculated as (Berger, 2003; Brusewitz, 1986)

$$H = -\sum_{k=1}^{L} P_k \log_2 P_k, \tag{5.89}$$

$$P_k = \int_{x_k}^{x_{k+1}} p(x)dx.$$

Rate-distortion theory states that the quantization error variance is expressed as (Berger, 2003):

$$\sigma_q^2 = f(R)\sigma_x^2, \tag{5.90}$$

where $f(R) = \gamma 2^{-2R}$ and the number of quantizer levels found as $L = 2^R$. The parameter γ, also called the *fudge factor*, depends on the pdf type of the information source.

The optimum allocation of the total bits Ramong multiple information sources (transform coefficients in transform coding (TC)) is an important task in lossy compression. Transform coefficient variances σ_k^2 (or eigenvalues λ_k) are quite uneven to achieve dimension reduction in TC. Therefore, an optimum bit allocation algorithm assigns bit rate R_k for quantization of coefficient θ_k in such a way that the resulting quantization errors for all coefficients are forced to be equal, $\sigma_{q_0}^2 = \sigma_{q_1}^2 = \ldots = \sigma_{q_{N-1}}^2$ (Akansu and Haddad, 1992). The number of levels for the kth quantizer, for coefficient θ_k, is found as $L_k = 2^{R_k}$. Optimally allocated bits R_k among multiple sources for the total bit budget R,with the assumption that all sources have the same pdf type, are calculated as (Akansu and Haddad, 1992)

$$R_k = R + \frac{1}{2}\log_2 \frac{\sigma_k^2}{\left(\prod_{i=0}^{N-1} \sigma_i^2\right)^{\frac{1}{N}}}, \tag{5.91}$$

where $R = \sum_{k=0}^{N-1} R_k$.

5.4.3 *Quantization of Eigen Subspace*

In TC, sparsity in transform coefficients is desired. In contrast, any sparse transform including KLT aims to sparse subspace (transform matrix) where basis vector components are interpreted as loading coefficients in some applications (Akansu and Torun, 2012; Bollen and Whaley, 2009; Choi and Varian, 2012; Mamaysky *et al.*, 2008; Ohlson and Rosenberg, 1982; Torun and Akansu, 2013). Quantization of a given subspace with an optimally designed single quantizer, Q, or a set of quantizers $\{Q_k\}$, is defined as

$$\widehat{\Phi} = Q(\Phi). \tag{5.92}$$

In this case, Q is a pdf-optimized midtread quantizer designed for the entire transform matrix. Then, transform coefficients are obtained by using the quantized matrix

$$\widehat{\theta} = \widehat{\Phi}\boldsymbol{x}. \tag{5.93}$$

Unlike in TC, coefficients are not quantized in sparse representation methods. Instead, transform coefficients of the quantized subspace for a given signal vector are obtained. As in TC, quantization error equals to reconstruction error, both in mse. Mean-squared quantization error is expressed as

$$\sigma_{q,S}^2 = \frac{1}{N^2} \sum_{k=0}^{N-1} \widetilde{\phi_k}^{\mathbf{T}} \widetilde{\phi_k}, \tag{5.94}$$

where $\widetilde{\phi_k} = \phi_k - \widehat{\phi_k}$.

5.4.4 pdf of Eigenvector

We model the probability density of eigenvector components in order to design pdf-optimized quantizers to sparse them. Each eigenvector of an AR(1) process is generated by a sinusoidal function as expressed in (5.27). The pdf, with arbitrary support, of a continuous sinusoidal function is modeled as (Balakrishnan and Nevzorov, 2004; Hejn *et al.*, 1998):

$$p(x) = \frac{1}{\pi\sqrt{(x-a)\ (b-x)}}, \tag{5.95}$$

where a and b define the support, $a \leq x \leq b$. The cumulative distribution function (cdf) of such a function type is of arcsine distribution and expressed as

$$P(x) = \frac{2}{\pi}\arcsin\left(\sqrt{\frac{x-a}{b-a}}\right). \tag{5.96}$$

Mean and variance of the arcsine distribution are calculated as

$$\mu = \frac{a+b}{2}, \tag{5.97}$$

$$\sigma^2 = \frac{(b-a)^2}{8}. \tag{5.98}$$

The pdf of arcsine distribution is symmetric and U-shaped. The arcsine distribution with $a = 0$ and $b = 1$, namely standard arcsine distribution, is a special case of the beta distribution with the parameters $\alpha = \beta = 0.5$. Figure 5.6 shows the pdf of arcsine distribution with parameters $a = -0.0854$ and $b = 0.0854$. Log-concavity of a pdf $p(x)$ is the sufficient condition for the uniqueness of a pdf-optimized quantizer. However, arcsine distribution type has the log-convex property (Bagnoli and Bergstrom, 2005). It is stated in Yee (2010) that for exponential sources and the sources with strictly log convex pdfs, the quantizer intervals (bins) and their bin representation (quanta) values are globally optimum and unique. Therefore, pdf-optimized

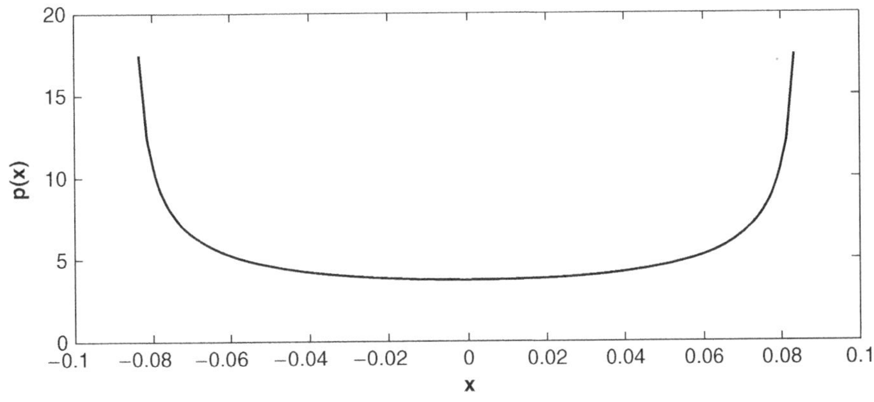

Figure 5.6 Probability density function of arcsine distribution for $a = -0.0854$ and $b = 0.0854$. Loadings of a second PC for an AR(1) signal source with $\rho = 0.9$ and $N = 256$ are fitted to arcsine distribution by finding minimum and maximum values in the PC.

quantizers can be designed for arcsine distribution (Lloyd, 1982; Max, 1960). The second principal component, ϕ_1, of an AR(1) source for $\rho = 0.9$ and size of $N = 256$ is shown to be fit by arcsine distribution with $a = \min(\phi_1)$ and $b = \max(\phi_1)$, respectively. Minimum and maximum valued components of the k*th* eigenvector depend on ρ, ω_k, and N as stated in (5.27). In order to maintain equal distortion levels among quantizers to sparse eigenvectors, we calculated optimal intervals for zero-zones of pdf-optimized midtread quantizers. Thus, most of the small valued eigenvector components are likely to be quantized as zero.

Figure 5.7a and Figure 5.7b display the normalized histograms of the first and second eigenvector components (PC1 and PC2 loading coefficients) for an AR(1) process with $\rho = 0.9$ and $N = 1024$. The value of N is selected large enough to generate proper histograms. The intervals of the histograms, Δ_k, are set as

$$\Delta_k = \frac{\max(\phi_k) - \min(\phi_k)}{N}, \tag{5.99}$$

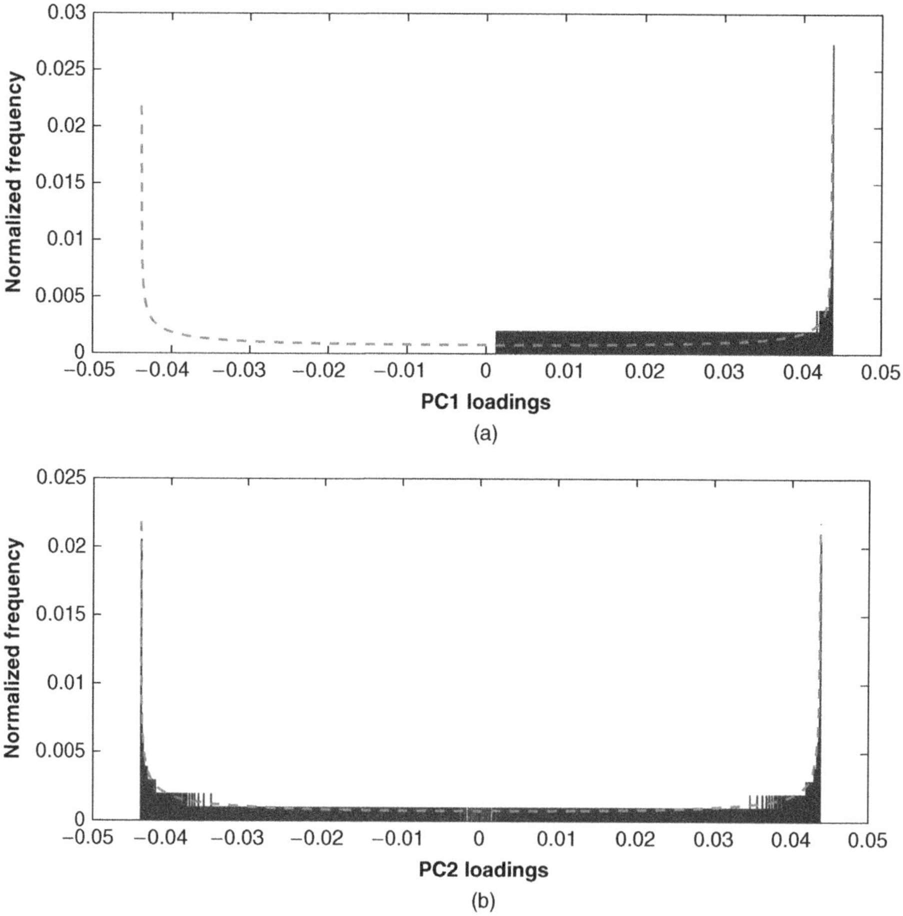

Figure 5.7 Normalized histograms of (a) PC1 and (b) PC2 loadings for an AR(1) signal source with $\rho = 0.9$ and $N = 1024$. The dashed lines in each histogram show the probability that is calculated by integrating an arcsine pdf for each bin interval.

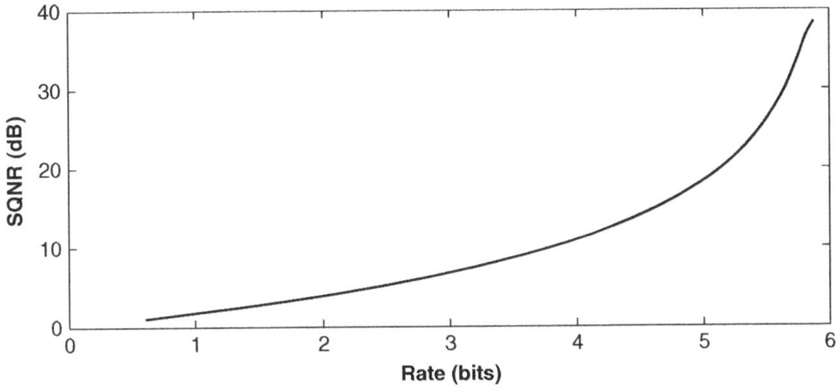

Figure 5.8 Rate (bits)-distortion (SQNR) performance of zero mean and unit variance arcsine pdf-optimized quantizer for $L = 65$ bins. The distortion level is increased by combining multiple bins around zero in a larger zero-zone.

where ϕ_k is the kth eigenvector. The dashed lines in each normalized histogram show the probability that is calculated by integrating the pdf of arcsine distribution in (5.95) for each bin interval. The histogram displayed in Figure 5.7a has only one side of the arcsine pdf, as expected from (5.27). In contrast, Figure 5.7b displays the histogram with a complete arcsine pdf shape. These figures confirm the arcsine distribution type for eigenvector components of an AR(1) process.

Now, we investigate the rate-distortion performance of an arcsine pdf-optimized zero-zone quantizer. Rate of quantizer output is calculated by using first-order entropy as defined in (5.89). Distortion caused by the quantizer is calculated in mse and represented in SQNR as defined in (5.87). Figure 5.8 displays rate-distortion performance of such a quantizer with $L = 65$. In this figure, distortion level is increased by increasing the zero-zone of the quantizer for more sparsity where rate decreases accordingly. One can design a quantizer with zero-zone for each eigenvector (PC) or for the entire eigenmatrix to achieve the desired level of matrix sparsity (Lloyd, 1982; Max, 1960).

5.4.5 Sparse KLT Method

In this subsection, we present in detail a simple method to sparse eigen subspace for an AR(1) process through a design example. The relevant parameter values for the sparse KLT (SKLT) example considered are tabulated in Table 5.1. The steps of design are summarized as follows.

1. First-order correlation coefficient ρ is calculated from the available dataset as described in (5.81), and $\mathbf{R}_x$ is constructed. Assume that $\rho = 0.9$ for the given example with $N = 256$.
2. Eigenvalues $\{\lambda_k\}$ and corresponding eigenvectors $\{\phi_k\}$ of $\mathbf{R}_x$ are calculated from (5.28) and (5.27), respectively. Eigenvalues of the first 16 eigenvectors (PCs) are listed in Table 5.1. These eigenvectors explain 68.28% of the total variance. Values of $\{\omega_k\}$ used to calculate each eigenvalue and corresponding eigenvector are also shown in Table 5.1. Note that $\eta_E(L)$ of (5.24) and the explained variance used in Table 5.1 measure energy compaction efficiency of a subspace.

3. PC loading coefficients (eigenvector components) are fitted to arcsine distribution by calculating $\{a_k = \min(\phi_k)\}\forall k$ and $\{b_k = \max(\phi_k)\}\forall k$. Then, variances $\left\{\sigma_k^2 = \frac{(b_k-a_k)^2}{8}\right\}\forall k$, are calculated by using (5.98). Table 5.1 also tabulates $\{a_k\}$, $\{b_k\}$, and $\{\sigma_k^2\}$ of eigenvectors.
4. For a given total rate R (desired level of sparsity), $\{R_k\}$ are calculated by plugging $\{\sigma_k^2\}$ in the optimum bit allocation equation given in (5.91). Then, quantizer levels $\{L_k\}$ are calculated as $\{L_k = 2^{R_k}\}\forall k$ and rounded up to the closest odd integer number. R is the sparsity tuning parameter of SKLT. As in all of the sparse PCA methods, R for a given sparsity has to be determined with cross-validation. Table 5.1 displays calculated rates and quantizer levels for the total rate of $R = 5.7$.
5. For this design example, a $L = 65$ level pdf-optimized zero-zone quantizer of arcsine distribution with zero mean and unit variance is used as the starting point. Then, several adjacent bins around zero are combined to adjust the zero-zone for the desired sparsity level. For the kth eigenvector, a predesigned $L = 65$ level pdf-optimized the zero-zone quantizer is converted to an $L_k \leq L$ level zero-zone quantizer.
6. PC loadings (eigenvector components) are normalized to have zero mean and unit variance, $\left\{\phi_k = \frac{(\phi_k - mean(\phi_k))}{std(\phi_k)}\right\}\forall k$, where *mean* and *std* are the mean and standard deviation of eigenvector components, respectively. Quantized (sparsed) eigenvectors are generated by applying quantization on eigenvectors of the original eigen subspace $\{\widehat{\phi_k} = Q_k(\phi_k)\}\forall k$. The number of zero components or sparsity level (S) of quantized PCs for this example are also given in Table 5.1.

Remark 5.1 *The number of bins for a predesigned pdf-optimized quantizer is selected based on the quantization noise and implementation cost. The increase in signal-to-quantization noise (SQNR) of a pdf-optimized zero-zone quantizer optimized for arcsine pdf with $L > 65$ is found not to be that significant.*

Table 5.1 Relevant parameter values of SKLT example for the first 16 PCs of an AR(1) source with $\rho = 0.9$ and $N = 256$. They explain 68.28% of the total variance.

	ω	λ	a	b	σ^2	R	L	S
PC1	0.0114	18.77	−0.0853	0.0853	0.0036	5.6546	51	26
PC2	0.0229	18.14	−0.0853	0.0853	0.0036	5.6563	51	28
PC3	0.0344	17.17	−0.0856	0.0856	0.0037	5.6588	51	40
PC4	0.0459	15.97	−0.0857	0.0857	0.0037	5.6620	51	34
PC5	0.0575	14.64	−0.0860	0.0860	0.0037	5.6655	51	36
PC6	0.0691	13.29	−0.0862	0.0862	0.0037	5.6691	51	38
PC7	0.0808	11.97	−0.0864	0.0864	0.0037	5.6725	51	42
PC8	0.0925	10.73	−0.0866	0.0866	0.0037	5.6754	51	42
PC9	0.1043	9.60	−0.0868	0.0868	0.0038	5.6790	51	40
PC10	0.1162	8.58	−0.0869	0.0869	0.0038	5.6819	51	36
PC11	0.1281	7.67	−0.0871	0.0870	0.0038	5.6835	51	44
PC12	0.1400	6.88	−0.0872	0.0872	0.0038	5.6866	53	36
PC13	0.1520	6.17	−0.0873	0.0873	0.0038	5.6881	53	36
PC14	0.1640	5.56	−0.0874	0.0874	0.0038	5.6902	53	36
PC15	0.1760	5.02	−0.0875	0.0875	0.0038	5.6915	53	36
PC16	0.1881	4.55	−0.0876	0.0876	0.0038	5.6930	53	36

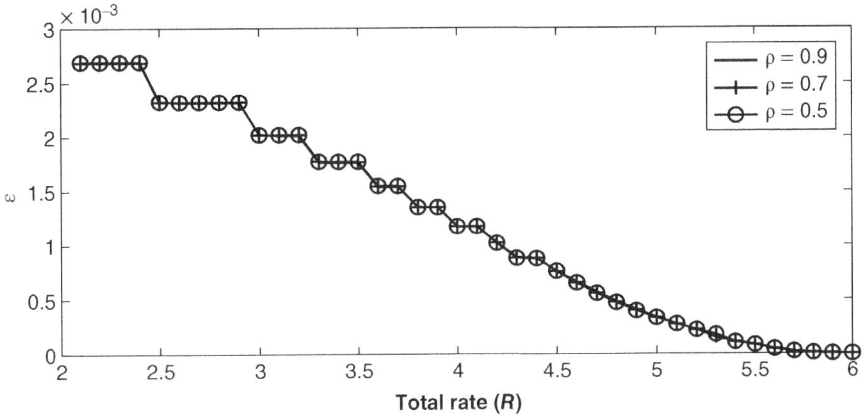

Figure 5.9 Orthogonality imperfectness-rate (sparsity) trade-off for sparse eigen subspaces of three AR(1) sources with $N = 256$.

Sparsity achieved by quantization of PCs leads to orthogonality imperfectness. We present orthogonality imperfectness ϵ in mse with respect to allowable total rate R (desired sparsity level) for various AR(1) sources defined as

$$\epsilon = \frac{1}{N^2} \sum_{i=0}^{N-1} \sum_{i=0}^{N-1} [\mathbf{I}(i,j) - \mathbf{K}(i,j)]^2, \tag{5.100}$$

where $\mathbf{I}$ is the $N \times N$ identity matrix; and $\mathbf{K} = \mathbf{A}\mathbf{A}^{*\mathbf{T}}$.

Figure 5.9 displays the trade-off between subspace sparsity and loss of orthogonality for various AR(1) sources and $N = 256$. It is observed from Figure 5.9 that the orthogonality imperfectness decreases almost linearly with increasing R, as expected.

5.4.6 *Sparsity Performance*

Let us compare performances of the presented SKLT method with the ST (Zou *et al.*, 2006), SPCA (Zou *et al.*, 2006), DSPCA (d'Aspremont *et al.*, 2007), and SPC (d'Aspremont *et al.*, 2007) methods for the AR(1) process, and also for an empirical correlation matrix of stock returns in the NASDAQ-100 index in this subsection. In order to provide a fair comparison, sparsity levels of all methods considered here are tuned in such a way that compared PCs have almost the same number of nonzero components. In most cases, the number of nonzero components of each PC in the SKLT method are kept slightly lower than the others in order to show the method's merit under mildly disadvantageous test conditions.

5.4.6.1 Eigen Subspace Sparsity for the AR(1) Process

The sparsity imposed on PCs may degrade the explained variance described in d'Aspremont *et al.* (2007). The explained variances of the PCs are calculated as $\{\lambda_k = \sigma_k^2 = \phi_k^{\mathrm{T}} \mathbf{R}_x \phi_k\} \forall k$, where ϕ_k is the k*th* eigenvector for a given $\mathbf{R}_x$. For the sparsed PCs, new explained variances (eigenvalues) are calculated as $\{\widehat{\lambda}_k = \widehat{\sigma_k^2} = \widehat{\phi_k}^{\mathbf{T}} \mathbf{R}_x \widehat{\phi_k}\} \forall k$, where $\widehat{\phi_k}$ is the k*th* sparse eigenvector. Then, the percentage of explained variance loss (VL) as a performance metric is defined as $\{V_k = \frac{(\lambda_k - \widehat{\lambda}_k)}{\lambda_k} \times 100\} \forall k$. Similarly, the cumulative explained variance loss of first L PCs is defined as $C_L = \sum_{k=1}^{N} \lambda_k - \sum_{k=1}^{L} \widehat{\lambda}_k$. In addition, we also used a nonsparsity (NS) performance metric for comparison. It is defined as the percentage of nonzero components in a given sparsed eigenvector. Thus, the performance is measured as the variance loss for the given NS level (d'Aspremont *et al.*, 2007; Zou and Hastie, 2005; Zou *et al.*, 2006). We are unable to provide their comparative rate-distortion performance due to the lack of models to generate sparse PCs for all methods reported here.

Figure 5.10a displays the VL measurements of sparsed first PCs generated by SKLT, SPCA, SPC, ST, and DSPCA methods with respect to NS for an AR(1) source with $\rho = 0.9$ and $N = 256$. For SKLT, an $L = 65$ level quantizer optimized for arcsine pdf with zero mean and

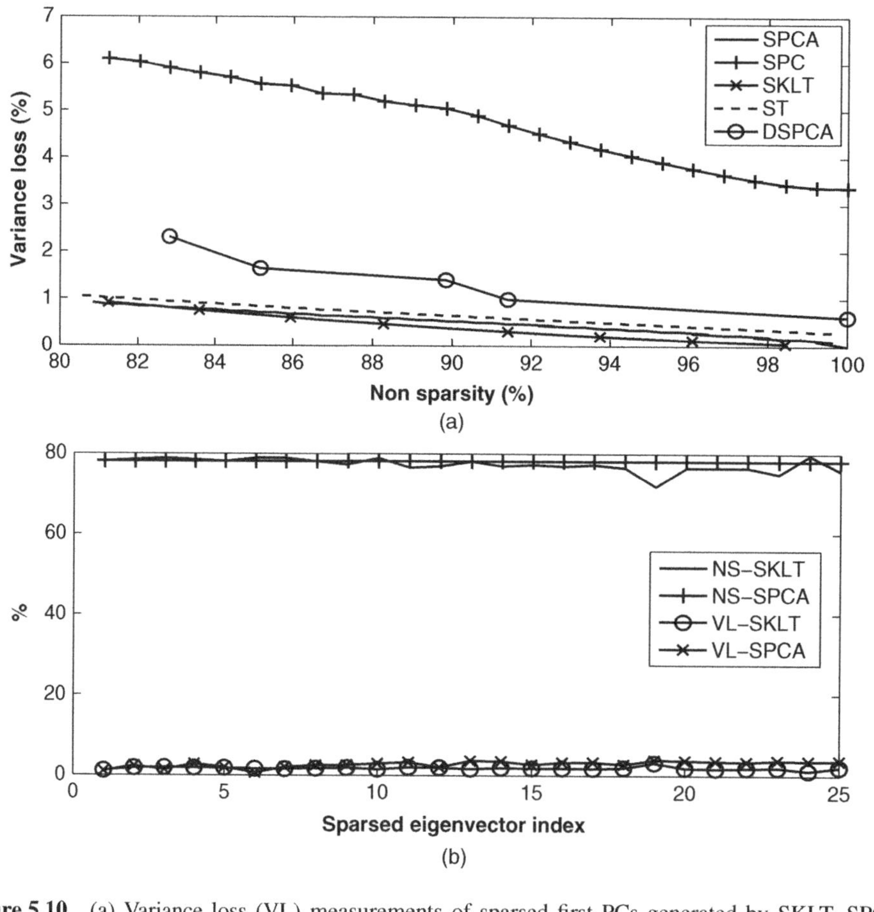

Figure 5.10 (a) Variance loss (VL) measurements of sparsed first PCs generated by SKLT, SPCA, SPC, ST, and DSPCA methods with respect to nonsparsity (NS) for an AR(1) source with $\rho = 0.9$ and $N = 256$; (b) NS and VL measurements of sparsed eigenvectors for an AR(1) source with $\rho = 0.9$ and $N = 256$ generated by the SKLT method and SPCA algorithm.

unit variance is used as the initial quantizer. The zero-zone width of the initial quantizer is adjusted for required sparsity, as explained in this chapter. Then, the generated quantizer is employed. Figure 5.10a shows that SKLT offers less variance loss than the other methods. SPCA provides competitive performance to SKLT. Figure 5.10b displays NS and VL performance comparisons of sparse PCs generated by SKLT and by SPCA for the same AR(1) process. The original eigenvectors that explain 90% of the total variance are selected for sparsity comparison. Figure 5.10b shows that the VL performance of SKLT is slightly better than that of SPCA. Note that the NS of SKLT is slightly lower than that of SPCA in this comparison.

5.4.6.2 Eigen Subspace Sparsity for the NASDAQ-100 Index

We present an example to sparse an eigen subspace that leads to the creation of the corresponding sparse eigenportfolios (Akansu and Torun, 2015). Eigendecomposition of empirical correlation matrices is a popular mathematical tool in finance employed for various tasks including eigenfiltering of measurement noise and creation of eigenportfolios for baskets of stocks (Akansu and Torun, 2012; Markowitz, 1959; Torun *et al.*, 2011). An empirical correlation matrix for the end-of-day (EOD) stock returns for the NASDAQ-100 index with $W = 30$-day time window ending on April 9, 2014, is measured (Torun *et al.*, 2011). The vector of 100 stock returns in the NASDAQ-100 index at time n is created as (Akansu and Torun, 2012):

$$\mathbf{r}(n) = [r_k(n)];\, k = 1, 2, \ldots, 100. \tag{5.101}$$

The empirical correlation matrix of returns at time n is expressed as

$$\mathbf{R}_E(n) \triangleq [E\{\mathbf{r}(n)\mathbf{r}^T(n)\}] = [R_{k,l}(n)], \tag{5.102}$$

$$= \begin{bmatrix} R_{1,1}(n) & R_{1,2}(n) & \cdots & R_{1,100}(n) \\ R_{2,1}(n) & R_{2,2}(n) & \cdots & R_{2,100}(n) \\ \vdots & \vdots & \ddots & \vdots \\ R_{100,1}(n) & R_{100,2}(n) & \cdots & R_{100,100}(n) \end{bmatrix},$$

where the matrix elements

$$R_{k,l}(n) = E\{r_k(n)r_l(n)\} = \frac{1}{W}\sum_{m=0}^{W-1} r_k(n-m)r_l(n-m)$$

represent measured pairwise correlations for an observation window of W samples. The returns are normalized to be zero mean and unit variance, and $\mathbf{R}_E(n)$ is a real, symmetric, and positive definite matrix. Now, we introduce eigendecomposition of $\mathbf{R}_E$ as follows:

$$\mathbf{R}_E(n) = \mathbf{A}_{KLT}^T \mathbf{\Lambda} \mathbf{A}_{KLT} = \sum_{k=1}^{N} \lambda_k \boldsymbol{\phi}_k \boldsymbol{\phi}_k^T, \tag{5.103}$$

where $\{\lambda_k, \boldsymbol{\phi}_k\}$ are eigenvalue–eigenvector pairs (Akansu and Torun, 2012).

The component values of eigenvector $\{\boldsymbol{\phi}_k\}$ are repurposed as the capital allocation coefficients to create the kth eigenportfolio for a group of stocks where the resulting coefficients $\{\theta_k\}$ are pairwise uncorrelated. These coefficients represent eigenportfolio returns. Eigenportfolios are used in various investment and trading strategies (Chamberlain and Rothschild, 1983). It is

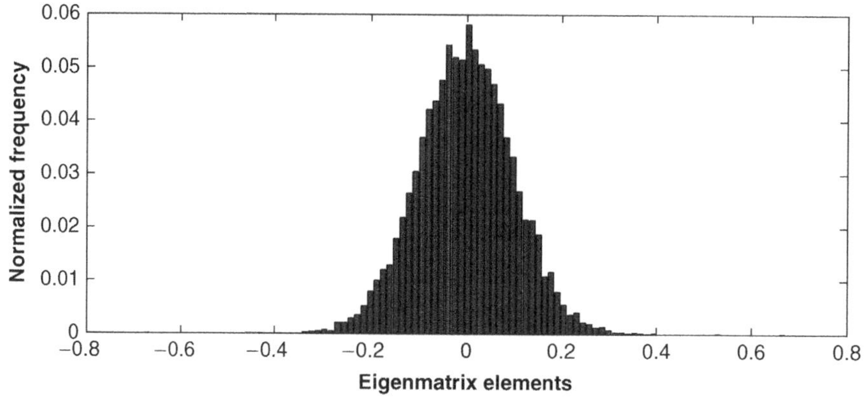

Figure 5.11 Normalized histogram of eigenmatrix elements for an empirical correlation matrix of end-of-day (EOD) returns for 100 stocks in the NASDAQ-100 index. $W = 30$-day measurement window ending on April 9, 2014.

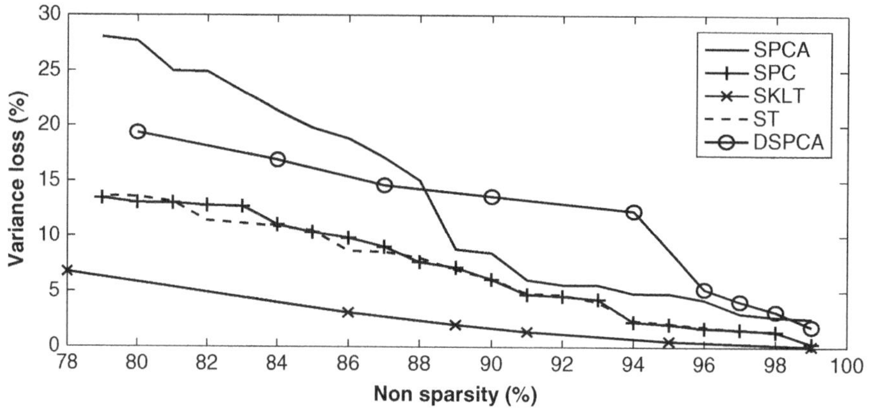

Figure 5.12 VL measurements of sparsed first PCs generated by SKLT, SPCA, SPC, ST, and DSPCA methods with respect to NS for an empirical correlation matrix of EOD returns for 100 stocks in the NASDAQ-100 index with $W = 30$-day measurement window ending on April 9, 2014.

required to buy and sell certain stocks in the amounts defined by the loading (capital allocation) coefficients in order to build and rebalance eigenportfolios in time for the targeted risk levels. Some of the loading coefficients may have relatively small values where their trading cost becomes a practical concern for portfolio managers. Therefore, sparsing eigen subspace of an empirical correlation matrix $\mathbf{R}_E(n)$ may offer cost reductions in desired trading activity. In contrast, although theoretically appealing, the optimization algorithms like SPCA, DSPCA, and SPC with constraints for forced sparsity (cardinality reduction of a set) may substantially alter intrinsic structures of original eigenportfolios. Therefore, such a sparse representation might cause a significant deviation from the measured empirical correlation matrix. Hence, the use of eigenportfolios generated by a sparsity-constrained optimization in a trading strategy may lead to poor performance.

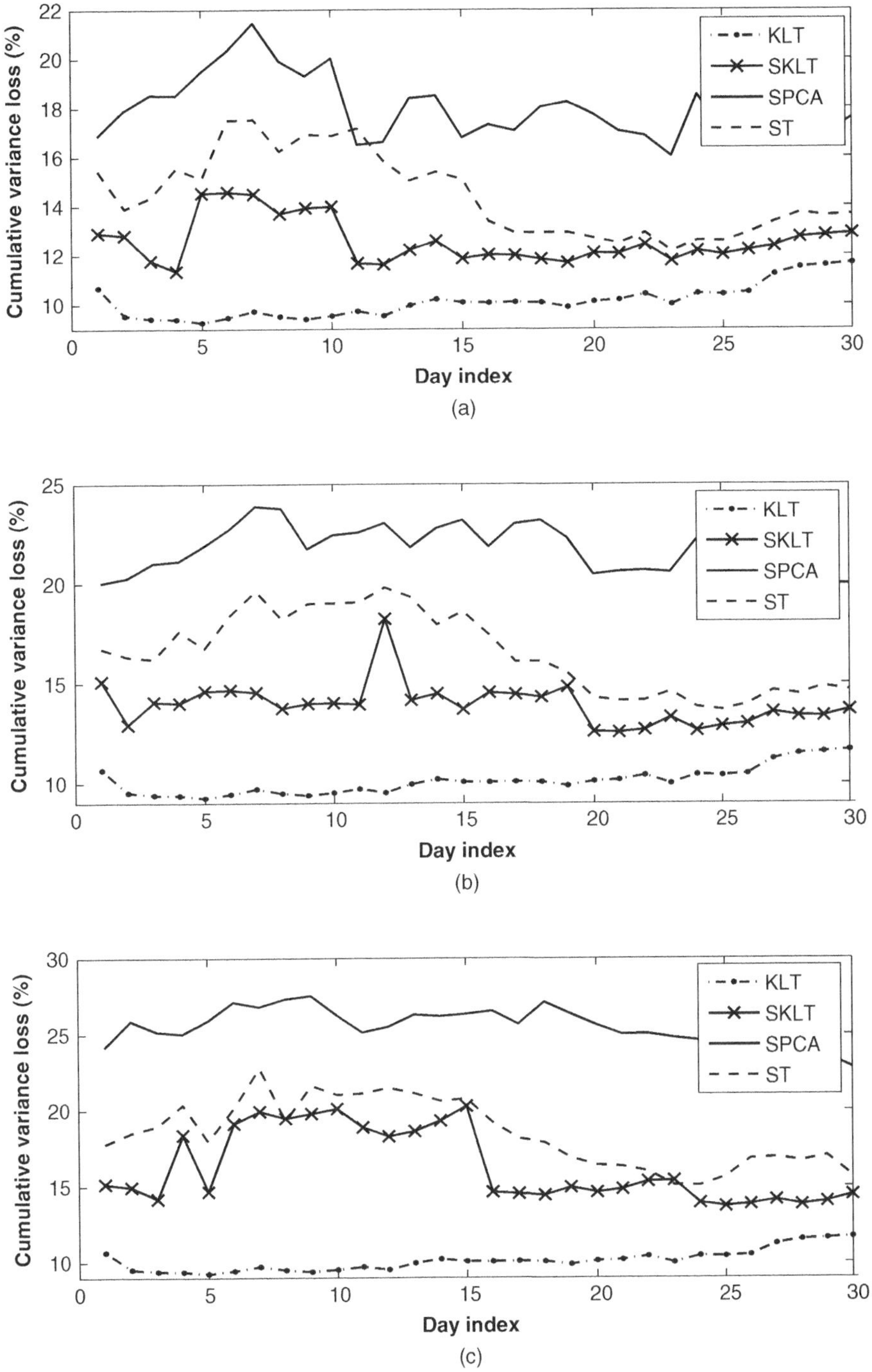

Figure 5.13 Cumulative explained variance loss with $L = 16$ generated daily from an empirical correlation matrix of EOD returns between April 9, 2014, and May 22, 2014, for 100 stocks in the NASDAQ-100 index by using KLT, SKLT, SPCA, and ST methods. NS levels of 85%, 80%, and 75% for all PCs are forced in (a), (b), and (c), respectively, using $W = 30$ days.

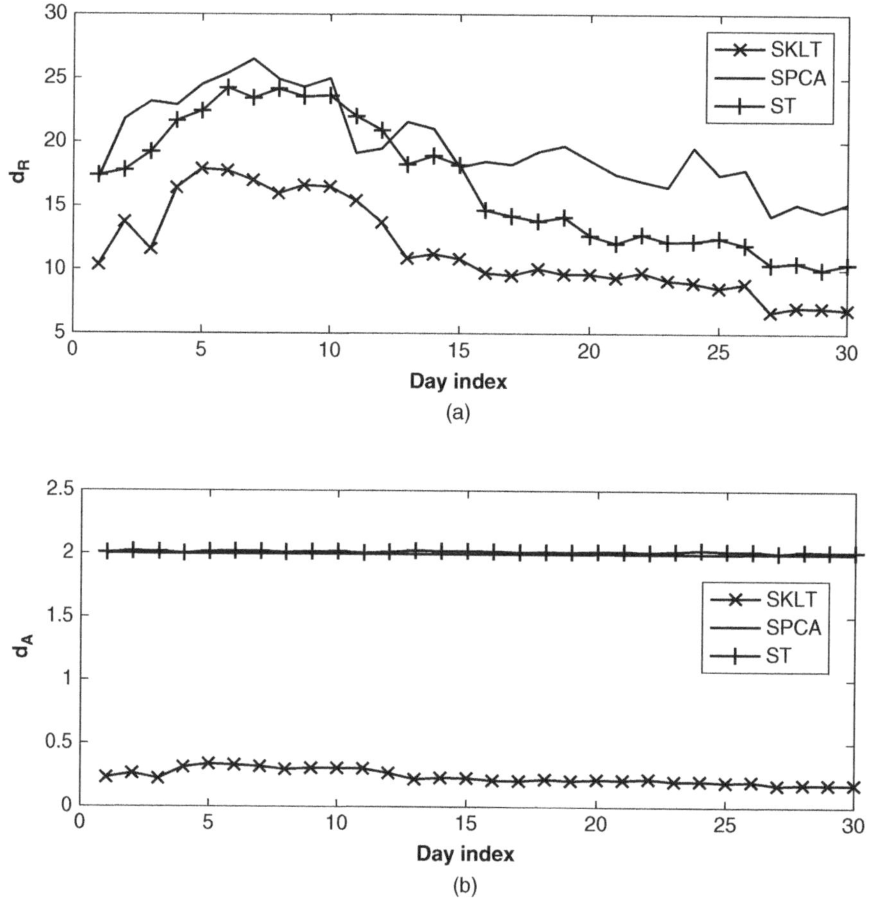

Figure 5.14 (a) $d_{\mathbf{R}}$ and (b) $d_{\mathbf{A}}$ of sparse eigen subspaces generated daily from an empirical correlation matrix of EOD returns between April 9, 2014, and May 22, 2014, for 100 stocks in the NASDAQ-100 index by using SKLT, SPCA, and ST methods, respectively. NS level of 85% for all PCs is forced with $W = 30$ days.

For simplicity, a single quantizer in the SKLT method is utilized to sparse the entire eigenmatrix $\mathbf{A}_{\mathrm{KLT}}$. It is optimized for the histogram of its elements as displayed in Figure 5.11. It is observed to be a Gaussian pdf. Figure 5.12 displays the VL measurements of sparsed first PCs generated by SKLT, SPCA, SPC, ST, and DSPCA methods with respect to NS. It is shown that SKLT offers less variance loss than other methods considered in this chapter. Figure 5.13 displays the cumulative explained variance loss with $L = 16$ generated daily from an empirical correlation matrix of EOD returns between April 9, 2014, and May 22, 2014, for 100 stocks in the NASDAQ-100 index by using KLT, SKLT, SPCA, and ST methods. The measurement window of the last 30 days, $W = 30$, is used. NS levels of 85%, 80%, and 75% for each PC are forced in experiments as displayed in Figure 5.13a, Figure 5.13b, and Figure 5.13c, respectively. The SKLT method consistently outperforms the others for this time-varying scenario.

The difference between the original $\mathbf{R}_E(n)$ and the modified correlation matrix $\widehat{\mathbf{R}_E(n)}$ due to sparsed eigenvectors is defined as

$$d_{\mathbf{R}} = \left\| \mathbf{R}_E(n) - \widehat{\mathbf{R}_E(n)} \right\|_2, \tag{5.104}$$

where $\|.\|_2$ is the norm-2 of a matrix. Hence, the distance between the original and the sparsed eigenmatrices is expressed as

$$d_{\mathbf{A}} = \left\| \mathbf{A}_{KLT} - \widehat{\mathbf{A}_{KLT}} \right\|_2. \tag{5.105}$$

Figure 5.14a and Figure 5.14b display the $d_{\mathbf{R}}$ and $d_{\mathbf{A}}$ of sparse eigen subspaces generated daily from an empirical correlation matrix of EOD returns between April 9, 2014, and May 22, 2014, for 100 stocks in the NASDAQ-100 index by using SKLT, SPCA, and ST methods, respectively. The NS level of 85% for all PCs is forced with $W = 30$ days. These performance measures highlight that the SKLT method sparses eigen subspace of the NASDAQ-100 index better than the SPCA and ST methods. Moreover, the SKLT does not force an alteration of the actual covariance structure like other methods.

5.5 Conclusions

Closed-form expressions for KLT kernel (eigenvectors) and corresponding transform coefficient variances (eigenvalues) of the AR(1) process were reported in the literature (Ray and Driver, 1970). However, they require solving a transcendental tangent equation (5.29). Mathematical steps leading to equations (5.27), (5.28), and (5.29) are discussed in detail, following the methodology used for a continuous stochastic process with exponential autocorrelation function (Davenport and Root, 1958; Pugachev, 1959a,b; Wilkinson, 1965). Then, a simple and fast method to find the roots of a transcendental equation is employed to derive the roots of (5.29) explicitly. That derivation made it possible to express the $N \times N$ KLT kernel and corresponding coefficient variances in explicit form, leading to extremely fast KLT implementations for processes that can be modeled with AR(1) process. The technique is shown to be more efficient than the D&Q algorithm (Golub and Loan, 1996).

The constrained optimization algorithms to generate sparse PCs do not guarantee a good performance for an arbitrary covariance matrix due to the nonconvex nature of the problem. The SKLT method to sparse subspaces is presented in this chapter. It utilizes the mathematical framework developed for transform coding in rate-distortion theory. The sparsity performance comparisons demonstrate the superiority of SKLT over the popular algorithms known in the literature.

References

Akansu, A.N. and Haddad, R.A. (1992). *Multiresolution signal decomposition: transforms, subbands, and wavelets*. New York: Academic Press.

Akansu, A.N. and Torun, M.U. (2012). Toeplitz approximation to empirical correlation matrix of asset returns: a signal processing perspective. *Journal of Selected Topics in Signal Processing*, 6 (4), 319–326.

Akansu, A.N. and Torun, M.U. (2015). *A primer for financial engineering: financial signal processing and electronic trading*. New York: Academic Press-Elsevier.

Allen, M. and Isaacson, E. (1997). *Numerical analysis for applied science*. New York: John Wiley & Sons.

Atal, B. and Hanauer, S. (1971). Speech analysis and synthesis by linear prediction of the speech wave. *The Journal of the Acoustical Society of America*, 50, 637.

Bagnoli, M. and Bergstrom, T. (2005). Log-concave probability and its applications. *Economic Theory*, 26 (2), 445–469.

Balakrishnan, N. and Nevzorov, V.B. (2004). *A primer on statistical distributions*. Hoboken, NJ: John Wiley & Sons.

Berger, T. (2003). *Rate-distortion theory*. Hoboken, NJ: John Wiley and Sons.

Bertero, M. and Boccacci, P. (1998). *Introduction to inverse problems in imaging*. London: Institute of Physics Publishing.

Bollen, N.P.B. and Whaley, R.E. (2009). Hedge fund risk dynamics: implications for performance appraisal. *The Journal of Finance*, 64 (2), 985–1035.

Brodie, J., Daubechies, I., De Mol, C., Giannone D. and Loris, I. (2009). Sparse and stable Markowitz portfolios. *Proceedings of the National Academy of Sciences*, 106 (30), 12267–12272.

Brown, J. and Churchill, R. (2009). *Complex variables and applications*. New York: McGraw-Hill.

Brusewitz, H. (1986). Quantization with entropy constraint and bounds to the rate distortion function, ser. Trita-TTT. *Telecommunication Theory, Electrical Engineering*, 8605, 28–29.

Cadima, J. and Jolliffe, I.T. (1995). Loading and correlations in the interpretation of principle components. *Journal of Applied Statistics*, 22 (2), 203–214.

Chamberlain, G. and Rothschild, M. (1983) Arbitrage, factor structure and mean-variance analysis on large asset markets. *Econometrica*, 51 (5), 1281–1304.

Choi, H. and Varian, H. (2012). Predicting the present with Google Trends. *Economic Record*, 88, 2–9.

Clarke, R.J. (1985). *Transform coding of images*. New York: Academic Press.

d'Aspremont, A. El Ghaoui, L. Jordan, M.I. and Lanckriet, G.R.G. (2007). A direct formulation for sparse PCA using semidefinite programming. *SIAM Review*, 49 (3), 434–448.

Davenport, W.B. and Root, W.L. (1958). *An introduction to the theory of random signals and noise*. New York: McGraw-Hill.

Doob, J. (1942). The Brownian movement and stochastic equations. *Annals of Mathematics*, 43 (2), 351–369.

Engl, H.W. Hanke, M. and Neubauer, A. (1996). *Regularization of inverse problems*. New York: Kluwer Academic.

Golub, G.H. and Loan, C.F.V. (1996). *Matrix computations*. Baltimore: Johns Hopkins University Press.

Gonzales, C. and Akansu, A.N. (1997). A very efficient low-bit-rate subband image/video codec using shift-only PR-QMF and zero-zone linear quantizers. *Proceedings of the IEEE International Conference on Acoustics, Speech and Signal Processing*, 4, 2993–2996.

Hajnal, A. and Juhasz, I. (1983). Remarks on the cardinality of compact spaces and their Lindelof subspaces. *Proceedings of the American Mathematical Society*, 51 (5), 146–148.

Hejn, K., Pacut, A. and Kramarski, L. (1998). The effective resolution measurements in scope of sine-fit test. *IEEE Transactions on Instrumentation and Measurement*, 47 (1), 45–50.

Jayant, N.S. and Noll, P. (1984). *Digital coding of waveforms: principles and applications to speech and video*. Englewood Cliffs, NJ: Prentice Hall Professional Technical Reference.

Jolliffe, I.T. (2002). *Principal component analysis*. New York: Springer-Verlag.

Kay, S. (1988). *Modern spectral estimation: theory and application*. Upper Saddle River, NJ: Prentice Hall.

Lloyd, S. (1982). Least squares quantization in PCM. *IEEE Transactions on Information Theory*, 28 (2), 129–137.

Luck, R. and Stevens, J. (2002). Explicit solutions for transcendental equations. *SIAM Review*, 44, 227–233.

Mamaysky, H. Spiegel, M. and Zhang, H. (2008). Estimating the dynamics of mutual fund alphas and betas. *Review of Financial Studies*, 21, 233–264.

Markowitz, H.M. (1959). *Portfolio selection: efficient diversification of investments*. New York: John Wiley & Sons.

Max, J. (1960). Quantizing for minimum distortion. *IRE Transactions on Information Theory*, 6 (1), 7–12.

Ohlson, J. and Rosenberg, B. (1982). Systematic risk of the CRSP equal-weighted common stock index: a history estimated by stochastic-parameter regression. *The Journal of Business*, 55 (1), 121–145.

Papoulis, A. (1991). *Probability, random variables, and stochastic processes*. New York: McGraw-Hill.

Pugachev, V. (1959a). A method for the determination of the eigenvalues and eigenfunctions of a certain class of linear integral equations. *Journal of Applied Mathematics and Mechanics* (translation of the Russian journal *Prikladnaya Matematika i Mekhanika*), 23 (3), 527–533.

Pugachev, V. (1959b). A method of solving the basic integral equation of statistical theory of optimum systems in finite form. *Journal of Applied Mathematics and Mechanics (translation of the Russian journal Prikladnaya Matematika i Mekhanika)*, 23 (1), 3–14.

Ray, W. and Driver, R. (1970). Further decomposition of the Karhunen-Loeve series representation of a stationary random process. *IEEE Transactions on Information Theory*, 16 (6), 663–668.

Shen, H. and Huang, J.Z. (2008). Sparse principal component analysis via regularized low rank matrix approximation. *Journal of Multivariate Analysis*, 99 (6), 1015–1034.

Sriperumbudur, B.K. Torres, D.A. and Lanckriet, G.R.G. The sparse eigenvalue problem. http://citeseerx.ist.psu.edu/viewdoc/download?doi=10.1.1.243.8515&rep=rep1&type=pdf

Strang, G. (1986). *Introduction to applied mathematics*. Wellesley, MA: Wellesley-Cambridge Press.

Tibshirani, R. (1996). Regression shrinkage and selection via the lasso. *Journal of the Royal Statistical Society, Series B*, 58, 267–288.

Torun, M.U. and Akansu, A.N. (2013). An efficient method to derive explicit KLT kernel for first-order autoregressive discrete process. *IEEE Transactions on Signal Processing*, 61 (15), 3944–3953.

Torun, M.U., Akansu, A.N. and Avellaneda, M. (2011). Portfolio risk in multiple frequencies. *IEEE Signal Processing Magazine, Special Issue on Signal Processing for Financial Applications*, 28 (5), 61–71.

Trendafilov, N. Jolliffe, I.T. and Uddin, M. (2003). A modified principal component technique based on the LASSO. *Journal of Computational and Graphical Statistics*, 12, 531–547.

Uhlenbeck, G. and Ornstein, L. (1930). On the theory of the Brownian motion. *Physical Review*, 36, 823–841.

Van Trees, H.L. (2001). *Detection, estimation, and modulation theory*. New York: John Wiley & Sons.

Wilkinson, J. (1965). *The algebraic eigenvalue problem*. Oxford: Oxford University Press.

Witten, D.M. Tibshirani, R. and Hastie, T. (2009). A penalized matrix decomposition, with applications to sparse principal components and canonical correlation analysis. *Biostatistics*, 10 (3), 515.

Yee, V.B. (2010). Studies on the asymptotic behavior of parameters in optimal scalar quantization. Ph.D. dissertation, The University of Michigan.

Zass, R. and Shashua, A. (2006). Nonnegative sparse PCA. In *Advances in* neural information processing systems (ed. Scholkopf, B., Platt, J. and Hoffman, T.). Cambridge, MA: MIT Press, 1561–1568.

Zou, H. and Hastie, T. (2005). Regularization and variable selection via the elastic net. *Journal of the Royal Statistical Society: Series B (Statistical Methodology)*, 67 (2), 301–320.

Zou, H. Hastie, T. and Tibshirani, R. (2006). Sparse principal component analysis. *Journal of Computational and Graphical Statistics*, 15, 262–286.

6

Approaches to High-Dimensional Covariance and Precision Matrix Estimations

Jianqing Fan[1], Yuan Liao[2] and Han Liu[3]
[1] *Bendheim Center for Finance, Princeton University, USA*
[2] *Department of Mathematics, University of Maryland, USA*
[3] *Department of Operations Research and Financial Engineering, Princeton University, USA*

6.1 Introduction

Large covariance and precision (inverse covariance) matrix estimations have become fundamental problems in multivariate analysis that find applications in many fields, ranging from economics and finance to biology, social networks, and health sciences. When the dimension of the covariance matrix is large, the estimation problem is generally challenging. It is well-known that the sample covariance based on the observed data is singular when the dimension is larger than the sample size. In addition, the aggregation of a huge amount of estimation errors can make considerable adverse impacts on the estimation's accuracy. Therefore, estimating large covariance and precision matrices has attracted rapidly growing research attention in the past decade. Many regularized methods have been developed: see Bickel and Levina (2008), El Karoui (2008), Friedman *et al.* (2008), Fryzlewicz (2013), Han *et al.* (2012), Lam and Fan (2009), Ledoit and Wolf (2003), Pourahmadi (2013), Ravikumar *et al.*, (2011b), Xue and Zou (2012), among others.

One of the commonly used approaches to estimating large matrices is to assume the covariance matrix to be sparse, that is, many off-diagonal components are either zero or nearly so. This effectively reduces the total number of parameters to estimate. However, such a sparsity assumption is restrictive in many applications. For example, financial returns depend on the common risk factors, housing prices depend on the economic health, and gene expressions can

Financial Signal Processing and Machine Learning, First Edition.
Edited by Ali N. Akansu, Sanjeev R. Kulkarni and Dmitry Malioutov.
© 2016 John Wiley & Sons, Ltd. Published 2016 by John Wiley & Sons, Ltd.

be stimulated by cytokines. Moreover, in many applications, it is more natural to assume that the precision matrix is sparse instead (e.g., in Gaussian graphical models).

In this chapter, we introduce several recent developments for estimating large covariance and precision matrices without assuming the covariance matrix to be sparse. One of the selected approaches assumes the precision matrix to be sparse and applies column-wise penalization for estimations. This method efficiently estimates the precision matrix in Gaussian graphical models. The other method is based on high-dimensional factor analysis. Both methods will be discussed in Sections 6.2 and 6.3, and are computationally more efficient than the existing ones based on penalized maximum likelihood estimation. We present several applications of these methods, including graph estimation for gene expression data, and several financial applications. In particular, we shall see that estimating covariance matrices of high-dimensional asset excess returns plays a central role in applications of portfolio allocations and in risk management.

In Section 6.4, we provide a detailed description of the so-called factor pricing model, which is one of the most fundamental results in finance. It postulates how financial returns are related to market risks, and has many important practical applications, including portfolio selection, fund performance evaluation, and corporate budgeting. In the model, the excess returns can be represented by a factor model. We shall also study a problem of testing "mean–variance efficiency." In such a testing problem, most of the existing methods are based on the Wald statistic, which has two main difficulties when the number of assets is large. First, the Wald statistic depends on estimating a large inverse covariance matrix, which is a challenging problem in a data-rich environment. Second, it suffers from a lower power in a high-dimensional, low-sample-size situation. To address the problem, we introduce a new test, called the *power enhancement test*, which aims to enhance the power of the usual Wald test.

In Section 6.5, we will present recent developments of efficient estimations in panel data models. As we shall illustrate, the usual principal components method for estimating the factor models is not statistically efficient since it treats the idiosyncratic errors as both cross-sectionally independent and homoscedastic. In contrast, using a consistent high-dimensional precision covariance estimator can potentially improve the estimation efficiency. We shall conclude in Section 6.6.

Throughout the chapter, we shall use $\|\mathbf{A}\|_2$ and $\|\mathbf{A}\|_F$ as the operator and Frobenius norms of a matrix $\mathbf{A}$. We use $\|\mathbf{v}\|$ to denote the Euclidean norm of a vector $\mathbf{v}$.

6.2 Covariance Estimation via Factor Analysis

Suppose we observe a set of stationary data $\{\boldsymbol{Y}_t\}_{t=1}^T$, where each $\boldsymbol{Y}_t = (Y_{1t}, \cdots, Y_{N,t})'$ is a high-dimensional vector; here, T and N respectively denote the sample size and the dimension. We aim to estimate the covariance matrix of $\boldsymbol{Y}_t$: $\boldsymbol{\Sigma} = \mathrm{Cov}(\boldsymbol{Y}_t)$, and its inverse $\boldsymbol{\Sigma}^{-1}$, which are assumed to be independent of t. This section introduces a method of estimating $\boldsymbol{\Sigma}$ and its inverse via factor analysis. In many applications, the cross-sectional units often depend on a few common factors. Fan *et al.* (2008) tackled the covariance estimation problem by considering the following factor model:

$$Y_{it} = \mathbf{b}_i' \boldsymbol{f}_t + u_{it}. \tag{6.1}$$

where Y_{it} is the observed response for the ith ($i = 1, \ldots, N$) individual at time $t = 1, \ldots, T$; $\mathbf{b}_i$ is a vector of factor loadings; $\boldsymbol{f}_t$ is a $K \times 1$ vector of common factors; and u_{it} is the error term, usually called *idiosyncratic component*, uncorrelated with $\boldsymbol{f}_t$. In fact, factor analysis has long been employed in financial studies, where Y_{it} often represents the excess returns of the ith asset (or stock) on time t. The literature includes, for instance, Campbell *et al.* (1997), Chamberlain and Rothschild (1983), Fama and French (1992). It is also commonly used in macroeconomics for forecasting diffusion indices (e.g., Stock and Watson, (2002).

The factor model (6.1) can be put in a matrix form as

$$\boldsymbol{Y}_t = \mathbf{B}\boldsymbol{f}_t + \boldsymbol{u}_t. \tag{6.2}$$

where $\mathbf{B} = (\mathbf{b}_1, \dots, \mathbf{b}_N)'$ and $\boldsymbol{u}_t = (u_{1t}, \dots, u_{Nt})'$. We are interested in $\boldsymbol{\Sigma}$, the $N \times N$ covariance matrix of $\boldsymbol{Y}_t$, and its inverse $\boldsymbol{\Theta} = \boldsymbol{\Sigma}^{-1}$, which are assumed to be time-invariant. Under model (6.1), $\boldsymbol{\Sigma}$ is given by

$$\boldsymbol{\Sigma} = \mathbf{B}\mathrm{Cov}(\boldsymbol{f}_t)\mathbf{B}' + \boldsymbol{\Sigma}_u, \tag{6.3}$$

where $\boldsymbol{\Sigma}_u = (\sigma_{u,ij})_{N\times N}$ is the covariance matrix of $\boldsymbol{u}_t$. Estimating the covariance matrix $\boldsymbol{\Sigma}_u$ of the idiosyncratic components $\{\boldsymbol{u}_t\}$ is also important for statistical inferences. For example, it is needed for large sample inference of the unknown factors and their loadings and for testing the capital asset pricing model (Sentana, 2009).

In the decomposition (6.3), it is natural to consider the *conditional sparsity*: given the common factors, most of the remaining outcomes are mutually weakly correlated. This gives rise to the approximate factor model (e.g., Chamberlain and Rothschild, 1983), in which $\boldsymbol{\Sigma}_u$ is a sparse covariance but not necessarily diagonal, and for some $q \in [0, 1)$,

$$m_N = \max_{i\leq N} \sum_{j\leq N} |\sigma_{u,ij}|^q \tag{6.4}$$

does not grow too fast as $N \to \infty$. When $q = 0$, m_N measures the maximum number of non zero components in each row.

We would like to emphasize that model (6.3) is related to but different from the problem recently studied in the literature on "low-rank plus sparse representation". In fact, the "low rank plus sparse" representation of (6.3) holds on the population covariance matrix, whereas the model considered by Candès *et al.* (2011) and Chandrasekaran *et al.* (2010) considered such a representation on the data matrix. As there is no $\boldsymbol{\Sigma}$ to estimate, their goal is limited to producing a low-rank plus sparse matrix decomposition of the data matrix, which corresponds to the identifiability issue of our study, and does not involve estimation or inference. In contrast, our ultimate goal is to estimate the population covariance matrices as well as the precision matrices. Our consistency result on $\boldsymbol{\Sigma}_u$ demonstrates that the decomposition (6.3) is identifiable, and hence our results also shed the light of the "surprising phenomenon" of Candès *et al.* (2011) that one can separate fully a sparse matrix from a low-rank matrix when only the sum of these two components is available.

Moreover, note that in financial applications, the common factors $\boldsymbol{f}_t$ are sometimes known, as in Fama and French (1992). In other applications, however, the common factors may be unknown and need to be inferred. Interestingly, asymptotic analysis shows that as the dimensionality grows fast enough (relative to the sample size), the effect of estimating the unknown factors is negligible, and the covariance matrices of $\boldsymbol{Y}_t$ and $\boldsymbol{u}_t$ and their inverses can be estimated as if the factors were known (Fan *et al.*, 2013).

We now divide our discussions into two cases: models with known factors and models with unknown factors.

6.2.1 Known Factors

When the factors are observable, one can estimate $\mathbf{B}$ by the ordinary least squares (OLS): $\hat{\mathbf{B}} = (\hat{\mathbf{b}}_1, \dots, \hat{\mathbf{b}}_N)'$, where,

$$\hat{\mathbf{b}}_i = \arg\min_{\mathbf{b}_i} \frac{1}{T} \sum_{t=1}^{T} (Y_{it} - \mathbf{b}_i' \boldsymbol{f}_t)^2, \quad i = 1, \dots, N.$$

The residuals are obtained using the plug-in method: $\hat{u}_{it} = Y_{it} - \hat{\mathbf{b}}_i' \boldsymbol{f}_t$.

Denote by $\hat{\boldsymbol{u}}_t = (\hat{u}_{1t}, \dots, \hat{u}_{pt})'$. We then construct the residual covariance matrix as:

$$\mathbf{S}_u = \frac{1}{T} \sum_{t=1}^{T} \hat{\boldsymbol{u}}_t \hat{\boldsymbol{u}}_t' = (s_{u,ij}).$$

Now we apply thresholding on $\mathbf{S}_u$. Define

$$\hat{\boldsymbol{\Sigma}}_u = (\hat{\sigma}_{ij})_{p \times p}, \hat{\sigma}_{ij}^{\mathcal{T}} = \begin{cases} s_{u,ii}, & i = j; \\ th(s_{u,ij}) I(|s_{u,ij}| \geq \tau_{ij}), & i \neq j. \end{cases} \tag{6.5}$$

where $th(\cdot)$ is a generalized shrinkage function of Antoniadis and Fan (2001), employed by Rothman *et al.* (2009) and Cai and Liu (2011), and $\tau_{ij} > 0$ is an entry-dependent threshold. In particular, the hard-thresholding rule $th(x) = xI(|x| \geq \tau_{ij})$ (Bickel and Levina, 2008) and the constant thresholding parameter $\tau_{ij} = \delta$ are allowed. In practice, it is more desirable to have τ_{ij} be entry-adaptive. An example of the threshold is

$$\tau_{ij} = \omega_T (s_{u,ii} s_{u,jj})^{1/2}, \quad \text{for a given}\, \omega_T > 0 \tag{6.6}$$

This corresponds to applying the thresholding with parameter ω_T to the correlation matrix of $\mathbf{S}_u$. Cai and Liu (2011) discussed an alternative type of "adaptive threshold." Moreover, we take ω_T to be: some $C > 0$,

$$\omega_T = C \sqrt{\frac{\log\ N}{T}},$$

which is a proper threshold level to overrides the estimation errors.

The covariance matrix $\mathrm{Cov}(\boldsymbol{f}_t)$ can be estimated by the sample covariance matrix

$$\widehat{\mathrm{Cov}}(\boldsymbol{f}_t) = T^{-1}\mathbf{F}'\mathbf{F} - T^{-2}\mathbf{F}'\mathbf{1}\mathbf{1}'\mathbf{F},$$

where $\mathbf{F}' = (\boldsymbol{f}_1, \dots, \boldsymbol{f}_T)$, and $\mathbf{1}$ is a T-dimensional column vector of ones. Therefore, we obtain a substitution estimator (Fan *et al.*, 2011):

$$\hat{\boldsymbol{\Sigma}} = \hat{\mathbf{B}} \widehat{\mathrm{Cov}}(\boldsymbol{f}_t) \hat{\mathbf{B}}' + \hat{\boldsymbol{\Sigma}}_u. \tag{6.7}$$

By the Sherman–Morrison–Woodbury formula,

$$\boldsymbol{\Sigma}^{-1} = \boldsymbol{\Sigma}_u^{-1} - \boldsymbol{\Sigma}_u^{-1}\mathbf{B}[\mathrm{Cov}(\boldsymbol{f}_t)^{-1} + \mathbf{B}'\boldsymbol{\Sigma}_u^{-1}\mathbf{B}]^{-1}\mathbf{B}'\boldsymbol{\Sigma}_u^{-1},$$

which is estimated by

$$\hat{\boldsymbol{\Sigma}}^{-1} = \hat{\boldsymbol{\Sigma}}_u^{-1} - \hat{\boldsymbol{\Sigma}}_u^{-1}\hat{\mathbf{B}}[\widehat{\mathrm{Cov}}(\boldsymbol{f}_t)^{-1} + \hat{\mathbf{B}}'\hat{\boldsymbol{\Sigma}}_u^{-1}\hat{\mathbf{B}}]^{-1}\hat{\mathbf{B}}'\hat{\boldsymbol{\Sigma}}_u^{-1}. \tag{6.8}$$

6.2.2 Unknown Factors

When factors are unknown, Fan *et al.* (2013) proposed a nonparametric estimator of $\mathbf{\Sigma}$ based on the principal component analysis. Let $\hat{\lambda}_1 \geq \hat{\lambda}_2 \geq \cdots \geq \hat{\lambda}_N$ be the ordered eigenvalues of the sample covariance matrix $\mathbf{S}$ of $\boldsymbol{Y}_t$ and $\{\hat{\xi}_i\}_{i=1}^N$ be their corresponding eigenvectors. Then the sample covariance has the following spectral decomposition:

$$\mathbf{S} = \sum_{i=1}^{K} \hat{\lambda}_i \hat{\xi}_i \hat{\xi}_i' + \boldsymbol{Q},$$

where $\boldsymbol{Q} = \sum_{i=K+1}^N \hat{\lambda}_i \hat{\xi}_i \hat{\xi}_i'$ is called the *principal orthogonal complement*, and K is the number of common factors. We can apply thresholding on $\boldsymbol{Q}$ as in (6.5) and (6.6). Denote the thresholded $\boldsymbol{Q}$ by $\hat{\mathbf{\Sigma}}_u$. Note that the threshold value in (6.6) now becomes, for some $C > 0$

$$\omega_T = C\left(\sqrt{\frac{\log\ N}{T}} + \frac{1}{\sqrt{N}}\right).$$

The estimator of $\mathbf{\Sigma}$ is then defined as:

$$\hat{\mathbf{\Sigma}}_K = \sum_{i=1}^{K} \hat{\lambda}_i \hat{\xi}_i \hat{\xi}_i' + \hat{\mathbf{\Sigma}}_u. \tag{6.9}$$

This estimator is called the principal orthogonal complement thresholding (POET) estimator. It is obtained by thresholding the remaining components of the sample covariance matrix, after taking out the first K principal components. One of the attractiveness of POET is that it is optimization-free, and hence is computationally appealing.

The POET (6.9) has an equivalent representation using a constrained least-squares method. The least-squares method seeks for $\hat{\mathbf{B}} = (\hat{\mathbf{b}}_1, \ldots, \hat{\mathbf{b}}_N)'$ and $\hat{\mathbf{F}}' = (\hat{f}_1, \ldots, \hat{f}_T)$ such that

$$(\hat{\mathbf{B}}, \hat{\mathbf{F}}) = \arg \min_{\mathbf{b}_i \in \mathbb{R}^K, f_t \in \mathbb{R}^K} \sum_{i=1}^{N} \sum_{t=1}^{T} (Y_{it} - \mathbf{b}_i' f_t)^2, \tag{6.10}$$

subject to the normalization

$$\frac{1}{T} \sum_{t=1}^{T} f_t f_t' = \mathbf{I}_K, \text{ and } \frac{1}{N} \sum_{i=1}^{N} \mathbf{b}_i \mathbf{b}_i' \text{ is diagonal.} \tag{6.11}$$

Putting it in a matrix form, the optimization problem can be written as

$$\arg \min_{\mathbf{B},\mathbf{F}} \|\mathbf{Y}' - \mathbf{B}\mathbf{F}'\|_F^2$$

$$T^{-1}\mathbf{F}'\mathbf{F} = \mathbf{I}_K, \mathbf{B}'\mathbf{B} \text{ is diagonal.} \tag{6.12}$$

where $\mathbf{Y}' = (\boldsymbol{Y}_1, \ldots, \boldsymbol{Y}_T)$ and $\mathbf{F}' = (f_1, \cdots, f_T)$. For each given $\mathbf{F}$, the least-squares estimator of $\mathbf{B}$ is $\hat{\mathbf{B}} = T^{-1}\mathbf{Y}'\mathbf{F}$, using the constraint (6.11) on the factors. Substituting this into (6.12), the objective function now becomes $\|\mathbf{Y}' - T^{-1}\mathbf{Y}'\mathbf{F}\mathbf{F}'\|_F^2 = \text{tr}[(\mathbf{I}_T - T^{-1}\mathbf{F}\mathbf{F}')\mathbf{Y}\mathbf{Y}']$. The minimizer is now clear: the columns of $\hat{\mathbf{F}}/\sqrt{T}$ are the eigenvectors corresponding to the K largest eigenvalues of the $T \times T$ matrix $\mathbf{Y}\mathbf{Y}'$ and $\hat{\mathbf{B}} = T^{-1}\mathbf{Y}'\hat{\mathbf{F}}$ (see e.g., Stock and Watson, 2002). The residual is given by $\hat{u}_{it} = Y_{it} - \hat{\mathbf{b}}_i'\hat{f}_t$, based on which we can construct the sample covariance matrix of $\boldsymbol{\Sigma}_u$. Then apply the thresholding to obtain $\hat{\boldsymbol{\Sigma}}_u$. The covariance of $\boldsymbol{Y}_t$ is then estimated by $\hat{\mathbf{B}}\hat{\mathbf{B}}' + \hat{\boldsymbol{\Sigma}}_u$. It can be proved that the estimator in (6.9) satisfies:

$$\hat{\boldsymbol{\Sigma}}_K = \hat{\mathbf{B}}\hat{\mathbf{B}}' + \hat{\boldsymbol{\Sigma}}_u.$$

Several methods have been proposed to consistently estimate the number of factors. For instance, Bai and Ng (2002) proposed to use:

$$\hat{K} = \arg \min_{0 \le k \le M} \frac{1}{N} \mathrm{tr}\left(\sum_{j=k+1}^{N} \hat{\lambda}_j \hat{\xi}_j \hat{\xi}_j' \right) + \frac{k(N+T)}{NT} \log \left(\frac{NT}{N+T} \right), \tag{6.13}$$

where M is a prescribed upper bound. The literature also includes, Ahn and Horenstein (2013), Alessi *et al.* (2010), Hallin and Liška (2007), Kapetanios (2010), among others. Numerical studies in Fan *et al.* (2013) showed that the covariance estimator is robust to overestimating K. Therefore, in practice, we can also choose a relatively large number for K. Consistency can still be guaranteed.

6.2.3 Choosing the Threshold

Recall that the threshold value ω_T depends on a user-specific constant C. In practice, we need to choose C to maintain the positive definiteness of the estimated covariances for any given finite sample. To do so, write the error covariance estimator as $\hat{\boldsymbol{\Sigma}}_u(C)$, which depends on C via the threshold. We choose C in the range where $\lambda_{\min}(\hat{\boldsymbol{\Sigma}}_u) > 0$. Define

$$C_{\min} = \inf \{C > 0 : \lambda_{\min}(\hat{\boldsymbol{\Sigma}}_u(M)) > 0, \forall M > C\}. \tag{6.14}$$

When C is sufficiently large, the estimator becomes diagonal, while its minimum eigenvalue must retain strictly positive. Thus, $C_{\min}$ is well defined and for all $C > C_{\min}$, $\hat{\boldsymbol{\Sigma}}_u(C)$ is positive definite under finite sample. We can obtain $C_{\min}$ by solving $\lambda_{\min}(\hat{\boldsymbol{\Sigma}}_u(C)) = 0, C \neq 0$. We can also approximate $C_{\min}$ by plotting $\lambda_{\min}(\hat{\boldsymbol{\Sigma}}_u(C))$ as a function of C, as illustrated in Figure 6.1. In practice, we can choose C in the range $(C_{\min} + \epsilon, M)$ for a small ϵ and large enough M. Choosing the threshold in a range to guarantee the finite-sample positive definiteness has also been previously suggested by Fryzlewicz (2013).

6.2.4 Asymptotic Results

Under regularity conditions (e.g., strong mixing, exponential-tail distributions), Fan *et al.* (2011, 2013) showed that for the error covariance estimator, assuming $\omega_T^{1-q} m_N = o(1)$,

$$\|\hat{\boldsymbol{\Sigma}}_u - \boldsymbol{\Sigma}_u\|_2 = O_P(\omega_T^{1-q} m_N),$$

and

$$\|\hat{\boldsymbol{\Sigma}}_u^{-1} - \boldsymbol{\Sigma}_u^{-1}\|_2 = O_P(\omega_T^{1-q} m_N).$$

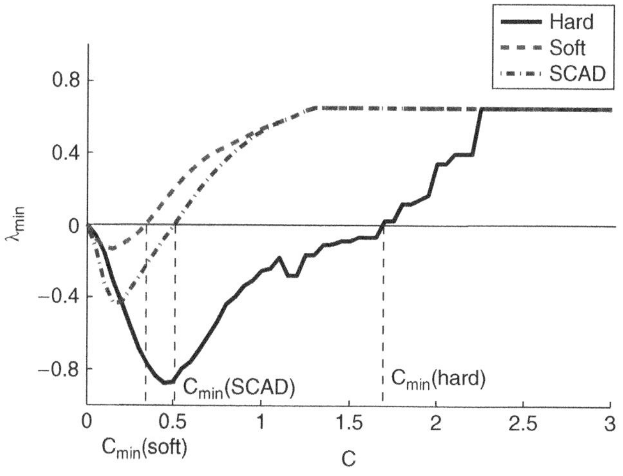

Figure 6.1 Minimum eigenvalue of $\hat{\boldsymbol{\Sigma}}_u(C)$ as a function of C for three choices of thresholding rules. Adapted from Fan *et al.* (2013).

Here $q \in [0, 1)$ quantifies the level of sparsity as defined in (6.4), and ω_T is given by: for some $C > 0$, when factors are known,

$$\omega_T = \sqrt{\frac{\log N}{T}}$$

when factors are unknown,

$$\omega_T = \sqrt{\frac{\log N}{T}} + \frac{1}{\sqrt{N}}.$$

The dimension N is allowed to grow exponentially fast in T.

As for the convergence of $\hat{\boldsymbol{\Sigma}}$, because the first K eigenvalues of $\boldsymbol{\Sigma}$ grow with N, one can hardly estimate $\boldsymbol{\Sigma}$ with satisfactory accuracy in either the operator norm or the Frobenius norm. This problem arises not from the limitation of any estimation method, but due to the nature of the high-dimensional factor model. We illustrate this in the following example.

Example 6.1 *Consider a simplified case where we know* $\mathbf{b}_i = (1, 0, \ldots, 0)'$ *for each* $i = 1, \ldots, N$, $\boldsymbol{\Sigma}_u = \mathbf{I}$, *and* $\{f_t\}_{t=1}^T$ *are observable. Then when estimating* $\boldsymbol{\Sigma}$, *we only need to estimate Cov(f) using the sample covariance matrix* $\widehat{Cov}(f_t)$, *and obtain an estimator for* $\boldsymbol{\Sigma}$:

$$\hat{\boldsymbol{\Sigma}} = \mathbf{B}\widehat{Cov}(f_t)\mathbf{B}' + \mathbf{I}.$$

Simple calculations yield to

$$\|\hat{\boldsymbol{\Sigma}} - \boldsymbol{\Sigma}\|_2 = |\frac{1}{T}\sum_{t=1}^{T}(f_{1t} - \bar{f}_1)^2 - Var\,(f_{1t})| \cdot \|\mathbf{1}_N\mathbf{1}_N'\|_2,$$

where $\mathbf{1}_N$ *denotes the N-dimensional column vector of ones with* $\|\mathbf{1}_N\mathbf{1}_N'\|_2 = N$. *Therefore, due to the central limit theorem employed on* $\frac{1}{\sqrt{T}}\sum_{t=1}^{T}(f_{1t} - \bar{f}_1)^2 - Var\,(f_{1t})$, $\frac{\sqrt{T}}{N}\|\hat{\boldsymbol{\Sigma}} - \boldsymbol{\Sigma}\|_2$ *is asymptotically normal. Hence* $\|\hat{\boldsymbol{\Sigma}} - \boldsymbol{\Sigma}\|_2$ *diverges if* $N \gg \sqrt{T}$, *even for such a simplified toy model.*

As we have seen from the above example, the small error of estimating Var (f_{1t}) is substantially amplified due to the presence of $\|\mathbf{1}_N \mathbf{1}'_N\|_2$; the latter in fact determines the size of the largest eigenvalue of $\boldsymbol{\Sigma}$. We further illustrate this phenomenon in the following example.

Example 6.2 *Consider an ideal case where we know the spectrum except for the first eigenvector of* $\boldsymbol{\Sigma}$. *Let* $\{\lambda_j, \xi_j\}_{j=1}^N$ *be the eigenvalues and vectors, and assume that the largest eigenvalue* $\lambda_1 \geq cN$ *for some* $c > 0$. *Let* $\hat{\xi}_1$ *be the estimated first eigenvector, and define the covariance estimator* $\hat{\boldsymbol{\Sigma}} = \lambda_1 \hat{\xi}_1 \hat{\xi}'_1 + \sum_{j=2}^N \lambda_j \xi_j \xi'_j$. *Assume that* $\hat{\xi}_1$ *is a good estimator in the sense that* $\|\hat{\xi}_1 - \xi_1\|^2 = O_P(T^{-1})$. *However,*

$$\|\hat{\boldsymbol{\Sigma}} - \boldsymbol{\Sigma}\|_2 = \|\lambda_1(\hat{\xi}_1 \hat{\xi}'_1 - \xi_1 \xi'_1)\|_2 = \lambda_1 O_P(\|\hat{\xi} - \xi\|) = O_P(\lambda_1 T^{-1/2}),$$

which can diverge when $T = O(N^2)$.

On the other hand, we can estimate the precision matrix with a satisfactory rate under the operator norm. The intuition follows from the fact that $\boldsymbol{\Sigma}^{-1}$ has bounded eigenvalues. Let $\hat{\boldsymbol{\Sigma}}^{-1}$ denote the inverse of the POET estimator. Fan *et al.* (2013) showed that $\hat{\boldsymbol{\Sigma}}^{-1}$ has the same rate of convergence as that of $\boldsymbol{\Sigma}_u^{-1}$. Specifically,

$$\|\hat{\boldsymbol{\Sigma}}^{-1} - \boldsymbol{\Sigma}^{-1}\|_2 = O_P(\omega_T^{1-q} m_N).$$

Comparing the rates of convergence of known and unknown factors, we see that when the common factors are unobservable, the rate of convergence has an additional term $m_N/N^{(1-q)/2}$, coming from the impact of estimating the unknown factors. This impact vanishes when $N \log N \gg T$, in which case the minimax rate as in Cai and Zhou (2010) is achieved. As N increases, more information about the common factors is collected, which results in more accurate estimation of the common factors $\{\boldsymbol{f}_t\}_{t=1}^T$. Then the rates of convergence in both observable factor and unobservable factor cases are the same.

6.2.5 A Numerical Illustration

We now illustrate the above theoretical results by using a simple three-factor model with a sparse error covariance matrix. The distribution of the data-generating process is taken from Fan *et al.* (2013) (Section 6.7). Specifically, we simulated from a standard Fama–French three-factor model. The factor loadings are drawn from a trivariate normal distribution $\mathbf{b}_i = (b_{1i}, b_{2i}, b_{3i})' \sim N_3(\boldsymbol{\mu}_B, \boldsymbol{\Sigma}_B)$, and $\boldsymbol{f}_t$ follows a vector autoregression of the first order (VAR(1)) model $\boldsymbol{f}_t = \boldsymbol{\mu} + \boldsymbol{\Phi} \boldsymbol{f}_{t-1} + \boldsymbol{\epsilon}_t$. To make the simulation more realistic, model parameters are calibrated from the real data on annualized returns of 100 industrial portfolios, obtained from the website of Kenneth French. As there are three common factors, the largest three eigenvalues of $\boldsymbol{\Sigma}$ are of the same order as $\sum_{i=1}^N b_{ji}^2, j = 1, 2, 3$, which are approximately $O(N)$, and grow linearly with N.

We generate a sparse covariance matrix $\boldsymbol{\Sigma}_u$ of the form: $\boldsymbol{\Sigma}_u = \boldsymbol{D}\boldsymbol{\Sigma}_0\boldsymbol{D}$. Here, $\boldsymbol{\Sigma}_0$ is the error correlation matrix, and $\boldsymbol{D}$ is the diagonal matrix of the standard deviations of the errors. We

Table 6.1 Mean and covariance matrix used to generate $\mathbf{b}_i$

μ_B	Σ_B		
0.0047	0.0767	−0.00004	0.0087
0.0007	−0.00004	0.0841	0.0013
−1.8078	0.0087	0.0013	0.1649

Table 6.2 Parameters of $\boldsymbol{f}_t$ generating process

μ	$\mathrm{Cov}(\boldsymbol{f}_t)$			$\boldsymbol{\Phi}$		
−0.0050	1.0037	0.0011	−0.0009	−0.0712	0.0468	0.1413
0.0335	0.0011	0.9999	0.0042	−0.0764	−0.0008	0.0646
−0.0756	−0.0009	0.0042	0.9973	0.0195	−0.0071	−0.0544

set $\boldsymbol{D} = \mathrm{diag}(\sigma_1, \dots, \sigma_p)$, where each σ_i is generated independently from a gamma distribution $G(\alpha, \beta)$, and α and β are chosen to match the sample mean and sample standard deviation of the standard deviations of the errors. The off-diagonal entries of $\boldsymbol{\Sigma}_0$ are generated independently from a normal distribution, with mean and standard deviation equal to the sample mean and sample standard deviation of the sample correlations among the estimated residuals. We then employ hard thresholding to make $\boldsymbol{\Sigma}_0$ sparse, where the threshold is found as the smallest constant that provides the positive definiteness of $\boldsymbol{\Sigma}_0$.

For the simulation, we fix $T = 300$, and let N increase from 20 to 600 in increments of 20. We plot the averages and standard deviations of the distance from $\hat{\boldsymbol{\Sigma}}$ and $\mathbf{S}$ to the true covariance matrix $\boldsymbol{\Sigma}$, under the norm $\|\mathbf{A}\|_\Sigma = \frac{1}{N}\|\boldsymbol{\Sigma}^{-1/2}\mathbf{A}\boldsymbol{\Sigma}^{-1/2}\|_F$ (recall that $\mathbf{S}$ denotes the sample covariance). It is easy to see that

$$\|\hat{\boldsymbol{\Sigma}} - \boldsymbol{\Sigma}\|_\Sigma = \frac{1}{N}\|\boldsymbol{\Sigma}^{-1/2}\hat{\boldsymbol{\Sigma}}\boldsymbol{\Sigma}^{-1/2} - \mathbf{I}\|_F,$$

which resembles the relative errors. We also plot the means and standard deviations of the distances from $\hat{\boldsymbol{\Sigma}}^{-1}$ and $\mathbf{S}^{-1}$ to $\boldsymbol{\Sigma}^{-1}$ under the spectral norm. Due to invertibility, the operator norm for $\mathbf{S}^{-1}$ is plotted only up to $N = 280$.

We observe that the unobservable factor model performs just as well as the estimator if the factors are known. The cost of not knowing the factors is negligible when N is large enough. As we can see from Figure 6.2, the impact decreases quickly. In addition, when estimating $\boldsymbol{\Sigma}^{-1}$, it is hard to distinguish the estimators with known and unknown factors, whose performances are quite stable compared to that of the sample covariance matrix. Intuitively, as the dimension increases, more information about the common factors becomes available, which helps infer the unknown factors. Indeed, as is shown in Bai (2003) and Fan *et al.* (2014a), the principal components method can estimate the unknown factors at a rate of:

$$\frac{1}{T}\sum_{t=1}^{T}\|\hat{f}_t - f_t\|^2 = O_P\left(\frac{1}{T^2} + \frac{1}{N}\right).$$

Hence, as long as N is relatively large, $\boldsymbol{f}_t$ can be estimated pretty accurately.

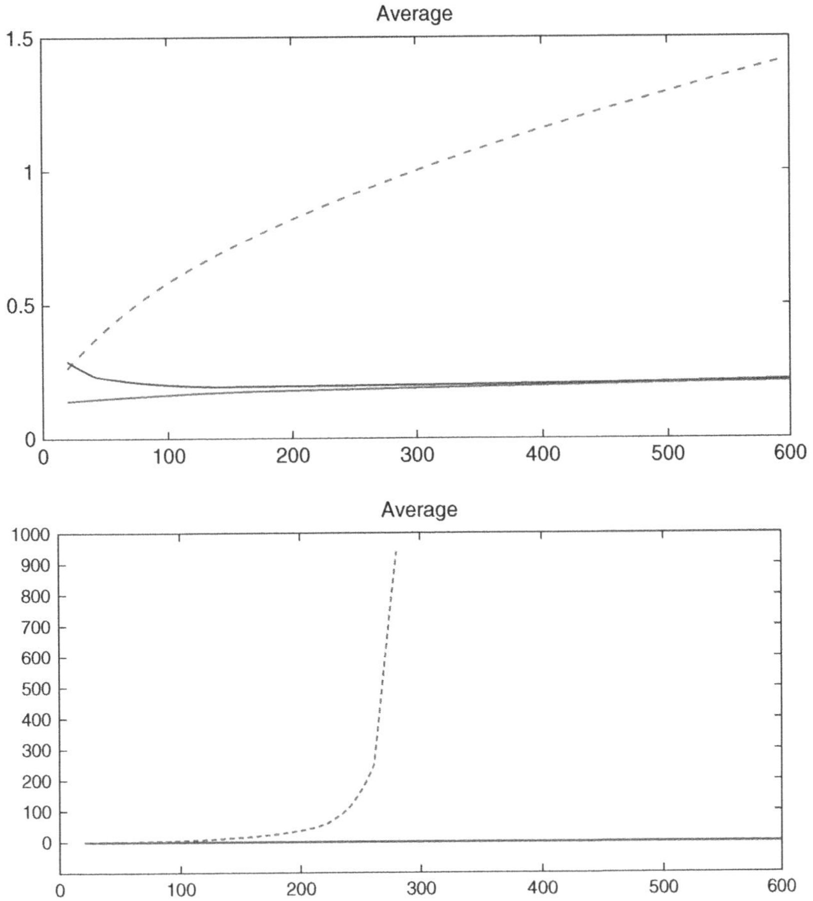

Figure 6.2 Averages of $N^{-1}\|\mathbf{\Sigma}^{-1/2}\hat{\mathbf{\Sigma}}\mathbf{\Sigma}^{-1/2}-\mathbf{I}\|_F$ (left panel) and $\|\hat{\mathbf{\Sigma}}^{-1}-\mathbf{\Sigma}^{-1}\|_2$ (right panel) with known factors (solid red curve), unknown factors (solid blue curve), and sample covariance (dashed curve) over 200 simulations, as a function of the dimensionality N. Taken from Fan *et al.* (2013).

6.3 Precision Matrix Estimation and Graphical Models

Let $\boldsymbol{Y}_1,\cdots,\boldsymbol{Y}_T$ be T data points from an N-dimensional random vector $\boldsymbol{Y}=(Y_1,\ldots,Y_N)'$ with $\boldsymbol{Y}\sim\mathcal{N}_N(\mathbf{0},\mathbf{\Sigma})$. We denote the precision matrix $\mathbf{\Theta}:=\mathbf{\Sigma}^{-1}$ and define an undirected graph $G=(V,E)$ based on the sparsity pattern of $\mathbf{\Theta}$: let $V=\{1,\cdots,N\}$ be the node set corresponding to the N variables in $\boldsymbol{Y}$, an edge $(j,k)\in E$ if and only $\mathbf{\Theta}_{jk}\neq 0$.

As we will explain in Section 6.3.1, the graph G describes the conditional independence relationships between $Y_1,\ldots,Y_N$: that is, letting $\boldsymbol{Y}_{\backslash\{j,k\}}:=\{Y_\ell:\quad \ell\neq j,k\}$, then Y_j is independent of Y_k given $\boldsymbol{Y}_{\backslash\{j,k\}}$ if and only if $(j,k)\notin E$.

In high-dimensional settings where $N\gg T$, we assume that many entries of $\mathbf{\Theta}$ are zero (or, in other words, the graph G is sparse). The problem of estimating a large sparse precision matrix $\mathbf{\Theta}$ is called *covariance selection* (Dempster, 1972).

6.3.1 Column-wise Precision Matrix Estimation

A natural approach for estimating $\mathbf{\Theta}$ is by penalizing the likelihood using the L_1-penalty (Banerjee *et al.*, 2008; Friedman *et al.*, 2008; Yuan and Lin, 2007). To further reduce the estimation bias, Jalali *et al.* (2012), Lam and Fan (2009), Shen *et al.*, (2012) propose either greedy algorithms or nonconvex penalties for sparse precision matrix estimation. Under certain conditions, Ravikumar *et al.* (2011a), Rothman *et al.* (2008), Wainwright (2009), Zhao and Yu (2006), Zou (2006), study the theoretical properties of the penalized likelihood methods.

Another approach is to estimate $\mathbf{\Theta}$ in a column-by-column fashion. For this, Yuan (2010) and Cai *et al.* (2011) propose the graphical Dantzig selector and CLIME, respectively, which can be solved by linear programming. More recently, Liu and Luo (2012) and Sun and Zhang (2012) have proposed the SCIO and scaled-lasso methods. Compared to the penalized likelihood methods, the column-by-column estimation methods are computationally simpler and are more amenable to theoretical analysis.

In the rest of this chapter, we explain the main ideas of the column-by-column precision matrix estimation methods. We start with an introduction of notations. Letting $\mathbf{v} := (v_1, \cdots, v_N)' \in \mathbb{R}^N$ and $I(\cdot)$ be the indicator function, for $0 < q < \infty$, we define

$$\|\mathbf{v}\|_q := \left(\sum_{j=1}^{N} |v_j|^q\right)^{1/q}, \|\mathbf{v}\|_0 := \sum_{j=1}^{N} I(v_j \neq 0), \quad \text{and} \quad \|\mathbf{v}\|_\infty := \max_j |v_j|.$$

Let $\mathbf{A} \in \mathbb{R}^{N\times N}$ be a symmetric matrix and $I, J \subset \{1, \cdots, N\}$ be two sets. Denote by $\mathbf{A}_{I,J}$ the submatrix of $\mathbf{A}$ with rows and columns indexed by I and J. Letting $\mathbf{A}_{*j}$ be the jth column of $\mathbf{A}$ and $\mathbf{A}_{*\setminus j}$ be the submatrix of $\mathbf{A}$ with the jth column $\mathbf{A}_{*j}$ removed. We define the following matrix norms:

$$\|\mathbf{A}\|_q := \max_{\|\mathbf{v}\|_q=1} \|\mathbf{A}\mathbf{v}\|_q, \|\mathbf{A}\|_{\max} := \max_{jk} |\mathbf{A}_{jk}|, \quad \text{and} \quad \|\mathbf{A}\|_F = \left(\sum_{j,k} |\mathbf{A}_{jk}|^2\right)^{1/2}.$$

We also denote $\Lambda_{\max}(\mathbf{A})$ and $\Lambda_{\min}(\mathbf{A})$ to be the largest and smallest eigenvalues of $\mathbf{A}$.

The column-by-column precision matrix estimation method exploits the relationship between conditional distribution of multivariate Gaussian and linear regression. More specifically, letting $\boldsymbol{Y} \sim \mathcal{N}_N(\mathbf{0}, \mathbf{\Sigma})$, the conditional distribution of Y_j given $\boldsymbol{Y}_{\setminus j}$ satisfies

$$Y_j \mid \boldsymbol{Y}_{\setminus j} \sim \mathcal{N}_{N-1}(\mathbf{\Sigma}_{\setminus j,j}(\mathbf{\Sigma}_{\setminus j,\setminus j})^{-1}\boldsymbol{Y}_{\setminus j}, \mathbf{\Sigma}_{jj} - \mathbf{\Sigma}_{\setminus j,j}(\mathbf{\Sigma}_{\setminus j,\setminus j})^{-1}\mathbf{\Sigma}_{\setminus j,j}).$$

Let $\boldsymbol{\alpha}_j := (\mathbf{\Sigma}_{\setminus j,\setminus j})^{-1}\mathbf{\Sigma}_{\setminus j,j} \in \mathbb{R}^{N-1}$ and $\sigma_j^2 := \mathbf{\Sigma}_{jj} - \mathbf{\Sigma}_{\setminus j,j}(\mathbf{\Sigma}_{\setminus j,\setminus j})^{-1}\mathbf{\Sigma}_{\setminus j,j}$. We have

$$Y_j = \boldsymbol{\alpha}_j' \boldsymbol{Y}_{\setminus j} + \epsilon_j, \tag{6.15}$$

where $\epsilon_j \sim \mathcal{N}(0, \sigma_j^2)$ is independent of $\boldsymbol{Y}_{\setminus j}$. By the block matrix inversion formula, we have

$$\mathbf{\Theta}_{jj} = (\text{Var}\ (\epsilon_j))^{-1} = \sigma_j^{-2}, \tag{6.16}$$

$$\mathbf{\Theta}_{\setminus j,j} = -(\text{Var}\ (\epsilon_j))^{-1}\boldsymbol{\alpha}_j = -\sigma_j^{-2}\boldsymbol{\alpha}_j. \tag{6.17}$$

Therefore, we can recover $\mathbf{\Theta}$ in a column-by-column manner by regressing Y_j on $\boldsymbol{Y}_{\setminus j}$ for $j = 1, 2, \cdots, N$. For example, let $\mathbf{Y} \in \mathbb{R}^{T\times N}$ be the data matrix. We denote by

$\boldsymbol{\alpha}_j := (\alpha_{j1}, \cdots, \alpha_{j(N-1)})' \in \mathbb{R}^{N-1}$. Meinshausen and Bühlmann (2006) propose to estimate each $\boldsymbol{\alpha}_j$ by solving the lasso regression:

$$\hat{\boldsymbol{\alpha}}_j = \arg \min_{\boldsymbol{\alpha}_j \in \mathbb{R}^{N-1}} \frac{1}{2T} \|\mathbf{Y}_{*j} - \mathbf{Y}_{*\setminus j} \boldsymbol{\alpha}_j\|_2^2 + \lambda_j \|\boldsymbol{\alpha}_j\|_1,$$

where λ_j is a tuning parameter. Once $\hat{\boldsymbol{\alpha}}_j$ is given, we get the neighborhood edges by reading out the nonzero coefficients of $\boldsymbol{\alpha}_j$. The final graph estimate $\hat{G}$ is obtained by either the "AND" or "OR" rule on combining the neighborhoods for all the N nodes. However, the neighborhood pursuit method of Meinshausen and Bühlmann (2006) only estimates the graph G but cannot estimate the inverse covariance matrix $\boldsymbol{\Theta}$.

To estimate $\boldsymbol{\Theta}$, Yuan (2010) proposes to estimate $\boldsymbol{\alpha}_j$ by solving the Dantzig selector:

$$\hat{\boldsymbol{\alpha}}_j = \arg \min_{\boldsymbol{\alpha}_j \in \mathbb{R}^{N-1}} \|\boldsymbol{\alpha}_j\|_1 \quad \text{subject to} \quad \|\mathbf{S}_{\setminus j,j} - \mathbf{S}_{\setminus j,\setminus j} \boldsymbol{\alpha}_j\|_\infty \le \gamma_j,$$

where $\mathbf{S} := T^{-1}\mathbf{Y}'\mathbf{Y}$ is the sample covariance matrix and γ_j is a tuning parameter. Once $\hat{\boldsymbol{\alpha}}_j$ is given, we can estimate σ_j^2 by $\hat{\sigma}_j^2 = [1 - 2\hat{\boldsymbol{\alpha}}_j' \mathbf{S}_{\setminus j,j} + \hat{\boldsymbol{\alpha}}_j' \mathbf{S}_{\setminus j,\setminus j} \hat{\boldsymbol{\alpha}}_j]^{-1}$. We then get the estimator $\hat{\boldsymbol{\Theta}}$ of $\boldsymbol{\Theta}$ by plugging $\hat{\boldsymbol{\alpha}}_j$ and $\hat{\sigma}_j^2$ into (6.16) and (6.17). Yuan (2010) analyzes the L_1-norm error $\|\hat{\boldsymbol{\Theta}} - \boldsymbol{\Theta}\|_1$ and shows its minimax optimality over certain model space.

In another work, Sun and Zhang (2012) propose to estimate $\boldsymbol{\alpha}_j$ and σ_j by solving a scaled-lasso problem:

$$\hat{\mathbf{b}}_j, \hat{\sigma}_j = \arg \min_{\mathbf{b}=(b_1,\cdots,b_N)',\sigma} \left\{ \frac{\mathbf{b}_j' \mathbf{S} \mathbf{b}_j}{2\sigma} + \frac{\sigma}{2} + \lambda \sum_{k=1}^{N} \mathbf{S}_{kk} |b_k| \quad \text{subject to} \quad b_j = -1 \right\}.$$

Once $\hat{\mathbf{b}}_j$ is obtained, $\boldsymbol{\alpha}_j = \hat{\mathbf{b}}_{\setminus j}$. Sun and Zhang (2012) provide the spectral-norm rate of convergence of the obtained precision matrix estimator.

Cai *et al.* (2011) proposes the CLIME estimator, which directly estimates the jth column of $\boldsymbol{\Theta}$ by solving

$$\hat{\boldsymbol{\Theta}}_{*j} = \arg \min_{\boldsymbol{\Theta}_{*j}} \|\boldsymbol{\Theta}_{*j}\|_1 \quad \text{subject to} \quad \|\mathbf{S}\boldsymbol{\Theta}_{*j} - \boldsymbol{e}_j\|_\infty \le \delta_j, \quad \text{for} \quad j = 1, \cdots, N,$$

where $\boldsymbol{e}_j$ is the jth canonical vector and δ_j is a tuning parameter. This optimization problem can be formulated into a linear program and has the potential to scale to large problems. In a closely related work of CLIME, Liu and Luo (2012) propose the SCIO estimator, which solves the jth column of $\boldsymbol{\Theta}$ by

$$\hat{\boldsymbol{\Theta}}_{*j} = \arg \min_{\boldsymbol{\Theta}_{*j}} \left\{ \frac{1}{2} \boldsymbol{\Theta}_{*j}' \mathbf{S} \boldsymbol{\Theta}_{*j} - \boldsymbol{e}_j' \boldsymbol{\Theta}_{*j} + \lambda_j \|\boldsymbol{\Theta}_{*j}\|_1 \right\}.$$

The SCIO estimator can be solved efficiently by the pathwise coordinate descent algorithm.

6.3.2 The Need for Tuning-insensitive Procedures

Most of the methods described in Section 6.3.1 require choosing some tuning parameters that control the bias–variance tradeoff. Their theoretical justifications are usually built on some theoretical choices of tuning parameters that cannot be implemented in practice. For example, in the neighborhood pursuit method and the graphical Dantzig selector, the tuning parameter λ_j and γ_j depend on σ_j^2, which is unknown. The tuning parameters of the CLIME and SCIO depend on $\|\boldsymbol{\Theta}\|_1$, which is unknown.

Choosing the regularization parameter in a data-dependent way remains an open problem. Popular techniques include the C_p-statistic, AIC (Akaike information criterion), BIC (Bayesian information criterion), extended BIC (Chen and Chen, 2008, 2012; Foygel and Drton, 2010), RIC (risk inflation criterion; Foster and George, 1994), cross validation, and covariance penalization (Efron, 2004). Most of these methods require data splitting and have been only justified for low-dimensional settings. Some progress has been made recently on developing likelihood-free regularization selection techniques, including permutation methods (Boos *et al.*, 2009; Lysen, 2009; Wu *et al.*, 2007) and subsampling methods (Bach, 2008; Ben-david *et al.*, 2006; Lange *et al.*, 2004; Meinshausen and Bühlmann, 2010). Meinshausen and Bühlmann (2010), Bach (2008), and Liu *et al.* (2012) also propose to select the tuning parameters using subsampling. However, these subsampling-based methods are computationally expensive and still lack theoretical guarantees.

To handle the challenge of tuning parameter selection, we introduce a "tuning-insensitive" procedure for estimating the precision matrix of high-dimensional Gaussian graphical models. Our method, named TIGER (tuning-insensitive graph estimation and regression), is asymptotically tuning-free and only requires very few efforts to choose the regularization parameter in finite sample settings.

6.3.3 TIGER: A Tuning-insensitive Approach for Optimal Precision Matrix Estimation

The main idea of the TIGER method is to estimate the precision matrix $\boldsymbol{\Theta}$ in a column-by-column fashion. For each column, the computation is reduced to a sparse regression problem. This idea has been adopted by many methods described in Section 6.3.1. These methods differ from each other mainly by how they solve the sparse regression subproblem. Unlike these existing methods, TIGER solves this sparse regression problem using the SQRT-lasso (Belloni *et al.*, 2012).

The SQRT-lasso is a penalized optimization algorithm for solving high-dimensional linear regression problems. For a linear regression problem $\tilde{\boldsymbol{Y}} = \tilde{\mathbf{X}}\boldsymbol{\beta} + \boldsymbol{\epsilon}$, where $\tilde{\boldsymbol{Y}} \in \mathbb{R}^T$ is the response, $\tilde{\mathbf{X}} \in \mathbb{R}^{T\times N}$ is the design matrix, $\boldsymbol{\beta} \in \mathbb{R}^N$ is the vector of unknown coefficients, and $\boldsymbol{\epsilon} \in \mathbb{R}^T$ is the noise vector. The SQRT-lasso estimates $\boldsymbol{\beta}$ by solving

$$\hat{\boldsymbol{\beta}} = \arg \min_{\boldsymbol{\beta}\in\mathbb{R}^N} \left\{ \frac{1}{\sqrt{T}} \|\tilde{\boldsymbol{Y}} - \tilde{\mathbf{X}}\boldsymbol{\beta}\|_2 + \lambda\|\boldsymbol{\beta}\|_1 \right\},$$

where λ is the tuning parameter. It is shown in Belloni *et al.* (2012) that the choice of λ for the SQRT-lasso method is asymptotically universal and does not depend on any unknown parameter. In contrast, most other methods, including the lasso and Dantzig selector, rely heavily on a known standard deviation of the noise. Moreover, the SQRT-lasso method achieves near oracle performance for the estimation of $\boldsymbol{\beta}$.

In Liu and Wang (2012), they show that the objective function of the scaled-lasso can be viewed as a variational upper bound of the SQRT-lasso. Thus, the TIGER method is essentially equivalent to the method in Sun and Zhang (2012). However, the SQRT-lasso is more amenable to theoretical analysis and allows us to simultaneously establish optimal rates of convergence for the precision matrix estimation under many different norms.

Let $\hat{\boldsymbol{\Gamma}} := \mathrm{diag}(\mathbf{S})$ be an N-dimensional diagonal matrix with the diagonal elements the same as those in $\mathbf{S}$. Conditioned on the observed data $\boldsymbol{Y}_1, \cdots, \boldsymbol{Y}_T$, we define

$$\mathbf{Z} := (Z_1, \cdots, Z_N)' = \boldsymbol{Y}\hat{\boldsymbol{\Gamma}}^{-1/2}.$$

By (6.15), we have

$$Z_j \hat{\boldsymbol{\Gamma}}_{jj}^{1/2} = \boldsymbol{\alpha}_j' \mathbf{Z}_{\backslash j} \hat{\boldsymbol{\Gamma}}_{\backslash j,\backslash j}^{1/2} + \epsilon_j, \tag{6.18}$$

We define

$$\boldsymbol{\beta}_j := \hat{\boldsymbol{\Gamma}}_{\backslash j,\backslash j}^{1/2} \hat{\boldsymbol{\Gamma}}_{jj}^{-1/2} \boldsymbol{\alpha}_j \qquad \text{and} \qquad \tau_j^2 = \sigma_j^2 \hat{\boldsymbol{\Gamma}}_{jj}^{-1}.$$

Therefore, we have

$$Z_j = \boldsymbol{\beta}_j' \mathbf{Z}_{\backslash j} + \hat{\boldsymbol{\Gamma}}_{jj}^{-1/2} \epsilon_j. \tag{6.19}$$

We define $\hat{\mathbf{R}}$ to be the sample correlation matrix: $\hat{\mathbf{R}} := (\mathrm{diag}(\mathbf{S}))^{-1/2}\mathbf{S}(\mathrm{diag}(\mathbf{S}))^{-1/2}$. Motivated by the model in (6.19), we propose the following precision matrix estimator.

TIGER Algorithm

For $j = 1, \ldots, N$, we estimate the jth column of $\boldsymbol{\Theta}$ by solving:

$$\hat{\boldsymbol{\beta}}_j := \arg \min_{\boldsymbol{\beta}_j \in \mathbb{R}^{N-1}} \left\{ \sqrt{1 - 2\boldsymbol{\beta}'_j \hat{\mathbf{R}}_{\backslash j,j} + \boldsymbol{\beta}'_j \hat{\mathbf{R}}_{\backslash j,\backslash j} \boldsymbol{\beta}_j} + \lambda \|\boldsymbol{\beta}_j\|_1 \right\}, \tag{6.20}$$

$$\hat{\tau}_j := \sqrt{1 - 2\hat{\boldsymbol{\beta}}'_j \hat{\mathbf{R}}_{\backslash j,j} + \hat{\boldsymbol{\beta}}'_j \hat{\mathbf{R}}_{\backslash j,\backslash j} \hat{\boldsymbol{\beta}}_j}, \tag{6.21}$$

$$\hat{\boldsymbol{\Theta}}_{jj} = \hat{\tau}_j^{-2} \hat{\boldsymbol{\Gamma}}_{jj}^{-1} \text{ and } \hat{\boldsymbol{\Theta}}_{\backslash j,j} = -\hat{\tau}_j^{-2} \hat{\boldsymbol{\Gamma}}_{jj}^{-1/2} \hat{\boldsymbol{\Gamma}}_{\backslash j,\backslash j}^{-1/2} \hat{\boldsymbol{\beta}}_j.$$

For the estimator in (6.20), λ is a tuning parameter. In Section 6.3.4, we show that by choosing $\lambda = \pi\sqrt{\frac{\log N}{2T}}$, the obtained estimator achieves the optimal rates of convergence in the asymptotic setting. Therefore, the TIGER procedure is asymptotically tuning free. For finite samples, we set

$$\lambda := \zeta\pi\sqrt{\frac{\log N}{2T}} \tag{6.22}$$

with ζ chosen from a range $[\sqrt{2}/\pi, 2]$. Since the choice of ζ does not depend on any unknown parameters, we call the procedure *tuning-insensitive*. Practically, we found that simply setting $\zeta = 1$ gives satisfactory finite sample performance in most applications.

If a symmetric precision matrix estimate is preferred, we conduct the following correction: $\tilde{\mathbf{\Theta}}_{jk} = \min\{\hat{\mathbf{\Theta}}_{jk}, \hat{\mathbf{\Theta}}_{kj}\}$ for all $k \neq j$. Another symmetrization method is

$$\tilde{\mathbf{\Theta}} = \frac{\hat{\mathbf{\Theta}} + \hat{\mathbf{\Theta}}'}{2}.$$

As has been shown by Cai *et al.* (2011), if $\hat{\mathbf{\Theta}}$ is a good estimator, then $\tilde{\mathbf{\Theta}}$ will also be a good estimator: they achieve the same rates of convergence in the asymptotic settings.

Let $\mathbf{Z} \in \mathbb{R}^{T\times N}$ be the normalized data matrix, that is, $\mathbf{Z}_{*j} = \mathbf{Y}_{*j}\mathbf{\Sigma}_{jj}^{-1/2}$ for $j = 1, \cdots, N$. An equivalent form of (6.20) and (6.21) is

$$\hat{\boldsymbol{\beta}}_j = \arg \min_{\boldsymbol{\beta}_j\in\mathbb{R}^{N-1}} \left\{ \frac{1}{\sqrt{T}}\|\mathbf{Z}_{*j} - \mathbf{Z}_{*\backslash j}\boldsymbol{\beta}_j\|_2 + \lambda\|\boldsymbol{\beta}_j\|_1 \right\}, \tag{6.23}$$

$$\hat{\tau}_j = \frac{1}{\sqrt{T}}\|\mathbf{Z}_{*j} - \mathbf{Z}_{*\backslash j}\hat{\boldsymbol{\beta}}_j\|_2. \tag{6.24}$$

Once $\hat{\mathbf{\Theta}}$ is estimated, we can also estimate the graph $\hat{\mathrm{G}} := (V, \hat{\mathrm{E}})$ based on the sparsity pattern of $\hat{\mathbf{\Theta}}_{jk} \neq 0$.

6.3.4 Computation

Instead of directly solving (6.20) and (6.21), we consider the following optimization:

$$\hat{\boldsymbol{\beta}}_j, \hat{\tau}_j := \arg \min_{\boldsymbol{\beta}_j\in\mathbb{R}^{N-1},\tau_j\geq 0} \left\{ \frac{1 - 2\boldsymbol{\beta}_j'\hat{\mathbf{R}}_{\backslash j,j} + \boldsymbol{\beta}_j'\hat{\mathbf{R}}_{\backslash j,\backslash j}\boldsymbol{\beta}_j}{2\tau_j} + \frac{\tau_j}{2} + \lambda\|\boldsymbol{\beta}_j\|_1 \right\}, \tag{6.25}$$

Liu and Wang (2012) show that the solution to (6.20) and (6.21) is the same as that to (6.25). Equation (6.25) is jointly convex with respect to $\boldsymbol{\beta}_j$ and τ_j and can be solved by a coordinate-descent procedure. In the tth iteration, for a given $\tau_j^{(t)}$, we first solve a subproblem

$$\boldsymbol{\beta}_j^{(t+1)} := \arg \min_{\boldsymbol{\beta}_j\in\mathbb{R}^{N-1}} \left\{ \frac{1 - 2\boldsymbol{\beta}_j'\hat{\mathbf{R}}_{\backslash j,j} + \boldsymbol{\beta}_j'\hat{\mathbf{R}}_{\backslash j,\backslash j}\boldsymbol{\beta}_j}{2\tau_j^{(t)}} + \lambda\|\boldsymbol{\beta}_j\|_1 \right\},$$

This is a lasso problem and can be efficiently solved by the coordinate-descent algorithm developed by Friedman *et al.* (2007). Once $\boldsymbol{\beta}_j^{(t+1)}$ is obtained, we can calculate $\tau_j^{(t+1)}$ as

$$\tau_j^{(t+1)} = \sqrt{1 - 2(\boldsymbol{\beta}_j^{(t+1)})'\hat{\mathbf{R}}_{\backslash j,j} + (\boldsymbol{\beta}_j^{(t+1)})'\hat{\mathbf{R}}_{\backslash j,\backslash j}(\boldsymbol{\beta}_j^{(t+1)})}.$$

We iterate these two steps until the algorithm converges.

6.3.5 Theoretical Properties of TIGER

Liu and Wang (2012) establish the rates of convergence of the TIGER estimator $\hat{\mathbf{\Theta}}$ to the true precision matrix $\mathbf{\Theta}$ under different norms. In particular, let $\|\mathbf{\Theta}\|_{\max} := \max_{jk} |\mathbf{\Theta}_{jk}|$ and

$\|\boldsymbol{\Theta}\|_1 := \max_j \sum_k |\boldsymbol{\Theta}_{jk}|$. Under the assumption that the condition number of $\boldsymbol{\Theta}$ is bounded by a constant, they establish the element-wise sup-norm rate of convergence:

$$\|\hat{\boldsymbol{\Theta}} - \boldsymbol{\Theta}\|_{\max} = O_P\left(\|\boldsymbol{\Theta}\|_1 \sqrt{\frac{\log N}{T}} \right). \tag{6.26}$$

Under mild conditions, the obtained rate in (6.26) is minimax optimal over the model class consisting of precision matrices with bounded condition numbers.

Let $I(\cdot)$ be the indicator function and $s := \sum_{j \neq k} I(\boldsymbol{\Theta}_{jk} \neq 0)$ be the number of nonzero off-diagonal elements of $\boldsymbol{\Theta}$. The result in (6.26) implies that the Frobenious norm error between $\hat{\boldsymbol{\Theta}}$ and $\boldsymbol{\Theta}$ satisfies:

$$\|\hat{\boldsymbol{\Theta}} - \boldsymbol{\Theta}\|_F := \sqrt{\sum_{i,j} |\hat{\boldsymbol{\Theta}}_{jk} - \boldsymbol{\Theta}_{jk}|^2} = O_P\left(\|\boldsymbol{\Theta}\|_1 \sqrt{\frac{(N+s)\log N}{T}} \right). \tag{6.27}$$

The rate in (6.27) is the minimax optimal rate for the Frobenious norm error in the same model class consisting of precision matrices with bounded condition numbers.

Let $\|\boldsymbol{\Theta}\|_2$ be the largest eigenvalue of $\boldsymbol{\Theta}$ (i.e., $\|\boldsymbol{\Theta}\|_2$ is the spectral norm of $\boldsymbol{\Theta}$) and $k := \max_{i=1,\cdots,N} \sum_j I(\boldsymbol{\Theta}_{ij} \neq 0)$. Liu and Wang (2010) also show that

$$\|\hat{\boldsymbol{\Theta}} - \boldsymbol{\Theta}\|_2 \leq \|\hat{\boldsymbol{\Theta}} - \boldsymbol{\Theta}\|_1 = O_P\left(k\|\boldsymbol{\Theta}\|_2 \sqrt{\frac{\log N}{T}} \right). \tag{6.28}$$

This spectral norm rate in (6.28) is also minimax optimal over the same model class as before.

6.3.6 *Applications to Modeling Stock Returns*

We apply the TIGER method to explore a stock price dataset collected from Yahoo! Finance (finance.yahoo.com). More specifically, the daily closing prices were obtained for 452 stocks that were consistently in the S&P 500 index between January 1, 2003, through January 1, 2011. This gives us altogether 2015 data points, and each data point corresponds to the vector of closing prices on a trading day. With $S_{t,j}$ denoting the closing price of stock j on day t, we consider the log-return variable $Y_{jt} = \log\ (S_{t,j}/S_{t-1,j})$ and build graphs over the indices j.

We Winsorize (or truncate) every stock so that its data points are within six times the mean absolute deviation from the sample average. In Figure 6.3, we show boxplots for 10 randomly chosen stocks. We see that the data contain outlier even after Winsorization; the reasons for these outliers include splits in a stock, which increase the number of shares. It is known that the log-return data are heavy-tailed. To suitably apply the TIGER method, we Gaussianize the marginal distribution of the data by the normal-score transformation. In Figure 6.3b, we compare the boxplots of the data before and after Gaussianization. We see that Gaussianization alleviates the effect of outliers.

In this analysis, we use the subset of the data between January 1, 2003, and January 1, 2008, before the onset of the financial crisis. The 452 stocks are categorized into 10 Global Industry Classification Standard (GICS) sectors, including `Consumer Discretionary` (70 stocks), `Consumer Staples` (35 stocks), `Energy` (37 stocks), `Financials` (74 stocks),

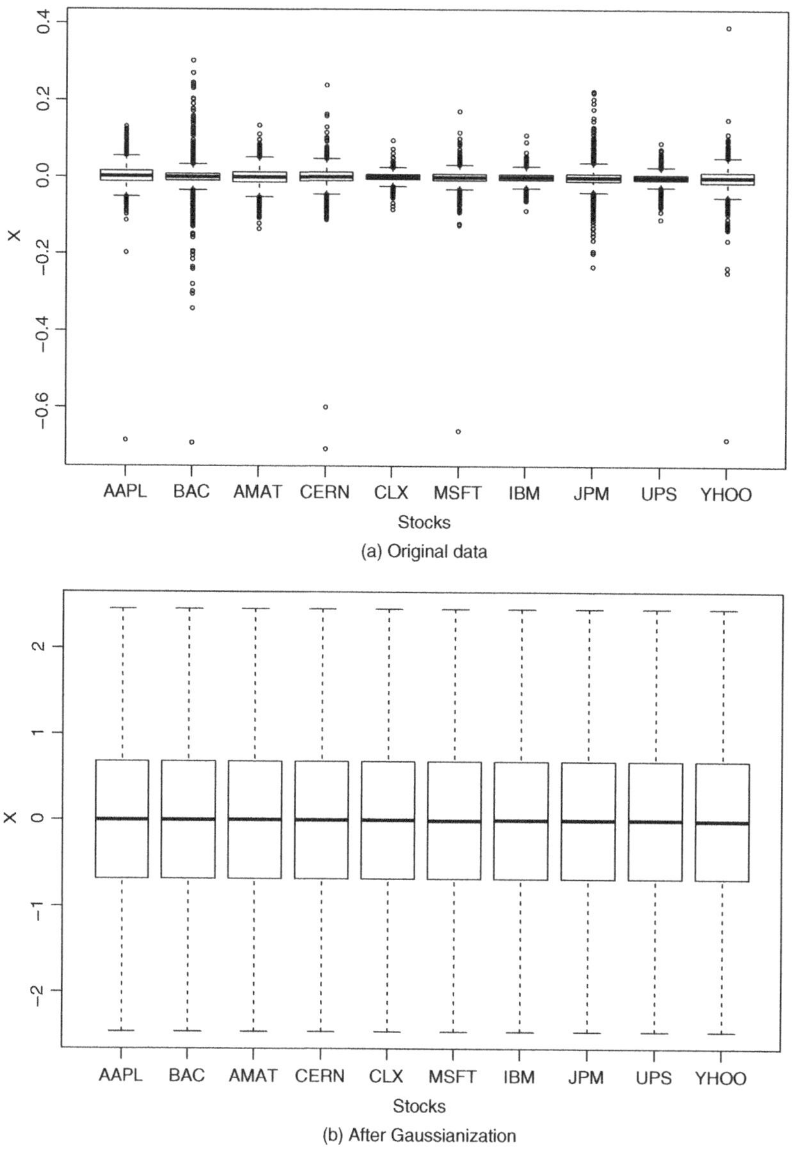

Figure 6.3 Boxplots of $Y_{jt} = \log\,(S_{t,j}/S_{t-1,j})$ for 10 stocks. As can be seen, the original data has many outliers, which is addressed by the normal-score transformation on the rescaled data (right).

`Health Care` (46 stocks), `Industrials` (59 stocks), `Information Technology` (64 stocks), `Materials` (29 stocks), `Telecommunications Services` (6 stocks), and `Utilities` (32 stocks). It is expected that stocks from the same GICS sectors should tend to be clustered together in the estimated graph, since stocks from the same GICS sector tend to interact more with each other. In Figure 6.4, the nodes are colored according to the GICS sector of the corresponding stock.

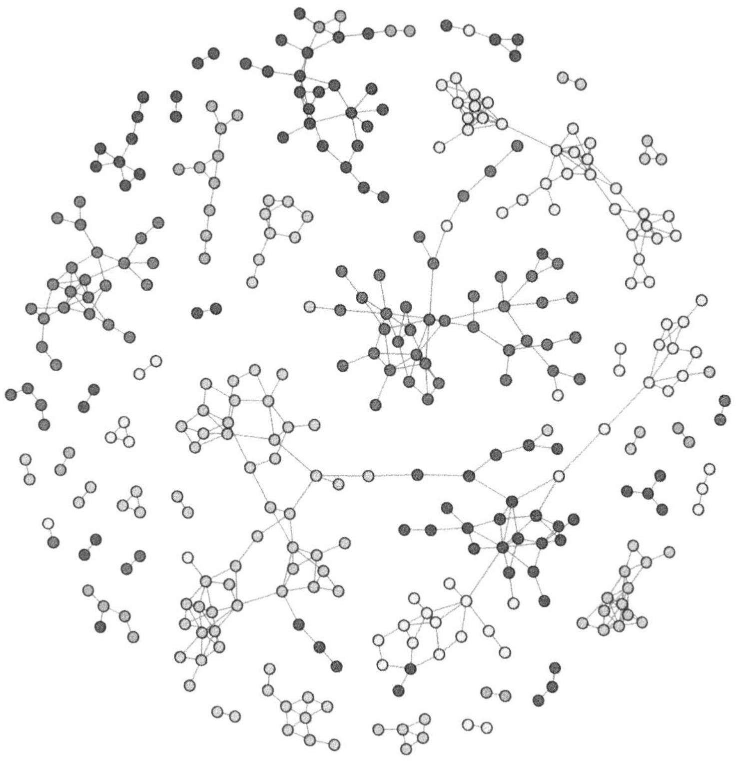

Figure 6.4 The estimated TIGER graph using the S&P 500 stock data from January 1, 2003, to January 1, 2008.

In Figure 6.4 we visualize the estimated graph using the TIGER method on the data from January 1, 2003, to January 1, 2008. There are altogether $T = 1257$ data points and $N = 452$ dimensions. Even though the TIGER procedure is asymptotically tuning-free, Liu and Wang (2010) show that a fine-tune step can further improve its finite sample performance. To fine-tune the tuning parameter, we adopt a variant of the stability selection method proposed by Meinshausen and Bühlmann (2010). As suggested in (6.22), we consider 10 equal-distance values of ζ chosen from a range $[\sqrt{2}/\pi, 2]$. We randomly sample 100 sub-datasets, each containing $B = \lfloor 10\sqrt{T} \rfloor = 320$ data points. On each of these 100 subsampled datasets, we estimate a TIGER graph for each tuning parameter. In the final graph shown in Figure 6.4, we use $\zeta = 1$, and an edge is present only if it appears more than 80% of the time among the 100 subsampled datasets (with all the singleton nodes removed).

From Figure 6.4, we see that stocks from the same GICS sectors are indeed close to each other in the graph. We refrain from drawing any hard conclusions about the effectiveness of the estimated TIGER graph—how it is used will depend on the application. One potential application of such a graph could be for portfolio optimization. When designing a portfolio, we may want to choose stocks with large graph distances to minimize the investment risk.

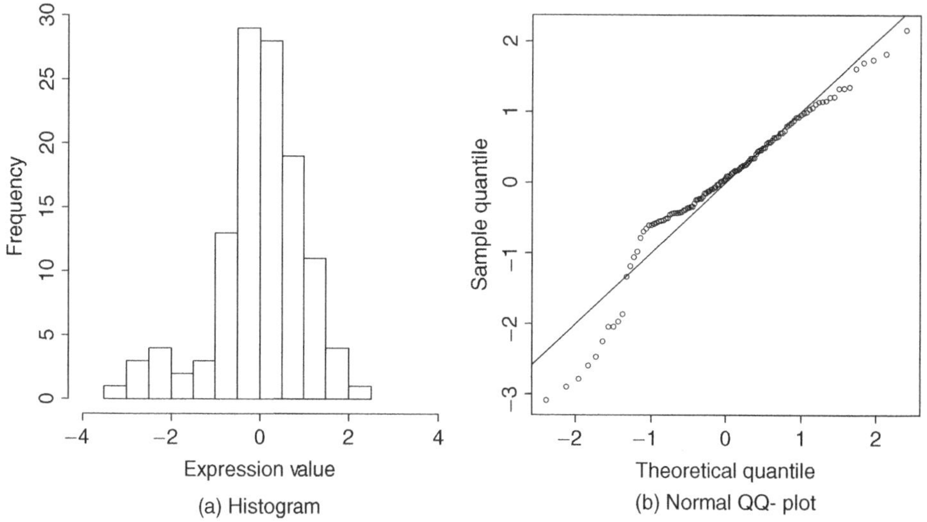

Figure 6.5 The histogram and normal QQ plots of the marginal expression levels of the gene MECPS. We see the data are not exactly Gaussian distributed. Adapted from Liu and Wang (2012).

6.3.7 Applications to Genomic Network

As discussed in this chapter, an important application of precision matrix estimation is to estimate high-dimensional graphical models. In this section, we apply the TIGER method on a gene expression dataset to reconstruct the conditional independence graph of the expression levels of 39 genes.

This dataset, which includes 118 gene expression arrays from *Arabidopsis thaliana*, originally appeared in Wille *et al.* (2004). Our analysis focuses on gene expression from 39 genes involved in two isoprenoid metabolic pathways: 16 from the mevalonate (MVA) pathway are located in the cytoplasm, 18 from the plastidial (MEP) pathway are located in the chloroplast, and 5 are located in the mitochondria. While the two pathways generally operate independently, crosstalk is known to happen (Wille *et al.* 2004). Our scientific goal is to recover the gene regulatory network, with special interest in crosstalk.

We first examine whether the data actually satisfy the Gaussian distribution assumption. In Figure 6.5, we plot the histogram and normal QQ plot of the expression levels of a gene named MECPS. From the histogram, we see the distribution is left-skewed compared to the Gaussian distribution. From the normal QQ plot, we see the empirical distribution has a heavier tail compared to Gaussian. To suitably apply the TIGER method on this dataset, we need to first transform the data so that its distribution is closer to Gaussian. Therefore, we Gaussianize the marginal expression values of each gene by converting them to the corresponding normal-scores. This is automatically done by the `huge.npn` function in the `R` package `huge` (Zhao *et al.*, 2012).

We apply the TIGER on the transformed data using the default tuning parameter $\zeta = \sqrt{2}/\pi$. The estimated network is shown in Figure 6.6. We note that the estimated network is very sparse with only 44 edges. Prior investigations suggest that the connections from genes AACT1 and HMGR2 to gene MECPS indicate a primary source of the crosstalk between the MEP and

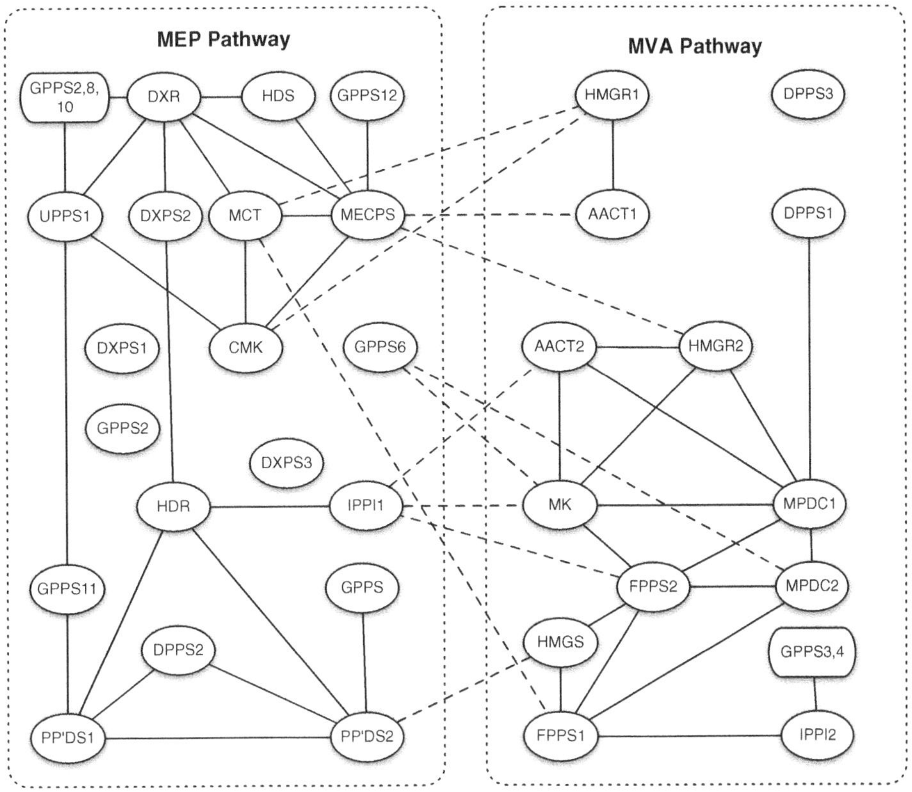

Figure 6.6 The estimated gene networks of the *Arabadopsis* dataset. The within-pathway edges are denoted by solid lines, and between-pathway edges are denoted by dashed lines. From Liu and Wang (2012).

MVA pathways, and these edges are presented in the estimated network. MECPS is clearly a hub gene for this pathway.

For the MEP pathway, the genes DXPS2, DXR, MCT, CMK, HDR, and MECPS are connected as in the true metabolic pathway. Similarly, for the MVA pathway, the genes AACT2, HMGR2, MK, MPDC1, MPDC2, FPPS1, and FPP2 are closely connected. Our analysis suggests 11 cross-pathway links. This is consistent to previous investigation in Wille *et al.* (2004). This result suggests that there might exist rich interpathway crosstalks.

6.4 Financial Applications

6.4.1 *Estimating Risks of Large Portfolios*

Estimating and assessing the risk of a large portfolio are important topics in financial econometrics and risk management. The risk of a given portfolio allocation vector $\mathbf{w}_N$ is conveniently measured by $(\mathbf{w}_N'\boldsymbol{\Sigma}\mathbf{w}_N)^{1/2}$, in which $\boldsymbol{\Sigma}$ is a volatility (covariance) matrix of the assets' returns. Often multiple portfolio risks are of interest, and hence it is essential to estimate the volatility matrix $\boldsymbol{\Sigma}$. On the other hand, assets' excess returns are often driven by a few common factors. Hence $\boldsymbol{\Sigma}$ can be estimated via factor analysis as described in Section 6.1.

Let$\{\boldsymbol{Y}_t\}_{t=1}^T$ be a strictly stationary time-series of an $N \times 1$ vector of observed asset returns and $\boldsymbol{\Sigma} = \text{Cov}(\boldsymbol{Y}_t)$. We assume that $\boldsymbol{Y}_t$ satisfies an approximate factor model:

$$\boldsymbol{Y}_t = \mathbf{B}\boldsymbol{f}_t + \boldsymbol{u}_t, t \leq T, \tag{6.29}$$

where $\mathbf{B}$ is an $N \times K$ matrix of factor loadings; $\boldsymbol{f}_t$ is a $K \times 1$ vector of common factors; and $\boldsymbol{u}_t$ is an $N \times 1$ vector of idiosyncratic error components. In contrast to N and T, here K is assumed to be fixed. The common factors may or may not be observable. For example, Fama and French (1993) identified three known factors that have successfully described the US stock market. In addition, macroeconomic and financial market variables have been thought to capture systematic risks as observable factors. On the other hand, in an empirical study, Bai and Ng (2002) determined two unobservable factors for stocks traded on the New York Stock Exchange during 1994–1998.

As described in Section 6.1, the factor model implies the following decomposition of $\boldsymbol{\Sigma}$:

$$\boldsymbol{\Sigma} = \mathbf{B}\text{Cov}(\boldsymbol{f}_t)\mathbf{B}' + \boldsymbol{\Sigma}_u. \tag{6.30}$$

In the case of observable factors, an estimator of $\boldsymbol{\Sigma}$ is constructed based on thresholding the covariance matrix of idiosyncratic errors, as in (6.7), denoted by $\hat{\boldsymbol{\Sigma}}_f$. In the case of unobservable factors, $\boldsymbol{\Sigma}$ can be estimated by POET as in (6.9), denoted by $\hat{\boldsymbol{\Sigma}}_P$. Because K, the number of factors, might also be unknown, this estimator uses a data-driven number of factors $\hat{K}$. Based on the factor analysis, the risk for a given portfolio $\mathbf{w}_N$ can be estimated by either $\sqrt{\mathbf{w}_N'\hat{\boldsymbol{\Sigma}}_f\mathbf{w}_N}$ or $\sqrt{\mathbf{w}_N'\hat{\boldsymbol{\Sigma}}_P\mathbf{w}_N}$, depending on whether$\boldsymbol{f}_t$ is observable.

6.4.1.1 Estimating a Minimum Variance Portfolio

There are also many methods proposed to choose data-dependent portfolios. For instance, estimated portfolio vectors can arise when the ideal portfolio $\mathbf{w}_N$ depends on the inverse of the large covariance $\boldsymbol{\Sigma}$ (Markowitz, 1952), by consistently estimating $\boldsymbol{\Sigma}^{-1}$. Studying the effects of estimating $\boldsymbol{\Sigma}$ is also important for portfolio allocations. In these problems, estimation errors in estimating $\boldsymbol{\Sigma}$ can have substantial implications (see discussions in El Karoui, 2010). For illustration, consider the following example of estimating the global minimum variance portfolio.

The *global minimum variance* portfolio is the solution to the problem:

$$\mathbf{w}_N^{gmv} = \arg\ \min_{\mathbf{w}}(\mathbf{w}'\boldsymbol{\Sigma}\mathbf{w}), \ \text{ such that } \mathbf{w}'\mathbf{e} = 1$$

where $\mathbf{e} = (1, \ldots, 1)$, yielding $\mathbf{w}_N^{gmv} = \boldsymbol{\Sigma}^{-1}\mathbf{e}/(\mathbf{e}'\boldsymbol{\Sigma}^{-1}\mathbf{e})$. Although this portfolio does not belong to the efficient frontier, Jagannathan and Ma (2003) showed that its performance is comparable with those of other tangency portfolios.

The factor model yields a positive definite covariance estimator for $\mathbf{\Sigma}$, which then leads to a data-dependent portfolio:

$$\hat{\mathbf{w}}_N^{gmv} = \frac{\hat{\mathbf{\Sigma}}^{-1}\mathbf{e}}{\mathbf{e}'\hat{\mathbf{\Sigma}}^{-1}\mathbf{e}}, \quad \hat{\mathbf{\Sigma}}^{-1} = \begin{cases} \hat{\mathbf{\Sigma}}_f^{-1} & \text{known factors;} \\ \hat{\mathbf{\Sigma}}_P^{-1}, & \text{unknown factors} \end{cases}.$$

It can be shown that $\hat{\mathbf{w}}_N^{gmv}$ is L_1-consistent, in the sense that

$$\|\hat{\mathbf{w}}_N^{gmv} - \mathbf{w}_N^{gmv}\|_1 = o_P(1).$$

We refer to El Karoui (2010) and Ledoit and Wolf (2003) for further discussions on the effects of estimating large covariance matrices for portfolio selections.

6.4.1.2 Statistical Inference of the Risks

Confidence intervals of the true risk $\mathbf{w}_N'\mathbf{\Sigma}\mathbf{w}_N$ can be constructed based on the estimated risk $\mathbf{w}_N'\hat{\mathbf{\Sigma}}\mathbf{w}_N$, where $\hat{\mathbf{\Sigma}} = \hat{\mathbf{\Sigma}}_f$ or $\hat{\mathbf{\Sigma}}_P$, depending on whether the factors are known or not. Fan *et al.* (2014a) showed that, under some regularity conditions, respectively,

$$\left[\mathrm{Var}\left(\sum_{t=1}^{T}(\mathbf{w}_N'\mathbf{B}\boldsymbol{f}_t)^2\right)\right]^{-1/2} T\hat{\mathbf{w}}_N'(\hat{\mathbf{\Sigma}} - \mathbf{\Sigma})\hat{\mathbf{w}}_N \to^d \mathcal{N}(0,1), \quad \hat{\mathbf{\Sigma}} = \hat{\mathbf{\Sigma}}_f \text{or } \hat{\mathbf{\Sigma}}_P,$$

where $\hat{\mathbf{w}}_N$ is an L_1-consistent estimator of $\mathbf{w}_N$.

An important implication is that the asymptotic variance is the same regardless of whether the factors are observable or not. Therefore, the impact of estimating the unknown factors is asymptotically negligible. In addition, it can also be shown that the asymptotic variance is slightly smaller than that of $\mathbf{w}_N'\mathbf{S}\mathbf{w}_N$, the sample covariance-based risk estimator. The asymptotic variance $\mathrm{Var}\left(\sum_{t=1}^{T}(\mathbf{w}_N'\mathbf{B}\boldsymbol{f}_t)^2\right)$ can be consistently estimated, using the heteroscedasticity and autocorrelation consistent covariance estimator of Newey and West (1987) based on the truncated sum of estimated autocovariance functions. Therefore, the above limiting distributions can be employed to assess the uncertainty of the estimated risks by, for example, constructing asymptotic confidence intervals for $(\mathbf{w}_N'\mathbf{\Sigma}\mathbf{w}_N)^{1/2}$. Fan *et al.* (2014a) showed that the confidence interval is practically accurate even at the finite sample.

6.4.2 Large Panel Test of Factor Pricing Models

The content of this section is adapted from the recent work by Fan *et al.* (2014b), including graphs and tables. We consider a *factor-pricing model*, in which the excess return has the following decomposition:

$$Y_{it} = \alpha_i + \mathbf{b}_i'\boldsymbol{f}_t + u_{it}, \quad i = 1, \ldots, N, t = 1, \ldots, T. \tag{6.31}$$

In this subsection, we shall focus on the case in which $\boldsymbol{f}_t$'s are observable.

Let $\boldsymbol{\alpha} = (\alpha_1, \dots, \alpha_N)'$ be the vector of intercepts for all N financial assets. The key implication from the multifactor pricing theory is that $\boldsymbol{\alpha}$ should be zero, known as *mean-variance efficiency*, for any asset i. An important question is then if such a pricing theory can be validated by empirical data, namely whether the null hypothesis

$$H_0 : \boldsymbol{\alpha} = 0, \tag{6.32}$$

is consistent with empirical data.

Most of the existing tests to the problem (6.32) are based on the quadratic statistic $W = \hat{\boldsymbol{\alpha}}'\hat{\boldsymbol{\Sigma}}_u^{-1}\hat{\boldsymbol{\alpha}}$, where $\hat{\boldsymbol{\alpha}}$ is the OLS estimator for $\boldsymbol{\alpha}$, $\hat{\boldsymbol{\Sigma}}_u^{-1}$ is the estimated inverse of the error covariance, and a_T is a positive number that depends on the factors $\boldsymbol{f}_t$ only. Prominent examples are the test given by Gibbons *et al.* (1989), the GMM test in MacKinlay and Richardson (1991), and the likelihood ratio test in Beaulieu *et al.* (2007), all in quadratic forms. Recently, Pesaran and Yamagata (2012) studied the limiting theory of the normalized W assuming $\boldsymbol{\Sigma}_u^{-1}$ were known. They also considered a quadratic test where $\hat{\boldsymbol{\Sigma}}_u^{-1}$ is replaced with its diagonalized matrix.

There are, however, two main challenges in the quadratic statistic W. The first is that estimating $\boldsymbol{\Sigma}_u^{-1}$ is a challenging problem when $N > T$, as described previously. Secondly, even though $\boldsymbol{\Sigma}_u^{-1}$ were known, this test suffers from a lower power in a high-dimensional-low-sample-size situation, as we now explain.

For simplicity, let us temporarily assume that $\{\boldsymbol{u}_t\}_{t=1}^T$ are independent and identically distributed (i.i.d.) Gaussian and $\boldsymbol{\Sigma}_u = \mathrm{Cov}(\boldsymbol{u}_t)$ is known, where $\boldsymbol{u}_t = (u_{1t}, \dots, u_{Nt})$. Under H_0, W is χ^2_N distributed, with the critical value $\chi^2_{N,q}$, which is of order N, at significant level q. The test has no power at all when $T\boldsymbol{\alpha}'\boldsymbol{\Sigma}_u\boldsymbol{\alpha} = o(N)$ or $\|\boldsymbol{\alpha}\|^2 = o(N/T)$, assuming that $\boldsymbol{\Sigma}_u$ has bounded eigenvalues. This is not unusual for the high-dimension-low-sample-size situation we encounter, where there are thousands of assets to be tested over a relatively short time period (e.g. 60 monthly data). And it is especially the case when there are only a few significant alphas that arouse market inefficiency. By a similar argument, this problem can not be rescued by using any genuine quadratic statistic, which are powerful only when a non-negligible fraction of assets are mispriced. Indeed, the factor N above reflects the noise accumulation in estimating N parameters of $\boldsymbol{\alpha}$.

6.4.2.1 High-dimensional Wald Test

Suppose $\{\boldsymbol{u}_t\}$ is i.i.d. $\mathcal{N}(0, \boldsymbol{\Sigma}_u)$. Then as $N, T \to \infty$, Pesaran and Yamagata (2012) showed that

$$\frac{Ta\hat{\boldsymbol{\alpha}}'\boldsymbol{\Sigma}_u^{-1}\hat{\boldsymbol{\alpha}} - N}{\sqrt{2N}} \to^d \mathcal{N}(0, 1)$$

where $a = 1 - \frac{1}{T}\sum_t \boldsymbol{f}_t'(\frac{1}{T}\sum_t \boldsymbol{f}_t\boldsymbol{f}_t')^{-1}\frac{1}{T}\sum_t \boldsymbol{f}_t$. This normalized Wald test is infeasible unless $\boldsymbol{\Sigma}_u^{-1}$ is consistently estimable. Under the sparse assumption of $\boldsymbol{\Sigma}_u$, this can be achieved by thresholding estimation as previously described. Letting $\hat{\boldsymbol{\Sigma}}_u^{-1}$ be the thresholding estimator, then a feasible high-dimensional Wald test is

$$J_{sw} \equiv \frac{Ta\hat{\boldsymbol{\alpha}}'\hat{\boldsymbol{\Sigma}}_u^{-1}\hat{\boldsymbol{\alpha}} - N}{\sqrt{2N}}.$$

With further technical arguments (see Fan *et al.*, 2014b), it can be shown that $J_{sw} \to^d \mathcal{N}(0, 1)$. Note that it is very technically involved to show that substituting $\hat{\boldsymbol{\Sigma}}_u^{-1}$ for $\boldsymbol{\Sigma}_u^{-1}$ is asymptotically negligible when $N/T \to \infty$.

6.4.2.2 Power Enhancement Test

Traditional tests of factor pricing models are not powerful unless there are enough stocks that have nonvanishing alphas. Even if some individual assets are significantly mispriced, their nontrivial contributions to the test statistic are insufficient to reject the null hypothesis. This problem can be resolved by introducing a power enhancement component (PEM) J_0 to the normalized Wald statistic J_{sw}. The PEM J_0 is a screening statistic, designed to detect sparse alternatives with significant individual alphas.

Specifically, for some predetermined threshold value $\delta_T > 0$, define a set

$$\hat{S} = \left\{ j : \frac{|\hat{\alpha}_j|}{\hat{\sigma}_j} > \delta_T, j = 1, \dots, N \right\}, \tag{6.33}$$

where $\hat{\alpha}_j$ is the OLS estimator and $\hat{\sigma}_j^2 = \frac{1}{T}\sum_{t=1}^T \hat{u}_{jt}^2/a$ is T times the estimated variance of $\hat{\alpha}_j$, with $\hat{u}_{jt}$ being the regression residuals. Denote a subvector of $\hat{\boldsymbol{\alpha}}$ by

$$\hat{\boldsymbol{\alpha}}_{\hat{S}} = (\hat{\alpha}_j \, : \, j \in \hat{S}),$$

the screened-out alpha estimators, which can be interpreted as estimated alphas of mispriced stocks. Let $\hat{\boldsymbol{\Sigma}}_{\hat{S}}$ be the submatrix of $\hat{\boldsymbol{\Sigma}}_u$ formed by the rows and columns whose indices are in $\hat{S}$. So $\hat{\boldsymbol{\Sigma}}_{\hat{S}}/(Ta)$ is an estimated conditional covariance matrix of $\hat{\boldsymbol{\alpha}}_{\hat{S}}$, given the common factors and $\hat{S}$.

With the above notation, we define the screening statistic as

$$J_0 = \sqrt{N} Ta \hat{\boldsymbol{\alpha}}_{\hat{S}}' \hat{\boldsymbol{\Sigma}}_{\hat{S}}^{-1} \hat{\boldsymbol{\alpha}}_{\hat{S}}. \tag{6.34}$$

The choice of δ_T must suppress most of the noises, resulting in an empty set of $\hat{S}$ under the null hypothesis. On the other hand, δ_T cannot be too large to filter out important signals of alphas under the alternative. For this purpose, noting that the maximum noise level is $O_P(\sqrt{\log\ N/T})$, we let

$$\delta_T = \log\ (\log\ T)\sqrt{\frac{\log\ N}{T}}.$$

This is a high criticism test. When $N = 500$ and $T = 60$, $\delta_T = 3.514$. With this choice of δ_T, if we define, for $\sigma_j^2 = (\boldsymbol{\Sigma}_u)_{jj}/(1 - Ef_t'(Ef_t f_t')^{-1}Ef_t)$,

$$S = \left\{ j : \frac{|\alpha_j|}{\sigma_j} > 2\delta_T, j = 1, \dots, N \right\}, \tag{6.35}$$

then under mild conditions, $P(S \subset \hat{S}) \to 1$, with some additional conditions, $P(S = \hat{S}) \to 1$, and $\hat{\boldsymbol{\alpha}}_{\hat{S}}$ behaves like $\boldsymbol{\alpha}_S = (\boldsymbol{\alpha}_j \, : \, j \in S)$.

The power enhancement test is then defined to be

$$J_0 + J_{sw},$$

Table 6.3 Variable descriptive statistics for the Fama–French three-factor model (Adapted from Fan *et al.*, 2014b)

Variables	Mean	Std dev.	Median	Min	Max
N_τ	617.70	26.31	621	574	665
$\lvert\hat{S}\rvert_0$	5.49	5.48	4	0	37
$\lvert\bar{\alpha}\rvert_i^\tau(\%)$	0.9973	0.1630	0.9322	0.7899	1.3897
$\lvert\bar{\alpha}\rvert_{i\in\hat{S}}^\tau(\%)$	4.3003	0.9274	4.1056	1.7303	8.1299
p-value of J_{wi}	0.2844	0.2998	0.1811	0	0.9946
p-value of J_{sw}	0.1861	0.2947	0.0150	0	0.9926
p-value of PEM	0.1256	0.2602	0.0003	0	0.9836

whose detectable region is the union of those of J_0 and J_{sw}. Note that under the null hypothesis, $S = \emptyset$, so by the selection consistency, $J_0 = 0$ with probability approaching one. Thus, the null distribution of the power enhancement test is that of J_{sw}, which is standard normal. This means adding J_0 does not introduce asymptotic size distortion. On the other hand, since $J_0 \geq 0$, the power of $J_0 + J_{sw}$ is always enhanced. Fan *et al.* (2014b) showed that the test is consistent against the alternative as any subset of:

$$\{\boldsymbol{\alpha} \in \mathbb{R}^N : \max_{j\leq N} |\alpha_j| > 2\delta_T \max_{j\leq N} \sigma_j\} \cup \{\boldsymbol{\alpha} \in \mathbb{R}^N : \|\boldsymbol{\alpha}\|^2 \gg (N \log N)/T\}.$$

6.4.2.3 Empirical Study

We study monthly returns on all the S&P 500 constituents from the CRSP database for the period January 1980 to December 2012, during which a total of 1170 stocks have entered the index for our study. Testing of market efficiency is performed on a rolling window basis: for each month from December 1984 to December 2012. The test statistics are evaluated using the preceding 60 months' returns ($T = 60$). The panel at each testing month consists of stocks without missing observations in the past 5 years, which yields a cross-sectional dimension much larger than the time-series dimension ($N > T$). For testing months $\tau = 12/1984, \ldots, 12/2012$, we fit the Fama–French three-factor (FF-3) model:

$$r_{it}^\tau - r_{ft}^\tau = \alpha_i^\tau + \beta_{i,\,\mathrm{MKT}}^\tau(\mathrm{MKT}_t^\tau - r_{ft}^\tau) + \beta_{i,\,\mathrm{SMB}}^\tau \mathrm{SMB}_t^\tau + \beta_{i,\,\mathrm{HML}}^\tau \mathrm{HML}_t^\tau + u_{it}^\tau, \tag{6.36}$$

for $i = 1, \ldots, N_\tau$ and $t = \tau - 59, \ldots, \tau$, where r_{it} represents the return for stock i at month t; r_{ft} is the risk-free rate; and MKT, SMB, and HML constitute the FF-3 model's market, size, and value factors.

Table 6.3 summarizes descriptive statistics for different components and estimates in the model. On average, 618 stocks (which is more than 500 because we are recording stocks that have *ever* become the constituents of the index) enter the panel of the regression during each 5-year estimation window, of which 5.5 stocks are selected by $\hat{S}$. The threshold $\delta_T = \sqrt{\log N/T} \log(\log T)$ is about 0.45 on average, which changes as the panel size N changes for every window of estimation. The selected stocks have much larger alphas than other stocks do, as expected. As far as the signs of those alpha estimates are concerned, 61.84% of all the estimated alphas are positive, and 80.66% of all the selected alphas are positive. This

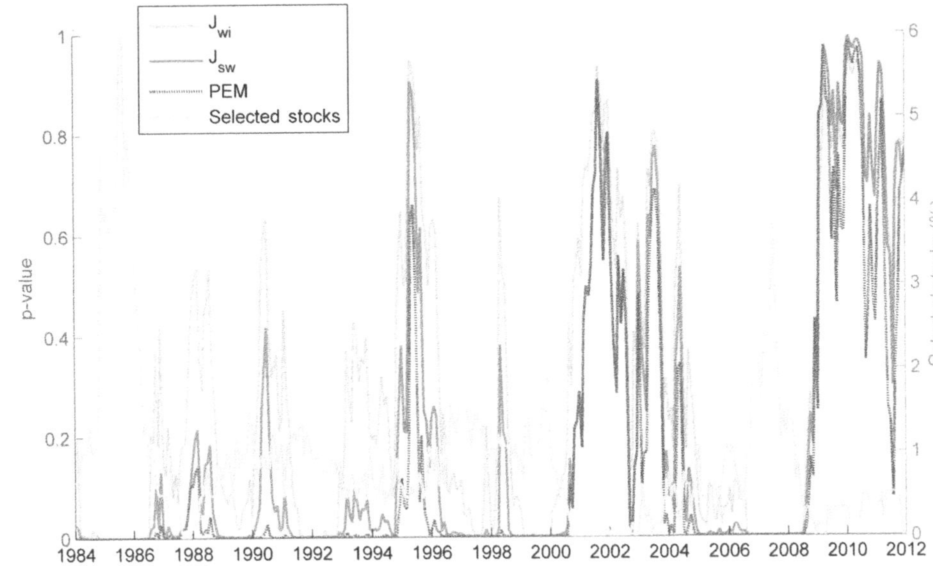

Figure 6.7 Dynamics of p-values and selected stocks (%, from Fan *et al.*, 2014b).

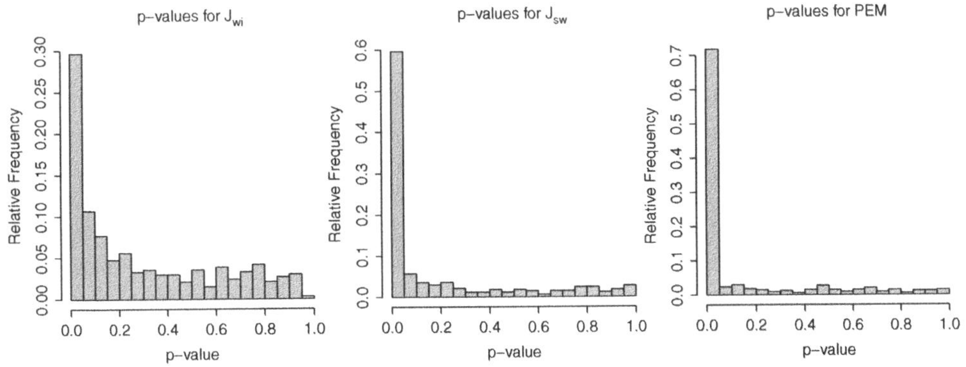

Figure 6.8 Histograms of p-values for J_{wi}, J_{sw}, and PEM (from Fan *et al.*, 2014b).

indicates that market inefficiency is primarily contributed by stocks with extra returns, instead of a large portion of stocks with small alphas, demonstrating the sparse alternatives. In addition, we notice that the p-values of the thresholded Wald test J_{sw} is generally smaller than that of the test J_{wi} given by Pesaran and Yamagata (2012).

We plot the running p-values of J_{wi}, J_{sw}, and the PEM test (augmented from J_{sw}) from December 1984 to December 2012. We also add the dynamics of the percentage of selected stocks ($|\hat{S}|_0/N$) to the plot, as shown in Figure 6.7. There is a strong negative correlation between the stock selection percentage and the p-values of these tests. This shows that the degree of market efficiency is influenced not only by the aggregation of alphas, but also by those extreme ones. We also observe that the p-value line of the PEM test lies beneath those of

J_{sw} and J_{wi} tests as a result of enhanced power, and hence it captures several important market disruptions ignored by the latter two (e.g. Black Monday in 1987, collapse of the Japanese bubble in late 1990, and the European sovereign debt crisis after 2010). Indeed, the null hypothesis of market efficiency is rejected by the PEM test at the 5% level during almost all financial crises, including major financial crises such as Black Wednesday in 1992, the Asian financial crisis in 1997, and the financial crisis in 2008, which are also detected by J_{sw} and J_{wi} tests. For 30%, 60%, and 72% of the study period, J_{wi}, J_{sw}, and the PEM test conclude that the market is inefficient, respectively. The histograms of the p-values of the three test statistics are displayed in Figure 6.8.

6.5 Statistical Inference in Panel Data Models

6.5.1 *Efficient Estimation in Pure Factor Models*

The sparse covariance estimation can also be employed to improve the estimation efficiency in factor models. Consider:

$$Y_{it} = \mathbf{b}_i'\boldsymbol{f}_t + u_{it}, i \leq N, t \leq T,$$

In the model, only Y_{it} is observable. In most literature, the factors and loadings are estimated via the principal components (PC) method, which solves a constraint minimization problem:

$$\min_{\mathbf{B},\boldsymbol{f}_t} \sum_{t=1}^{T} (\boldsymbol{Y}_t - \mathbf{B}\boldsymbol{f}_t)'(\boldsymbol{Y}_t - \mathbf{B}\boldsymbol{f}_t)$$

subject to some identifiability constraints so that the solution is unique. The PC method does not incorporate the error covariance $\boldsymbol{\Sigma}_u$, hence it essentially treats the error terms $\boldsymbol{u}_t$ as cross-sectionally homoscedastic and uncorrelated. It is well known that under either cross-sectional heteroscedasticity or correlations, the PC method is not efficient. On the other hand, when $\boldsymbol{\Sigma}_u$ is assumed to be sparse and estimated via thresholding, we can incorporate this covariance estimator into the estimation, and improve the estimation efficiency.

6.5.1.1 Weighted Principal Components

We can estimate the factors and loadings via the weighted least squares. For some $N \times N$ positive definite weight matrix $\mathbf{W}$, solve the following optimization problem:

$$\min_{\mathbf{B},\boldsymbol{f}_t} \sum_{t=1}^{T} (\boldsymbol{Y}_t - \mathbf{B}\boldsymbol{f}_t)'\mathbf{W}(\boldsymbol{Y}_t - \mathbf{B}\boldsymbol{f}_t),$$

subject to:

$$\frac{1}{T}\sum_{t=1}^{T} \boldsymbol{f}_t\boldsymbol{f}_t' = \mathbf{I}, \quad \mathbf{B}'\mathbf{W}\mathbf{B} \text{ is diagonal.}$$

Here, $\mathbf{W}$ can be either stochastic or deterministic. When $\mathbf{W}$ is stochastic, it can be understood as a consistent estimator of some deterministic matrix.

Solving the constrained optimization problem gives the WPC estimators: $\hat{\mathbf{b}}_j$ and $\hat{\boldsymbol{f}}_t$ are both $K \times 1$ vectors such that the columns of the $T \times K$ matrix $\hat{\mathbf{F}}/\sqrt{T} = (\hat{\boldsymbol{f}}_1, \ldots, \hat{\boldsymbol{f}}_T)'/\sqrt{T}$ are the eigenvectors corresponding to the largest K eigenvalues of $\mathbf{YWY}'$, and $\hat{\mathbf{B}} = T^{-1}\mathbf{Y}'\hat{\mathbf{F}} = (\hat{\mathbf{b}}_1, \ldots, \hat{\mathbf{b}}_N)'$. This method is called *weighted principal components* (WPC; see Bai and Liao, 2013), to distinguish from the traditional PC method that uses $\mathbf{W} = \mathbf{I}$. Note that PC does not encounter the problem of estimating large covariance matrices, and is not efficient when $\{u_{it}\}$'s are cross-sectionally correlated across i.

Bai and Liao (2013) studied the inferential theory of the WPC estimators. In particular, they showed that for the estimated common component, as $T, N \to \infty$,

$$\frac{\hat{\mathbf{b}}_i'\hat{\boldsymbol{f}}_t - \mathbf{b}_i'\boldsymbol{f}_t}{(\mathbf{b}_i'\Xi_{\mathbf{W}}\mathbf{b}_i/N + \boldsymbol{f}_t'\boldsymbol{\Omega}_i\boldsymbol{f}_t/T)^{1/2}} \to^d \mathcal{N}(0,1). \tag{6.37}$$

with $\Xi_{\mathbf{W}} = \boldsymbol{\Sigma}_\Lambda^{-1}\mathbf{B}'\mathbf{W}\boldsymbol{\Sigma}_u\mathbf{W}\mathbf{B}\boldsymbol{\Sigma}_\Lambda^{-1}/N$ and $\boldsymbol{\Omega}_i = \mathrm{Cov}(\boldsymbol{f}_t)^{-1}\Phi_i\mathrm{Cov}(\boldsymbol{f}_t)^{-1}$, where

$$\Phi_i = E(\boldsymbol{f}_t\boldsymbol{f}_t'u_{it}^2) + \sum_{t=1}^{\infty} E[(\boldsymbol{f}_1\boldsymbol{f}_{1+t}' + \boldsymbol{f}_{1+t}\boldsymbol{f}_1')u_{i1}u_{i,1+t}],$$

and $\boldsymbol{\Sigma}_\Lambda = \lim_{N\to\infty}\mathbf{B}'\mathbf{W}\mathbf{B}/N$, assumed to exist. Note that although the factors and loadings are not individually identifiable, $\hat{\mathbf{b}}_i'\hat{\boldsymbol{f}}_t$ can consistently estimate the common component $\mathbf{b}_i'\boldsymbol{f}_t$, without introducing a rotational transformation.

6.5.1.2 Optimal Weight Matrix

There are three interesting choices for the weight matrix $\mathbf{W}$. The most commonly seen weight is the identity matrix, which leads to the regular PC estimator. The second choice of the weight matrix takes $\mathbf{W} = \mathrm{diag}^{-1}\{\mathrm{Var}\,(u_{1t}), \ldots, \mathrm{Var}\,(u_{Nt})\}$. The third choice is the optimal weight. Note that the asymptotic variance of the estimated common component in (6.37) depends on $\mathbf{W}$ only through

$$\Xi_{\mathbf{W}} = \boldsymbol{\Sigma}_\Lambda^{-1}\mathbf{B}'\mathbf{W}\boldsymbol{\Sigma}_u\mathbf{W}\mathbf{B}\boldsymbol{\Sigma}_\Lambda^{-1}/N.$$

It is straightforward to show that when $\mathbf{W}^* = \boldsymbol{\Sigma}_u^{-1}$, the asymptotic variance is minimized, that is, for any positive definite matrix $\mathbf{W}$, $\Xi_{\mathbf{W}} - \Xi_{\mathbf{W}^*}$ is semipositive definite. In other words, the choice $\mathbf{W} = \boldsymbol{\Sigma}_u^{-1}$ as the weight matrix of the WPC estimator yields the minimum asymptotic variance of the estimated common component.

Table 6.4 gives the estimators and the corresponding weight matrix. The heteroscedastic WPC uses $\mathbf{W} = \mathbf{I}$, which takes into account the cross-sectional heteroscedasticity of $(u_{1t}, \ldots, u_{Nt})$, while the efficient WPC uses the optimal weight matrix $\boldsymbol{\Sigma}_u^{-1}$. Under the sparsity assumption, the optimal weight matrix can be estimated using the POET estimator as described in Section 6.3.

6.5.2 *Panel Data Model with Interactive Effects*

A closely related model is the panel data with a factor structure in the error term:

$$Y_{it} = \boldsymbol{x}_{it}'\beta + \varepsilon_{it}, \quad \varepsilon_{it} = \mathbf{b}_i'\boldsymbol{f}_t + u_{it}, \quad i \le N, t \le T, \tag{6.38}$$

Table 6.4 Three interesting choices of the weight matrix.

	Eigenvectors of	$\mathbf{W}$
Regular PC	$\mathbf{Y}\mathbf{Y}'$	$\mathbf{I}$
Heteroscedastic WPC	$\mathbf{Y}\mathrm{diag}(\boldsymbol{\Sigma}_u)^{-1}\mathbf{Y}'$	$\mathrm{diag}(\boldsymbol{\Sigma}_u)^{-1}$
Efficient WPC	$\mathbf{Y}\boldsymbol{\Sigma}_u^{-1}\mathbf{Y}'$	$\boldsymbol{\Sigma}_u^{-1}$

The estimated $\hat{\mathbf{F}}/\sqrt{T}$ is the eigenvectors of the largest r eigenvalues of $\mathbf{YWY}'$, and $\hat{\mathbf{B}} = T^{-1}\mathbf{Y}'\hat{\mathbf{F}}$.

Table 6.5 Canonical correlations for simulation study (from Bai and Liao, 2013)

		Loadings			Factors			$(\frac{1}{NT}\sum_{i,t}(\hat{\mathbf{b}}_i'\hat{f}_t - \mathbf{b}_i'f_t)^2)^{1/2}$		
T	N	PC	HWPC	EWPC	PC	HWPC	EWPC	PC	HWPC	EWPC
		(The larger the better)			(The larger the better)			(The smaller the better)		
100	80	0.433	0.545	0.631	0.427	0.551	0.652	0.570	0.540	0.496
100	150	0.613	0.761	0.807	0.661	0.835	0.902	0.385	0.346	0.307
100	200	0.751	0.797	0.822	0.827	0.882	0.924	0.333	0.312	0.284
150	100	0.380	0.558	0.738	0.371	0.557	0.749	0.443	0.394	0.334
150	200	0.836	0.865	0.885	0.853	0.897	0.942	0.313	0.276	0.240
150	300	0.882	0.892	0.901	0.927	0.946	0.973	0.257	0.243	0.222

The columns of loadings and factors report the canonical correlations.

where $\boldsymbol{x}_{it}$ is a $d \times 1$ vector of regressors; $\boldsymbol{\beta}$ is a $d \times 1$ vector of unknown coefficients. The regression noise ε_{it} has a factor structure with unknown loadings and factors, regarded as an *interactive effect* of the individual and time effects. In the model, the only observables are $(Y_{it}, \boldsymbol{x}_{it})$. This model has been considered by many researchers, such as Ahn *et al.* (2001), Pesaran (2006), Bai (2009), and Moon and Weidner (2010), and has broad applications in social sciences. For example, in the income studies, Y_{it} represents the income of individual i at age t, and $\boldsymbol{x}_{it}$ is a vector of observable characteristics that are associated with income. Here $\mathbf{b}_i$ represents a vector of unmeasured skills, such as innate ability, motivation, and hardworking; $\boldsymbol{f}_t$ is a vector of unobservable prices for the unmeasured skills, which can be time-varying.

The goal is to estimate the structural parameter $\boldsymbol{\beta}$, whose dimension is fixed. Because the regressor and factor can be correlated, simply regressing Y_{it} on $\boldsymbol{x}_{it}$ is not consistent. Let $\boldsymbol{X}_t = (\boldsymbol{x}_{1t}, \dots, \boldsymbol{x}_{Nt})'$. The least-squares estimator of $\boldsymbol{\beta}$ is

$$\hat{\boldsymbol{\beta}} = \arg\min \min_{\mathbf{B}, f_t} \sum_{t=1}^{T} (\boldsymbol{Y}_t - \boldsymbol{X}_t\boldsymbol{\beta} - \mathbf{B}\boldsymbol{f}_t)'\mathbf{W}(\boldsymbol{Y}_t - \boldsymbol{X}_t'\boldsymbol{\beta} - \mathbf{B}\boldsymbol{f}_t), \tag{6.39}$$

with a high-dimensional weight matrix $\mathbf{W}$. In particular, it allows a consistent estimator for $\boldsymbol{\Sigma}_u^{-1}$ as the optimal weight matrix, which takes into account both cross-sectional correlation and heteroscedasticity of u_{it} over i. The minimization is subjected to the constraint $\frac{1}{T}\sum_{t=1}^{T} \boldsymbol{f}_t\boldsymbol{f}_t'/T = \mathbf{I}$ and $\mathbf{B}'\mathbf{W}\mathbf{B}$ being diagonal.

The estimated $\boldsymbol{\beta}$ for each given $(\mathbf{B}, \{\boldsymbol{f}_t\})$ is simply

$$\boldsymbol{\beta}(\mathbf{B}, \{\boldsymbol{f}_t\}) = \left(\sum_{t=1}^{T} \boldsymbol{X}_t'\mathbf{W}\boldsymbol{X}_t\right)^{-1} \sum_{t=1}^{T} \boldsymbol{X}_t'\mathbf{W}(\boldsymbol{Y}_t - \mathbf{B}\boldsymbol{f}_t).$$

On the other hand, given $\boldsymbol{\beta}$, the variable $\boldsymbol{Y}_t - \boldsymbol{X}_t\boldsymbol{\beta}$ has a factor structure. Hence the estimated $(\mathbf{B},\boldsymbol{f}_t)$ are the weighted principal components estimators: let $\boldsymbol{X}(\hat{\boldsymbol{\beta}})$ be an $N \times T$ matrix $\boldsymbol{X}(\hat{\boldsymbol{\beta}}) = (\boldsymbol{X}_1\hat{\boldsymbol{\beta}}, \dots, \boldsymbol{X}_T\hat{\boldsymbol{\beta}})$. The columns of the $T \times r$ matrix $\hat{\mathbf{F}}/\sqrt{T} = (\hat{\boldsymbol{f}}_1, \dots, \hat{\boldsymbol{f}}_T)'/\sqrt{T}$ are the eigenvectors corresponding to the largest r eigenvalues of $(\mathbf{Y}' - \boldsymbol{X}(\hat{\boldsymbol{\beta}}))'\mathbf{W}(\mathbf{Y}' - \boldsymbol{X}(\hat{\boldsymbol{\beta}}))$, and $\hat{\mathbf{B}} = T^{-1}(\mathbf{Y}' - \boldsymbol{X}(\hat{\boldsymbol{\beta}}))\hat{\mathbf{F}}$. Therefore, given $(\mathbf{B},\boldsymbol{f}_t)$, we can estimate $\boldsymbol{\beta}$, and given $\boldsymbol{\beta}$, we can estimate $(\mathbf{B},\boldsymbol{f}_t)$. So $\hat{\boldsymbol{\beta}}$ can be simply obtained by iterations, with an initial value. The inversion $(\sum_{t=1}^{T} \boldsymbol{X}_t'\mathbf{W}\boldsymbol{X}_t)^{-1}$ does not update during iterations.

6.5.2.1 Optimal Weight Matrix

To present the inferential theory of $\hat{\boldsymbol{\beta}}$, additional notation are needed. Rearrange the design matrix

$$\boldsymbol{Z} = (\boldsymbol{X}_{11}, \dots, \boldsymbol{X}_{1T}, \boldsymbol{X}_{21}, \dots, \boldsymbol{X}_{2T}, \dots, \boldsymbol{X}_{N1}, \dots, \boldsymbol{X}_{NT})', NT \times \dim\ (\boldsymbol{\beta}).$$

Let

$$\mathbf{A}_\mathbf{W} = [\mathbf{W} - \mathbf{W}\mathbf{B}(\mathbf{B}'\mathbf{W}\mathbf{B})^{-1}\mathbf{B}'\mathbf{W}] \otimes (\mathbf{I} - \mathbf{F}(\mathbf{F}'\mathbf{F})^{-1}\mathbf{F}'/T).$$

Under regularity conditions, Bai and Liao (2013) showed that

$$\sqrt{NT}(\hat{\boldsymbol{\beta}} - \boldsymbol{\beta}) \to^{d} \mathcal{N}(0, \mathbf{V}_\mathbf{W}),$$

where, for $\boldsymbol{\Sigma}_u = \mathrm{Cov}(\boldsymbol{u}_t)$,

$$\mathbf{V}_\mathbf{W} = \mathrm{plim}_{N,T\to\infty}(\frac{1}{NT}\mathbf{Z}'\mathbf{A}_\mathbf{W}\mathbf{Z})^{-1}\frac{1}{NT}\mathbf{Z}'\mathbf{A}_\mathbf{W}(\boldsymbol{\Sigma}_u \otimes \mathbf{I})\mathbf{A}_\mathbf{W}\mathbf{Z}(\frac{1}{NT}\mathbf{Z}'\mathbf{A}_\mathbf{W}\mathbf{Z})^{-1}$$

assuming the right-hand side converges in probability.

It is not difficult to show that $\mathbf{W}^* = \boldsymbol{\Sigma}_u^{-1}$ is the optimal weight matrix, in the sense that $\mathbf{V}_\mathbf{W} - \mathbf{V}_{\mathbf{W}^*}$ is semipositive definite for all positive definite weight matrix $\mathbf{W}$. With $\mathbf{W} = \mathbf{W}^*$, the asymptotic variance of $\hat{\boldsymbol{\beta}}$ is

$$\mathbf{V}_{\mathbf{W}^*} = \mathrm{plim}_{N,T\to\infty} = (\frac{1}{NT}\mathbf{Z}'\mathbf{A}_{\mathbf{W}^*}\mathbf{Z})^{-1}.$$

Assuming $\boldsymbol{\Sigma}_u$ to be sparse, one can estimate $\mathbf{W}^*$ based on an initial estimator of $\boldsymbol{\beta}$. Specifically, define $\hat{\boldsymbol{\beta}}_0$ as in (6.39) with $\mathbf{W} = \mathbf{I}$, which is the estimator used in Bai (2009) and Moon and Weidner (2010). Apply the singular value decomposition to

$$\frac{1}{T}\sum_{t=1}^{T}(\boldsymbol{Y}_t - \boldsymbol{X}_t\hat{\boldsymbol{\beta}}_0)(\boldsymbol{Y}_t - \boldsymbol{X}_t\hat{\boldsymbol{\beta}}_0)' = \sum_{i=1}^{N} \nu_i\xi_i\xi_i',$$

where $(v_j, \xi_j)_{j=1}^N$ are the eigenvalues–eigenvectors of $\frac{1}{T}\sum_{t=1}^T (\mathbf{Y}_t - \mathbf{X}_t\hat{\boldsymbol{\beta}}_0)(\mathbf{Y}_t - \mathbf{X}_t\hat{\boldsymbol{\beta}}_0)'$ in a decreasing order such that $v_1 \geq v_2 \geq \ldots \geq v_N$. Then $\hat{\boldsymbol{\Sigma}}_u = (\hat{\Sigma}_{u,ij})_{N\times N}$,

$$\hat{\Sigma}_{u,ij} = \begin{cases} \tilde{R}_{ii}, & i = j \\ th_{ij}(\tilde{R}_{ij}), & i \neq j \end{cases}, \quad \tilde{R}_{ij} = \sum_{k=r+1}^{N} v_k \xi_{ki} \xi_{kj},$$

where $th_{ij}(\cdot)$ is the same thresholding function. The optimal weight matrix $\mathbf{W}^*$ can then be estimated by $\hat{\boldsymbol{\Sigma}}_u^{-1}$, and the resulting estimator $\hat{\boldsymbol{\beta}}$ achieves the asymptotic variance $\mathbf{V}_{\mathbf{W}^*}$.

6.5.3 *Numerical Illustrations*

We present a simple numerical example to compare the weighted principal components with the popular methods in the literature. The idiosyncratic error terms are generated as follows: let $\{\epsilon_{it}\}_{i\leq N, t\leq T}$ be i.i.d. $\mathcal{N}(0, 1)$ in both t, i. Let

$$u_{1t} = \epsilon_{1t}, \; u_{2t} = \epsilon_{2t} + a_1\epsilon_{1t}, \; u_{3t} = \epsilon_{3t} + a_2\epsilon_{2t} + b_1\epsilon_{1t},$$
$$u_{i+1,t} = \epsilon_{i+1,t} + a_i\epsilon_{it} + b_{i-1}\epsilon_{i-1,t} + c_{i-2}\epsilon_{i-2,t},$$

where $\{a_i, b_i, c_i\}_{i=1}^N$ are i.i.d. $\mathcal{N}(0, 1)$. Then $\boldsymbol{\Sigma}_u$ is a banded matrix, with both cross-sectional correlation and heteroscedasticity. Let the two factors $\{f_{1t}, f_{2t}\}$ be i.i.d. $\mathcal{N}(0, 1)$, and $\{b_{i,1}, b_{i,2}\}_{i\leq N}$ be uniform on $[0, 1]$.

6.5.3.1 Pure Factor Model

Consider the pure factor model $Y_{it} = b_{i1}f_{1,t} + b_{i,2}f_{2t} + u_{it}$. Estimators based on three weight matrices are compared: PC using $\mathbf{W} = \mathbf{I}$; HWPC using $\mathbf{W} = \text{diag}(\boldsymbol{\Sigma}_u)^{-1}$ and EWPC using $\mathbf{W} = \boldsymbol{\Sigma}_u^{-1}$. Here $\boldsymbol{\Sigma}_u$ is estimated using the POET estimator. The smallest canonical correlation (the larger the better) between the estimators and parameters are calculated, as an assessment of the estimation accuracy. The simulation is replicated 100 times, and the average canonical correlations are reported in Table 6.5. The mean squared error of the estimated common components are also compared.

We see that the estimation becomes more accurate when we increase the dimensionality. HWPC improves the regular PC, while the EWPC gives the best estimation results.

6.5.3.2 Interactive Effects

Adding a regression term, we consider the panel data model with interactive effect: $Y_{it} = \mathbf{x}_{it}'\boldsymbol{\beta} + b_{i1}f_{1,t} + b_{i,2}f_{2t} + u_{it}$, where the true $\boldsymbol{\beta} = (1, 3)'$. The regressors are generated to be dependent on $(\boldsymbol{f}_t, \mathbf{b}_i)$:

$$x_{it,1} = 2.5b_{i1}f_{1,t} - 0.2b_{i2}f_{2,t} - 1 + \eta_{it,1}, x_{it,2} = b_{i1}f_{1,t} - 2b_{i2}f_{2,t} + 1 + \eta_{it,2}$$

where $\eta_{it,1}$ and $\eta_{it,2}$ are independent i.i.d. standard normal.

Both methods, PC (Bai, 2009; Moon and Weidner, 2010) and WPC with $\mathbf{W} = \hat{\boldsymbol{\Sigma}}_u^{-1}$, are carried out to estimate $\boldsymbol{\beta}$ for the comparison. The simulation is replicated 100 times; results

Table 6.6 Method comparison for the panel data with interactive effects (from Bai and Liao, 2013)

		$\beta_1 = 1$				$\beta_2 = 3$			
		Mean		Normalized SE		Mean		Normalized SE	
T	N	WPC	PC	WPC	PC	WPC	PC	WPC	PC
100	100	1.002	1.010	0.550	1.418	3.000	3.003	0.416	1.353
100	150	1.003	1.007	0.681	1.626	2.999	3.000	0.611	1.683
100	200	1.002	1.005	0.631	1.800	3.000	3.000	0.774	1.752
150	100	1.003	1.006	0.772	1.399	3.000	2.999	0.714	1.458
150	150	1.001	1.005	0.359	1.318	3.000	3.001	0.408	1.379
150	200	1.001	1.003	0.547	1.566	3.000	3.000	0.602	1.762

"Mean" is the average of the estimators; "Normalized SE" is the standard error of the estimators multiplied by $\sqrt{NT}$.

are summarized in Table 6.6. We see that both methods are almost unbiased, while the efficient WPC indeed has significantly smaller standard errors than the regular PC method in the panel model with interactive effects.

6.6 Conclusions

Large covariance and precision (inverse covariance) matrix estimations have become fundamental problems in multivariate analysis that find applications in many fields, ranging from economics and finance to biology, social networks, and health sciences.

We introduce two efficient methods for estimating large covariance matrices and precision matrices. The introduced precision matrix estimator assumes the precision matrix to be sparse, which is immediately applicable for Gaussian graphical models. It is tuning-parameter insensitive, and simultaneously achieves the minimax optimal rates of convergence in precision matrix estimation under different matrix norms. On the other hand, the estimator based on factor analysis imposes a conditional sparsity assumption. Computationally, our procedures are significantly faster than existing methods. Both theoretical properties and numerical performances of these methods are presented and illustrated. In addition, we also discussed several financial applications of the proposed methods, including risk management, testing high-dimensional factor pricing models. We also illustrate how the proposed covariance estimators can be used to improve statistical efficiency in estimating factor models and panel data models.

References

Ahn, S., Lee, Y. and Schmidt, P. (2001). Gmm estimation of linear panel data models with time-varying individual effects. *Journal of Econometrics*, 101, 219–255.

Ahn S.C. and Horenstein, A.R. (2013). Eigenvalue ratio test for the number of factors. *Econometrica*, 81 (3), 1203–1227.

Alessi, L., Barigozzi, M. and Capasso, M. (2010). Improved penalization for determining the number of factors in approximate factor models. *Statistics & Probability Letters*, 80 (23), 1806–1813.

Antoniadis, A. and Fan, J. (2001). Regularization of wavelet approximations. *Journal of the American Statistical Association*, 96 (455), 939–955.

Bach, F.R. (2008). Bolasso: model consistent lasso estimation through the bootstrap. In Proceedings of the Twenty-fifth International Conference on Machine Learning (ICML). http://www.di.ens.fr/~fbach/fbach_bolasso_icml2008.pdf

Bai, J. (2003). Inferential theory for factor models of large dimensions. *Econometrica*, 71 (1), 135–171.

Bai, J. (2009). Panel data models with interactive fixed effects. *Econometrica*, 77, 1229–1279.

Bai, J. and Liao, Y. (2013). Statistical inferences using large estimated covariances for panel data and factor models. Technical report, University of Maryland.

Bai, J. and Ng, S. (2002). Determining the number of factors in approximate factor models. *Econometrica*, 70, 191–221.

Banerjee, O., El Ghaoui, L. and d'Aspremont, A. (2008). Model selection through sparse maximum likelihood estimation for multivariate Gaussian or binary data. *The Journal of Machine Learning Research*, 9, 485–516.

Beaulieu, M., Dufour, J. and Khalaf, L. (2007). Multivariate tests of mean-variance efficiency with possibly non-Gaussian errors: an exact simulation based approach. *Journal of Business and Economic Statistics*, 25, 398–410.

Belloni, A., Chernozhukov, V. and Wang, L. (2012). Square-root lasso: pivotal recovery of sparse signals via conic programming. *Biometrika*, 98, 791–806.

Ben-david, S., Luxburg, U.V. and Pal, D. (2006). A sober look at clustering stability. In *Proceedings of the Annual Conference of Learning Theory*. New York: Springer, pp. 5–19.

Bickel, P. and Levina, E. (2008). Covariance regularization by thresholding. *Annals of Statistics*, 36 (6), 2577–2604.

Boos, D.D., Stefanski, L.A. and Wu, Y. (2009). Fast FSR variable selection with applications to clinical trials. *Biometrics*, 65 (3), 692–700.

Cai, T. and Liu, W. (2011). Adaptive thresholding for sparse covariance matrix estimation. *Journal of the American Statistical Association*, 106 (494), 672–684.

Cai, T., Liu, W. and Luo, X. (2011). A constrained ℓ_1 minimization approach to sparse precision matrix estimation. *Journal of the American Statistical Association*, 106 (494), 594–607.

Campbell, J.Y., Lo, A.W.C., MacKinlay, A.C., *et al.* (1997). *The econometrics of financial markets*, vol. 2. Princeton, NJ: Princeton University Press.

Candès, E.J., Li, X., Ma, Y. and Wright, J. (2011). Robust principal component analysis? *Journal of the ACM (JACM)*, 58 (3), 11.

Chamberlain, G. and Rothschild, M. (1983). Arbitrage, factor structure and mean-variance analysis in large asset markets. *Econometrica*, 51, 1305–1324.

Chandrasekaran, V., Parrilo, P.A. and Willsky, A.S. (2010). Latent variable graphical model selection via convex optimization. In *48th Annual Allerton Conference on Communication, Control, and Computing (Allerton), 2010*. Piscataway, NJ: IEEE, pp. 1610–1613.

Chen, J. and Chen, Z. (2008). Extended Bayesian information criteria for model selection with large model spaces. *Biometrika*, 95 (3), 759–771.

Chen, J. and Chen, Z. (2012). Extended BIC for small-n-large-p sparse GLM. *Statistica Sinica*, 22, 555–574.

Dempster, A. (1972). Covariance selection. *Biometrics*, 28, 157–175.

Efron, B. (2004). The estimation of prediction error: covariance penalties and cross-validation. *Journal of the American Statistical Association*, 99, 619–632.

El Karoui, N. (2008). Operator norm consistent estimation of large dimensional sparse covariance matrices. *Annals of Statistics*, 36 (6), 2717–2756.

El Karoui, N. (2010). High-dimensionality effects in the Markowitz problem and other quadratic programs with linear constraints: risk underestimation. *Annals of Statistics*, 38 (6), 3487–3566.

Fama, E. and French, K. (1992). The cross-section of expected stock returns. *Journal of Finance*, 47, 427–465.

Fan, J., Fan, Y. and Lv, J. (2008). High dimensional covariance matrix estimation using a factor model. *Journal of Econometrics*, 147 (1), 186–197.

Fan, J., Liao, Y. and Mincheva, M. (2011). High dimensional covariance matrix estimation in approximate factor models. *Annals of Statistics*, 39, 3320–3356.

Fan, J., Liao, Y. and Mincheva, M. (2013). Large covariance estimation by thresholding principal orthogonal complements (with discussion). *Journal of the Royal Statistical Society, Series B*, 75, 603–680.

Fan, J., Liao, Y. and Shi, X. (2014a). *Risks of large portfolios*. Technical report. Princeton, NJ: Princeton University.

Fan, J., Liao, Y. and Yao, J. (2014b). Large panel test of factor pricing models. Technical report. Princeton, NJ: Princeton University.

Foster, D.P. and George, E.I. (1994). The risk inflation criterion for multiple regression. *Annals of Statistics*, 22 (4), 1947–1975.
Foygel, R. and Drton, M. (2010). Extended Bayesian information criteria for Gaussian graphical models. *Advances in Neural Information Processing Systems*, 23, 604–612.
Friedman, J., Hastie, T. and Tibshirani, R. (2007). Pathwise coordinate optimization. *Annals of Applied Statistics*, 1 (2), 302–332.
Friedman, J., Hastie, T. and Tibshirani, R. (2008). Sparse inverse covariance estimation with the graphical lasso. *Biostatistics*, 9 (3), 432–441.
Fryzlewicz, P. (2013). High-dimensional volatility matrix estimation via wavelets and thresholding. *Biometrika*, 100, 921–938.
Gibbons, M., Ross, S. and Shanken, J. (1989). A test of the efficiency of a given portfolio. *Econometrica*, 57, 1121–1152.
Hallin, M. and Liška, R. (2007). Determining the number of factors in the general dynamic factor model. *Journal of the American Statistical Association*, 102 (478), 603–617.
Han, F., Zhao, T. and Liu, H. (2012). *High dimensional nonparanormal discriminant analysis*. Technical report, Department of Biostatistics. Baltimore: Johns Hopkins Bloomberg School of Public Health.
Jagannathan, R. and Ma, T. (2003). Risk reduction in large portfolios: why imposing the wrong constraints helps. *Journal of Finance*, 58, 1651–1684.
Jalali, A., Johnson, C. and Ravikumar, P. (2012). High-dimensional sparse inverse covariance estimation using greedy methods. International Conference on Artificial Intelligence and Statistics.
Kapetanios, G. (2010). A testing procedure for determining the number of factors in approximate factor models with large datasets. *Journal of Business & Economic Statistics*, 28 (3), 397–409.
Lam, C. and Fan, J. (2009). Sparsistency and rates of convergence in large covariance matrix estimation. *Annals of Statistics*, 37 (6B), 4254.
Lange, T., Roth, V., Braun, M.L. and Buhmann, J.M. (2004). Stability-based validation of clustering solutions. *Neural Computation*, 16 (6), 1299–1323.
Ledoit, O. and Wolf, M. (2003). Improved estimation of the covariance matrix of stock returns with an application to portfolio selection. *Journal of Empirical Finance*, 10 (5), 603–621.
Liu, H., Roeder, K. and Wasserman, L. (2010). Stability approach to regularization selection (stars) for high dimensional graphical models Proceedings of the Twenty-Third Annual Conference on Neural Information Processing Systems (NIPS).
Liu, H. and Wang, L. (2012). TIGER: a tuning-insensitive approach for optimally estimating gaussian graphical models. Technical report, Department of Operations Research and Financial Engineering, Princeton University.
Liu, W. and Luo, X. (2012). High-dimensional sparse precision matrix estimation via sparse column inverse operator. arXiv/1203.3896.
Lysen, S. (2009). Permuted inclusion criterion: A variable selection technique. Publicly accessible Penn Dissertations. Paper 28.
MacKinlay, A. and Richardson, M. (1991). Using generalized method of moments to test mean-variance efficiency. *Journal of Finance*, 46, 511–527.
Markowitz, H. (1952). Portfolio selection. *Journal of Finance*, 7, 77–91.
Meinshausen, N. and Bühlmann, P. (2006). High dimensional graphs and variable selection with the lasso. *Annals of Statistics*, 34 (3), 1436–1462.
Meinshausen, N. and Bühlmann, P. (2010). Stability selection. *Journal of the Royal Statistical Society, Series B, Methodological*, 72, 417–473.
Moon, R. and Weidner, M. (2010). *Linear regression for panel with unknown number of factors as interactive fixed effects*. Technical report. Los Angeles: University of South California.
Newey, W. and West, K. (1987). A simple, positive semi-definite, heteroskedasticity and autocorrelation consistent covariance matrix. *Econometrica*, 55, 703–708.
Pesaran, H. (2006). Estimation and inference in large heterogeneous panels with a multifactor error structure. *Econometrica*, 74, 967–1012.
Pesaran, H. and Yamagata, T. (2012). *Testing capm with a large number of assets*. Technical report. Los Angeles: University of South California.
Pourahmadi, M. (2013). *High-dimensional covariance estimation: with high-dimensional data*. Hoboken, NJ: John Wiley & Sons.
Ravikumar, P., Wainwright, M.J., Raskutti, G. and Yu, B. (2011a). High-dimensional covariance estimation by minimizing ℓ_1-penalized log-determinant divergence. *Electronic Journal of Statistics*, 5, 935–980.

Ravikumar, P., Wainwright, M.J., Raskutti, G., Yu, B., *et al.* (2011b). High-dimensional covariance estimation by minimizing ℓ_1-penalized log-determinant divergence. *Electronic Journal of Statistics*, 5, 935–980.

Rothman, A.J., Bickel, P.J., Levina, E. and Zhu, J. (2008). Sparse permutation invariant covariance estimation. *Electronic Journal of Statistics* 2, 494–515.

Rothman, A.J., Levina, E. and Zhu, J. (2009). Generalized thresholding of large covariance matrices. *Journal of the American Statistical Association*, 104(485), 177–186.

Sentana, E. (2009). The econometrics of mean-variance efficiency tests: a survey. *Econometrics Journal*, 12, 65–101.

Shen, X., Pan, W. and Zhu, Y. (2012). Likelihood-based selection and sharp parameter estimation. *Journal of the American Statistical Association*, 107(497), 223–232.

Stock, J.H. and Watson, M.W. (2002). Forecasting using principal components from a large number of predictors. *Journal of the American Statistical Association*, 97(460), 1167–1179.

Sun, T. and Zhang, C.H. (2012). *Sparse matrix inversion with scaled lasso*. Technical report, Department of Statistics. New Brunswick, NJ: Rutgers University.

Wainwright, M. (2009). Sharp thresholds for high-dimensional and noisy sparsity recovery using constrained quadratic programming. *IEEE Transactions on Information Theory*, 55(5), 2183–2201.

Wille, A., Zimmermann, P., Vranova, E., Frholz, A., Laule, O., *et al.* (2004). Sparse graphical Gaussian modeling of the isoprenoid gene network in *Arabidopsis thaliana*. *Genome Biology*, 5 (11), R92.

Wu, Y., Boos, D.D. and Stefanski, L.A. (2007). Controlling variable selection by the addition of pseudovariables. *Journal of the American Statistical Association*, 102, 235–243.

Xue, L. and Zou, H. (2012). Positive-definite ℓ_1-penalized estimation of large covariance matrices. *Journal of the American Statistical Association*, 107, 1480–1491.

Yuan, M. (2010). High dimensional inverse covariance matrix estimation via linear programming. *Journal of Machine Learning Research*, 11(8), 2261–2286.

Yuan, M. and Lin, Y. (2007). Model selection and estimation in the Gaussian graphical model. *Biometrika*, 94 (1), 19–35.

Zhao, P. and Yu, B. (2006). On model selection consistency of lasso. *Journal of Machine Learning Research*, 7 (11), 2541–2563.

Zhao, T., Liu, H., Roeder, K., Lafferty, J. and Wasserman, L. (2012). The huge package for high-dimensional undirected graph estimation in *r*. *Journal of Machine Learning Research*, forthcoming.

Zou, H. (2006). The adaptive lasso and its oracle properties. *Journal of the American Statistical Association*, 101 (476), 1418–1429.

7

Stochastic Volatility

Modeling and Asymptotic Approaches to Option Pricing and Portfolio Selection

Matthew Lorig[1] and Ronnie Sircar[2]
[1] *University of Washington, USA*
[2] *Princeton University, USA*

7.1 Introduction

Understanding and measuring the inherent uncertainty in market volatility are crucial for portfolio optimization, risk management, and derivatives trading. The problem is made difficult since volatility is not directly observed. Rather, volatility is a statistic of the observable returns of, for example, a stock, and so estimates of it are at best noisy. Among the major empirical challenges have been separating contributions of diffusive and jump components of log returns, typical timescales of fluctuation, and memory effects. Until recently, data were limited to low frequencies, typically daily. The availability of high-frequency data over the past 20 years brings with it issues of deciphering market microstructure effects such as the bid–ask bounce, which contaminate the potential usefulness of such large datasets, and we refer to the recent book by Aït-Sahalia and Jacod (2014) for an overview of the difficulties.

The major problem that has been the driver of *stochastic volatility* models is the valuation and hedging of derivative securities. This market grew in large part from the landmark paper by Black and Scholes (1973), which showed how to value simple options contracts when volatility is constant. Even at the time of their paper, Black and Scholes realized that the constant volatility assumption was a strong idealization. In an empirical paper by Black and Scholes (1972), the authors tested their option price formulas and concluded: "we found that using past data to estimate the variance caused the model to overprice options on high-variance stocks and underprice options on low-variance stocks." Indeed the overwhelming evidence from time-series data reveals that volatility exhibits unpredictable variation. In addition, as we will describe

Financial Signal Processing and Machine Learning, First Edition.
Edited by Ali N. Akansu, Sanjeev R. Kulkarni and Dmitry Malioutov.
© 2016 John Wiley & Sons, Ltd. Published 2016 by John Wiley & Sons, Ltd.

in this chapter, option prices exhibit a significant departure from the Black–Scholes (constant volatility) theory, the *implied volatility skew*, which can be explained by allowing volatility to vary randomly in time.

A second important problem is portfolio optimization: namely, how to optimally invest capital between a risky stock and a riskless bank account. In a continuous time stochastic model with constant volatility, the pioneering work was by Robert Merton (Merton, 1969, 1971; reprinted in Merton, 1992). Since then, understanding the effect of volatility uncertainty, stochastic growth rate, transaction costs, price impact, illiquidity, and other frictions on the portfolio choice problem has generated considerable research. Here, we will focus on the effect of stochastic volatility and present some new results in Section 7.3.

The increased realism obtained by allowing volatility to be stochastic comes with increased computational difficulties. Moreover, there is no broad consensus concerning how to best model volatility. Here, we will discuss some computationally efficient approaches, focusing particularly on asymptotic approximations.

7.1.1 Options and Implied Volatility

The most liquidly traded derivatives contracts are call and put options, which give the option holder the right to buy (in the case of a call) or sell (in the case of a put) one unit of the underlying security at a *fixed strike price* K on a *fixed expiration date* T. Here we are focusing specifically on *European*-style options (i.e., no early exercise), which are typically traded on indices such as the S&P 500. If S_t represents the price of a stock or index at time t, then a European-style derivative has a payoff at time T, which is a function h of S_T. In the case of calls and puts, the payoff functions are $h(S) = (S - K)^+$ and $h(S) = (K - S)^+$, respectively.

7.1.1.1 Black–Scholes model

In the Black–Scholes model, the stock price S is a geometric Brownian motion described by the following stochastic differential equation (SDE):

$$\frac{\mathrm{d}S_t}{S_t} = \mu \, \mathrm{d}t + \sigma \, \mathrm{d}W_t, \tag{7.1}$$

where W is a standard Brownian motion with respect to a historical (or real-world, or physical) probability measure $\mathbb{P}$. Here, the parameters are the expected growth rate μ and the volatility σ, both assumed constant. The remarkable finding of Black and Scholes (1973) is that the no-arbitrage price of an option does not depend on μ, and so, to price an option, the only parameter that needs to be estimated from data is the volatility σ. Unless otherwise stated, we shall assume throughout this article that interest rates are zero.

It will be convenient to introduce the following notation:

$$\tau := T - t, \qquad x := \log S_t, \qquad k := \log K,$$

where t is the current time; and K and T are the strike and expiration date, respectively, of a call or put option. Then, for fixed (t, T, x, k), the Black–Scholes pricing formula for a call option with time to expiration $\tau > 0$ is given by

$$u^{\mathrm{BS}}(\sigma) := \mathrm{e}^{x}\mathcal{N}(d_{+}(\sigma)) - \mathrm{e}^{k}\mathcal{N}(d_{-}(\sigma)), \qquad d_{\pm}(\sigma) := \frac{1}{\sigma\sqrt{\tau}}\left(x - k \pm \frac{\sigma^{2}\tau}{2}\right), \tag{7.2}$$

where $\mathcal{N}$ is the CDF of a standard normal random variable, and we have stressed the volatility argument σ in the notation.

It turns out that the Black–Scholes price (7.2) can be expressed as the expected payoff of the option, but where the expectation is taken with respect to a different probability measure $\mathbb{Q}$ under which the stock price is a martingale (that is, it is a pure fluctuation process with no trend or growth rate). This means that there is a so-called risk-neutral world in which the stock price follows the dynamics

$$\frac{\mathrm{d}S_t}{S_t} = \sigma \,\mathrm{d}W_t^{\mathbb{Q}},$$

where $W^{\mathbb{Q}}$ is a Brownian motion under $\mathbb{Q}$, and the call option price (7.2) can be expressed as the conditional expectation

$$u^{\mathrm{BS}}(\sigma) = \mathbb{E}^{\mathbb{Q}}[(S_T - K)^{+} \mid \log S_t = x],$$

where $\mathbb{E}^{\mathbb{Q}}$ denotes that the expectation is taken under the probability measure $\mathbb{Q}$.

7.1.1.2 Implied Volatility

The *implied volatility* of a given call option with price u (which is either observed in the market or computed from a model) is the unique positive solution I of

$$u^{\mathrm{BS}}(I) = u. \tag{7.3}$$

It is the volatility parameter that has to be put into the Black–Scholes formula to match the observed price u.

Note that the implied volatility I depends implicitly on the maturity date T and the log strike k as the option price u will depend on these quantities. The map $(T, k) \mapsto I(T, k)$ is known as the *implied volatility surface*. If market option prices reflected Black–Scholes assumptions, I would be constant and equal to the stock's historical volatility σ. However, in equities data, the function $I(T, \cdot)$ exhibits downward-sloping behavior in k, whose slope varies with the option maturities T, as illustrated in Figure 7.1. This downward slope is known as the *implied volatility skew*.

These features of the implied volatility surface can be reproduced by enhancing the Black–Scholes model (7.1) with stochastic volatility and/or jumps. One focus of this chapter will be to survey some approaches taken to capturing the implied volatility skew.

7.1.2 Volatility Modeling

While the overwhelming evidence from time-series and option price data indicates that the volatility σ in (7.1) should be allowed to vary stochastically in time:

$$\frac{\mathrm{d}S_t}{S_t} = \mu \,\mathrm{d}t + \sigma_t \,\mathrm{d}W_t,$$

there is no consensus as to how exactly the (stochastic) volatility σ_t should be modeled.

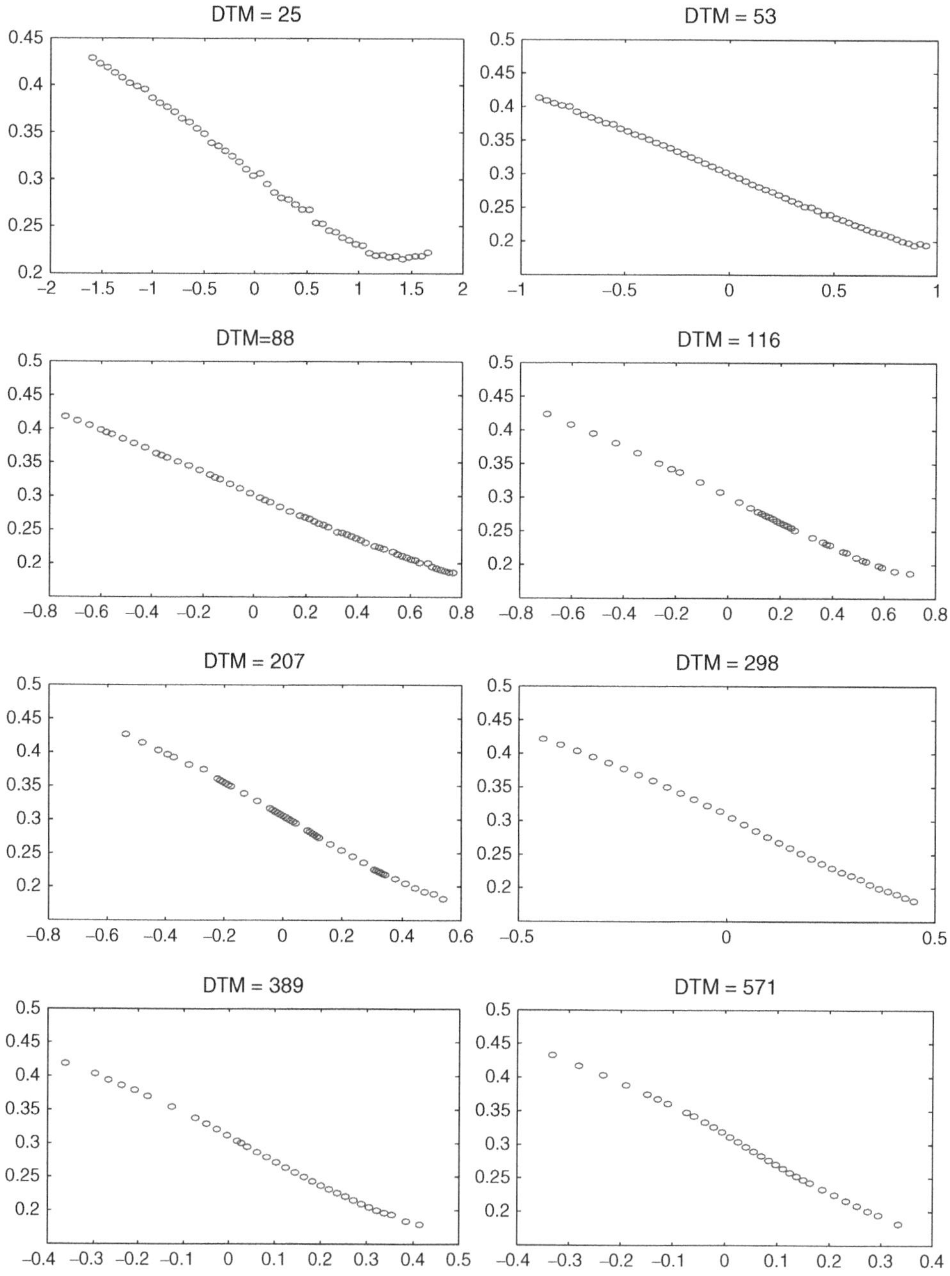

Figure 7.1 Implied volatility from S&P 500 index options on May 25, 2010, plotted as a function of log-moneyness to maturity ratio: $(k - x)/(T - t)$. DTM, days to maturity.

7.1.2.1 No-arbitrage Pricing and Risk-Neutral Measure

In standard option-pricing theory, it is assumed that markets do not admit arbitrage. No arbitrage pricing implies that all traded asset prices (after discounting) are martingales under some probability measure $\mathbb{Q}$, typically referred to as a *risk-neutral* measure. Consequently, the price u_t of an option at time t with payoff $\varphi(S_T)$ at time T is given by

$$u_t = \mathbb{E}^{\mathbb{Q}}[\varphi(S_T)|\mathcal{F}_t], \tag{7.4}$$

where $\mathcal{F}_t$ is the history of the market up to time t. Typically there are nontraded sources of randomness such as jumps or stochastic volatility. As a result, there exist infinitely many risk-neutral measures. The nonuniqueness of these measures is often referred to as *market incompleteness*, meaning not every derivative asset can be perfectly hedged. In practice, one assumes that the market has chosen a specific risk-neutral measure, which is consistent with observed option prices.

In what follows, we will model asset dynamics under a unique risk-neutral pricing measure $\mathbb{Q}$, which we assume has been chosen by the market. Under $\mathbb{Q}$, we have

$$S_t = \mathrm{e}^{X_t}, \quad \mathrm{d}X_t = -\frac{1}{2}\sigma_t^2 \mathrm{d}t + \sigma_t \mathrm{d}W_t^{\mathbb{Q}}, \tag{7.5}$$

which describes the dynamics of $X_t = \log\ S_t$. In the rest of this section, we review some of the most common models of volatility and discuss some of their advantages and disadvantages.

7.1.2.2 Local Volatility Models

In *local volatility* (LV) models, the volatility σ_t of the underlying is modeled as a deterministic function $\sigma(\cdot,\cdot)$ of time t, and the time-t value of the underlying X_t. That is,

$$\mathrm{d}X_t = -\tfrac{1}{2}\sigma^2(t, X_t)\ \mathrm{d}t + \sigma(t, X_t)\mathrm{d}W_t^{\mathbb{Q}}, \quad \text{(local volatility)}$$

Typically, one assumes that the function $\sigma(t,\cdot)$ increases as x decreases in order to capture the *leverage effect*, which refers to the tendency for the value of an asset to decrease as its volatility increases.

One advantage of local volatility models is that markets remain *complete*, meaning that derivatives written on S can be hedged perfectly – just as in the Black–Scholes model. While market completeness is convenient from a theoretical point of view, it is not necessarily a realistic property of financial markets. Indeed, if markets are complete, then one can ask: why do we need derivatives?

Another advantage of local volatility models is that they can provide a very tight fit to option prices quoted on the market. In fact, Dupire (1994) shows that there exists a local volatility model that can *exactly* match option prices quoted on the market and that there is an explicit formula for how to construct this model from observed call and put prices under the assumption that they can be interpolated across continuous strikes and maturities. However, a tight fit must be balanced with stability: local volatility models are notoriously bad at providing stability and typically need to be recalibrated hourly.

7.1.2.3 Stochastic Volatility Models

In a *stochastic volatility* (SV) model, promoted in the late 1980s by Hull and White (1987), Scott (1987), and Wiggins (1987), the volatility σ_t of the underlying is modeled as a deterministic function $\sigma(\cdot)$ of some auxiliary process Y, which is usually modeled as a diffusion:

$$\begin{aligned} \mathrm{d}X_t &= -\frac{1}{2}\sigma^2(Y_t)\,\mathrm{d}t + \sigma(Y_t)\,\mathrm{d}W_t^{\mathbb{Q}}, \\ \mathrm{d}Y_t &= \alpha(Y_t)\,\mathrm{d}t + \beta(Y_t)\,\mathrm{d}B_t^{\mathbb{Q}}, \qquad \text{(stochastic volatility)} \qquad (7.6) \\ \mathrm{d}\langle W^{\mathbb{Q}}, B^{\mathbb{Q}}\rangle_t &= \rho\,\mathrm{d}t, \end{aligned}$$

with $|\rho| < 1$. Here, $B^{\mathbb{Q}}$ is a Brownian motion that is correlated with $W^{\mathbb{Q}}$. One typically takes the correlation ρ to be negative in order to capture the empirical observation that when volatility goes up, stock prices tend to go down, which is called the *leverage effect*. In a single-factor stochastic volatility setting such as that described by (7.6), derivatives written on S cannot be perfectly hedged by continuously trading a bond and the underlying S alone. However, a derivative written on S can be perfectly replicated by continuously trading a bond, the underlying S, and a single option on S. Thus, assuming options can be traded continuously can *complete the market*. However, as transaction costs on options are much higher than on stocks, and as their liquidity is typically lower, this assumption typically is not made. Unlike the local volatility case, there is no explicit formula for constructing Y dynamics and a volatility function $\sigma(\cdot)$ so that model-induced option prices fit observed market prices exactly.

It is common to assume that the volatility driving process Y is mean-reverting, or *ergodic*, meaning there exists a distribution Π such that the ergodic theorem holds:

$$\lim_{t\to\infty} \frac{1}{t}\int_0^t g(Y_s)\mathrm{d}s = \int g(y)\Pi(\mathrm{d}y),$$

for all bounded functions g.

Equation (7.6) actually refers specifically to *one-factor* stochastic volatility models. One can always introduce another auxiliary process Z, and model the volatility σ_t as a function $\sigma(\cdot,\cdot)$ of both Y and Z. If S, Y, and Z are driven by three distinct Brownian motions, then continuously trading a bond, the underlying S, and two options on S would be required to perfectly hedge further options on S. *Multifactor* stochastic volatility models have the ability of fit option prices better than their one-factor counterparts. But, each additional factor of volatility brings with it additional computational challenges. Multifactor and multiscale stochastic volatility models are discussed at length in Fouque *et al.* (2011).

7.1.2.4 Local-Stochastic Volatility Models

As the name suggests, *local-stochastic volatility* (LSV) models combine features of both local volatility and stochastic volatility models by modeling the volatility σ_t as a function $\sigma(\cdot,\cdot,\cdot)$ of time t, the underlying X, and an auxiliary process Y (possibly multidimensional). For example,

$$\begin{aligned} \mathrm{d}X_t &= -\frac{1}{2}\sigma^2(t, X_t, Y_t)\,\mathrm{d}t + \sigma(t, X_t, Y_t)\,\mathrm{d}W_t^{\mathbb{Q}}, \\ \mathrm{d}Y_t &= f(t, X_t, Y_t)\,\mathrm{d}t + \beta(t, X_t, Y_t)\,\mathrm{d}B_t^{\mathbb{Q}}, \qquad \text{(local-stochastic volatility)} \\ \mathrm{d}\langle W^{\mathbb{Q}}, B^{\mathbb{Q}}\rangle_t &= \rho\,\mathrm{d}t, \qquad (7.7) \end{aligned}$$

where $|\rho| < 1$. Note that the class of models described by (7.7) nests all LV models and all (one-factor) SV models. However, while LSV models offer more modeling flexibility than LV or SV models separately, these models also present new computational challenges. Indeed, while there exist LV and SV models for which option prices can be computed in closed form (or semiclosed form, e.g., up to Fourier inversion), explicit formulas for option prices are available in an LSV setting only when $\rho = 0$. We remark that having a closed-form formula or a fast approximation is crucial for the inverse problem of calibrating an LSV model from observed option prices.

7.1.2.5 Models with Jumps

Some authors argue that diffusion models of the form (7.5) are not adequate to capture the complex dynamics of stock price processes because diffusion models do not allow stock prices to jump. Discontinuities in the stock price process can be modeled by adding a jump term $\mathrm{d}J_t$ to the process (7.5) as follows:

$$\mathrm{d}X_t = \left(-\frac{1}{2}\sigma_t^2 - \lambda \int (\mathrm{e}^z - 1)F(\mathrm{d}z)\right)\mathrm{d}t + \sigma \mathrm{d}W_t^{\mathbb{Q}} + \mathrm{d}J_t,$$

and this type of model dates back to Merton (1976). Here, jumps arrive as a Poisson process with intensity λ and have distribution F, and the drift of X is compensated to ensure that $S = \mathrm{e}^X$ is a martingale under $\mathbb{Q}$. As with SV models, jumps render a market incomplete. Adding jumps also to volatility can help fit the strong implied volatility smile that is commonly observed for short maturity options, and we refer to Bakshi *et al.* (1997) for an analysis.

7.1.2.6 ARCH and GARCH Models

Although our focus will be on the continuous-time models, it is worth mentioning that discrete-time models for stock returns are widely studied in econometrics literature. A large class of discrete-time models are the autoregressive conditional heteroscedasticity (ARCH) processes introduced by Engle (1982), later generalized under the name GARCH. The discrete-time models that are closest to the type of continuous-time stochastic volatility models (7.6) driven by diffusions are the EGARCH models developed in Nelson (1991, 1990). Those papers also discuss convergence of the discrete-time EGARCH process to an exponential Ornstein–Uhlenbeck continuous-time stochastic volatility model.

7.2 Asymptotic Regimes and Approximations

Given a model and its parameters, computing expectations of the form (7.4) one time is straightforward using Monte Carlo methods or a numerical solution of the associated pricing partial integro-differential equation (PIDE). However, when the computation is part of an iterative procedure to calibrate the model to the observed implied volatility surface, it becomes important to have a fast method of computing option prices or model-induced implied volatilities. As a result, a number of different efficient approximation methods have developed.

Broadly speaking, there are two methods of setting up asymptotic expansions for option pricing and implied volatility. In *contract asymptotics*, one considers extreme regimes specific to the option contract, in other words, large or small time-to-maturity $(T - t)$, or large or small strikes K. In the *model asymptotics* approach, one views the complicated incomplete market model as a perturbation around a more tractable model, often the Black–Scholes model (7.1).

7.2.1 Contract Asymptotics

The foundational papers in this approach appeared in 2004. The main result from Lee (2004), commonly referred to as the *moment formula*, relates the implied volatility slope at extreme strikes to the largest finite moment of the stock price. We refer to the recent book by Gulisashvili (2012) for an overview of this and related asymptotics.

The approach pioneered by Berestycki *et al.* (2004) uses large-deviation calculations for the short-time regime. The regime where the time-to-maturity is large is studied by Tehranchi (2009). However, for the rest of this chapter, we concentrate on model asymptotics as they are adaptable to other option contracts and, moreover, are amenable to nonlinear portfolio optimization problems, as we discuss in Section 7.3.

7.2.2 Model Asymptotics

We will present the analysis in terms of the log-stock price $X_t = \log S_t$, and consider a European option with payoff $\varphi(X_T)$ at time T, where $\varphi(x) = h(e^x)$. There may be other factors such as stochastic volatility driving the stock price dynamics, and we denote these by the (possibly multi-dimensional) process Y. If (X, Y) is (jointly) a Markov process, then the time t price $u(t, x, y)$ of the European option is an expectation of the form

$$u(t, x, y) = \mathbb{E}^{\mathbb{Q}}[\varphi(X_T)|X_t = x, Y_t = y].$$

Here, we are using the Markov property of (X, Y) to replace the filtration $\mathcal{F}_t$ in (7.4) with the time-t values of (X, Y).

Under mild conditions on the processes (X, Y) and the payoff function φ, the function $u(t, x, y)$ is sufficiently smooth to be the solution of the *Kolmogorov backward equation* (KBE)

$$(\partial_t + \mathcal{A}(t))u = 0, \qquad u(T, x) = \varphi(x), \tag{7.8}$$

where the operator $\mathcal{A}(t)$ is the *generator* of (X, Y) (which may have t-dependence from the coefficients of (X, Y)). The operator $\mathcal{A}(t)$ is, in general, a second-order partial integro-differential operator. Unfortunately, equation (7.8) rarely has a closed-form solution – especially when we include realistic features such as jumps and stochastic volatility. As such, one typically seeks an approximate solution to (7.8). We will discuss an approach to this using *perturbation theory* (also referred to as asymptotic analysis).

Perturbation theory is a classical tool developed to solve problems arising in physics and engineering. We describe its use here to find an approximate solution to (typically) a PIDE starting from the exact solution of a related PIDE. More specifically, suppose the integro-differential operator $\mathcal{A}(t)$ in (7.8) can be written in the form

$$\mathcal{A}(t) = \sum_{n=0}^{\infty} \varepsilon^n \mathcal{A}_n(t), \tag{7.9}$$

where each $\mathcal{A}_n(t)$ in the sequence $(\mathcal{A}_n(t))$ is an integro-differential operators, and $\varepsilon > 0$ is a (typically small) parameter. Formally, one seeks an approximate solution to (7.8) by expanding the function u as a power series in the parameter ε

$$u = \sum_{n=0}^{\infty} \varepsilon^n u_n. \tag{7.10}$$

Inserting the expansion (7.9) for $\mathcal{A}(t)$ and the expansion (7.10) for u into the PIDE (7.8), and collecting like powers of ε, one finds

$$\begin{aligned}
&\mathcal{O}(1): && (\partial_t + \mathcal{A}_0(t))u_0 = 0, && u_0(T,x,y) = \varphi(x),\\
&\mathcal{O}(\varepsilon): && (\partial_t + \mathcal{A}_0(t))u_1 = -\mathcal{A}_1(t)u_0, && u_n(T,x,y) = 0,\\
&\quad\vdots && \quad\vdots && \quad\vdots\\
&\mathcal{O}(\varepsilon^n): && (\partial_t + \mathcal{A}_0(t))u_n = -\sum_{k=1}^{\infty} \mathcal{A}_k(t)u_{n-k}, && u_n(T,x,y) = 0.
\end{aligned}$$

The approximating sequence of functions (u_n) is then found by solving the above nested sequence of PIDEs.

This method is most useful when the *fundamental solution* Γ_0 (also referred to as *Green's function*), corresponding to the operator $\mathcal{A}_0(t)$, is available in closed form. It is the solution of

$$(\partial_t + \mathcal{A}_0(t))\Gamma_0(t,x,y;T,\xi,\omega) = 0, \qquad \Gamma_0(T,x,y;T,\xi,\omega) = \delta(x-\xi)\delta(y-\omega),$$

where $\delta(\cdot)$ is the Dirac delta function (or point mass at zero).

Upon finding Γ_0, the approximating sequence of functions (u_n) can be written down directly:

$$u_0(t,x,y) = \mathcal{P}_0(t,T)\varphi(x) := \int \mathrm{d}\xi \mathrm{d}\omega\, \Gamma_0(t,x,y;T,\xi,\omega)\varphi(\xi), \tag{7.11}$$

$$u_n(t,x,y) = \int_t^T \mathrm{d}t_1 \mathcal{P}_0(t,t_1) \sum_{k=1}^{n} \mathcal{A}_k u_{n-k}(t_1,x,y). \tag{7.12}$$

where the operator $\mathcal{P}_0(t,T)$ is referred to as the *semigroup* generated by $\mathcal{A}_0(t)$.

Finding an appropriate decomposition of the generator $\mathcal{A}(t) = \sum_{n=0}^{\infty} \mathcal{A}_n(t)$ is a bit of an art. In general, the most appropriate decomposition will depend strongly on the underlying process X (from which $\mathcal{A}(t)$ is derived). As a starting point, it will help to identify operators $\mathcal{A}_0$ for which the fundamental solution Γ_0 can be written in closed form or semiclosed form, and we shall discuss some examples in Section 7.2.4.

7.2.3 *Implied Volatility Asymptotics*

Models are typically calibrated to implied volatilities rather than to prices directly. As such, it is useful to have closed-form approximations for model-induced implied volatilities. In this section, we will show how to translate an expansion for option prices into an expansion for implied volatilities. Throughout this section, we fix a model for $X = \log S$, a time t, a maturity date $T > t$, the initial values $X_t = x$, and a call option payoff $\varphi(X_T) = (\mathrm{e}^{X_T} - \mathrm{e}^k)^+$. Our goal is to find the implied volatility for *this particular call option*. To ease notation, we will suppress much of the dependence on (t, T, x, k). However, the reader should keep in mind that the implied volatility of the option under consideration *does* depend on (t, T, x, k), even if this is not explicitly indicated.

Assume that the option price u has an expansion of the form

$$u = u_0 + \sum_{n=1} \varepsilon^n u_n, \qquad \text{where} \qquad u_0 = u^{\mathrm{BS}}(\sigma_0), \text{ for some } \sigma_0 > 0. \tag{7.13}$$

We wish to find the implied volatility I corresponding to u, which is the unique positive solution of (7.3). To find the unknown implied volatility I, we expand it in powers of ε as follows:

$$I = I_0 + E^{\varepsilon}, \text{where} \quad E^{\varepsilon} = \sum_{n=1}^{\infty} \varepsilon^n I_n.$$

Expanding $u^{\mathrm{BS}}(I)$ about the point I_0, we find

$$\begin{aligned} u^{\mathrm{BS}}(I) &= u^{\mathrm{BS}}(I_0 + E^{\varepsilon}) \\ &= u^{\mathrm{BS}}(I_0) + \sum_{n=1}^{\infty} \frac{(E^{\varepsilon})^n}{n!} \partial_\sigma u^{\mathrm{BS}}(I_0) \\ &= u^{\mathrm{BS}}(I_0) + \varepsilon I_1 \partial_\sigma u^{\mathrm{BS}}(I_0) + \varepsilon^2 \left(I_2 \partial_\sigma + \frac{1}{2!} I_1^2 \partial_\sigma^2 \right) u^{\mathrm{BS}}(I_0) \\ &\quad + \varepsilon^3 \left(I_3 \partial_\sigma + \frac{1}{2!} 2 I_1 I_2 \partial_\sigma^2 + \frac{1}{3!} I_1^3 \partial_\sigma^3 \right) u^{\mathrm{BS}}(I_0) + \cdots . \end{aligned} \tag{7.14}$$

Inserting the expansion (7.13) for u and the expansion (7.14) for $u^{\mathrm{BS}}(I)$ into equation (7.3), and collecting like powers of ε, we obtain

$$\begin{aligned} \mathcal{O}(1) : &\qquad u_0 = u^{\mathrm{BS}}(I_0), \\ \mathcal{O}(\varepsilon) : &\qquad u_1 = I_1 \partial_\sigma u^{\mathrm{BS}}(I_0), \\ \mathcal{O}(\varepsilon^2) : &\qquad u_2 = \left(I_2 \partial_\sigma + \frac{1}{2!} I_1^2 \partial_\sigma^2 \right) u^{\mathrm{BS}}(I_0) \\ \mathcal{O}(\varepsilon^3) : &\qquad u_3 = \left(I_3 \partial_\sigma + \frac{1}{2!} 2 I_1 I_2 \partial_\sigma^2 + \frac{1}{3!} I_1^3 \partial_\sigma^3 \right) u^{\mathrm{BS}}(I_0). \end{aligned}$$

Using $u_0 = u^{\mathrm{BS}}(\sigma_0)$, we can solve for the sequence (I_n) recursively. We have

$$\mathcal{O}(1) : \qquad I_0 = \sigma_0, \tag{7.15}$$

$$\begin{aligned} \mathcal{O}(\varepsilon) : &\qquad I_1 = \frac{1}{\partial_\sigma u^{\mathrm{BS}}(I_0)} u_1, \\ \mathcal{O}(\varepsilon^2) : &\qquad I_2 = \frac{1}{\partial_\sigma u^{\mathrm{BS}}(I_0)} \left(u_2 - \left(\frac{1}{2!} I_1^2 \partial_\sigma^2 \right) u^{\mathrm{BS}}(I_0) \right) \\ \mathcal{O}(\varepsilon^3) : &\qquad I_3 = \frac{1}{\partial_\sigma u^{\mathrm{BS}}(I_0)} \left(u_3 - \left(\frac{1}{2!} 2 I_1 I_2 \partial_\sigma^2 + \frac{1}{3!} I_1^3 \partial_\sigma^3 \right) u^{\mathrm{BS}}(I_0) \right). \end{aligned} \tag{7.16}$$

Note that the implied volatility expansion involves both *model-independent* and *model-dependent* terms:

$$\text{model independent:} \qquad \frac{\partial_\sigma^n u^{\text{BS}}(I_0)}{\partial_\sigma u^{\text{BS}}(I_0)} \qquad \text{model dependent:} \qquad \frac{u_n}{\partial_\sigma u^{\text{BS}}(I_0)}. \tag{7.17}$$

The model-independent terms are always explicit. For example, using (7.2), a direct computation reveals

$$\frac{\partial_\sigma^2 u^{\text{BS}}(\sigma)}{\partial_\sigma u^{\text{BS}}(\sigma)} = \frac{(k-x)^2}{t\sigma^3} - \frac{t\sigma}{4}, \frac{\partial_\sigma^3 u^{\text{BS}}(\sigma)}{\partial_\sigma u^{\text{BS}}(\sigma)} = -\frac{t}{4} + \frac{(k-x)^4}{t^2\sigma^6} - \frac{3(k-x)^2}{t\sigma^4} - \frac{(k-x)^2}{2\sigma^2} + \frac{t^2\sigma^2}{16}.$$

Higher order terms are also explicit. Whether or not the model-dependent terms in (7.17) can be computed explicitly (i.e., meaning without numerical integration or special functions) depends on the specific form the sequence (u_n) takes.

7.2.4 Tractable Models

We say that a model X is *tractable* if its generator $\mathcal{A}_0(t)$ admits a closed-form (or semiclosed form) fundamental solution Γ_0. A large class of tractable models are the exponential Lévy models. In this class, a traded asset $S = e^X$ is described by the following Lévy–Itô SDE:

$$dX_t = \mu \, dt + \sigma dW_t^{\mathbb{Q}} + \int_{\mathbb{R}} z \, d\widetilde{N}_t(dz). \tag{7.18}$$

Here, $W^{\mathbb{Q}}$ is a standard Brownian motion and $\widetilde{N}$ is an independent compensated Poisson random measure:

$$d\widetilde{N}_t(dz) = dN_t(dz) - \nu(dz)dt.$$

The last term in (7.18) can be understood as follows: for any Borel set A, the process $N(A)$ is a Poisson process with intensity $\nu(A)$. Thus, the probability that X experiences a jump of size $z \in A$ in the time interval $[t, t + dt)$ is $\nu(A)dt$. In order for S to be a martingale, the drift μ must be given by

$$\mu = -\frac{1}{2}\sigma^2 - \int_{\mathbb{R}} \nu(dz)(e^z - 1 - z).$$

The generator $\mathcal{A}_0$ of X is given by

$$\mathcal{A}_0 = \mu\partial_x + \frac{1}{2}\sigma^2\partial_{xx}^2 + \int \nu(dz)(\theta_z - 1 - z\partial_x), \tag{7.19}$$

where the operator θ_z is a *shift operator*: $\theta_z f(x) = f(x+z)$.

When X has no jump component (i.e., when $\nu \equiv 0$), the generator $\mathcal{A}_0$ has a fundamental solution Γ_0 that can be written in closed form as a Gaussian kernel:

$$\Gamma_0(t, x; T, y) = \frac{1}{\sqrt{2\pi\sigma^2(T-t)}} \exp\left(-\frac{(x - y + \mu(T-t))^2}{2\sigma^2(T-t)}\right).$$

More generally, when jumps are present, the generator $\mathcal{A}_0$ has a fundamental solution Γ_0, which is available in semiclosed form as a Fourier transform:

$$\Gamma_0(t,x;T,y) = \frac{1}{2\pi}\int \mathrm{d}\xi\ \mathrm{e}^{\mathrm{i}\xi(x-y)+(T-t)\Phi_0(\xi)}, \tag{7.20}$$

where

$$\Phi_0(\xi) = \mathrm{i}\mu\xi - \frac{1}{2}\sigma^2\xi^2 + \int \nu(\mathrm{d}z)(\mathrm{e}^{\mathrm{i}\xi z} - 1 - \mathrm{i}\xi z).$$

The function Φ_0 is referred to as the *characteristic exponent* of X, since it satisfies

$$\mathbb{E}^{\mathbb{Q}}[\mathrm{e}^{\mathrm{i}\xi X_T}|X_t = x] = \mathrm{e}^{\mathrm{i}\xi x+(T-t)\Phi_0(\xi)}.$$

Using (7.11) and (7.20), we can express the action of the semigroup operator $\mathcal{P}_0(t,T)$ on a general function ψ as follows:

$$\begin{aligned}\mathcal{P}_0(t,T)\psi(x) &= \int \mathrm{d}y\ \Gamma_0(t,x;T,y)\psi(y) = \frac{1}{2\pi}\int \mathrm{d}y \int \mathrm{d}\xi\ \mathrm{e}^{i\xi(x-y)+(T-t)\Phi_0(\xi)}\psi(y)\\ &= \frac{1}{2\pi}\int \mathrm{d}\xi\ \mathrm{e}^{i\xi x+(T-t)\Phi_0(\xi)}\hat{\psi}(\xi),\end{aligned}$$

where $\hat{\psi}$ is the Fourier transform of ψ:

$$\hat{\psi}(\xi) := \int \mathrm{d}y\ \mathrm{e}^{-\mathrm{i}\xi y}\psi(y).$$

Because the semigroup operator $\mathcal{P}_0(t,T)$ corresponding to $\mathcal{A}_0$ is well understood in the exponential Lévy setting, if one can write the generator $\mathcal{A}$ of a process X as $\mathcal{A} = \sum_{n=0}^{\infty} \varepsilon^n \mathcal{A}_n$ with $\mathcal{A}_0$ given by (7.19), then one can use (7.11)–(7.12) to find approximate solutions to the full pricing PIDE $(\partial_t + \mathcal{A})u = 0$.

7.2.5 *Model Coefficient Polynomial Expansions*

Model coefficient expansions are developed in a series of papers by Lorig *et al.* (2014, 2015a, b, c, d). The authors' method (which we shall henceforth refer to as LPP) can be used to find closed-form asymptotic approximations for option prices and implied volatilities in a general d-dimensional Markov setting. Here, for simplicity, we focus on two simple cases: (i) general two-dimensional *local-stochastic volatility* (LSV) models, and (ii) general scalar Lévy-type models.

7.2.5.1 LSV models

We consider a general class of models, in which an asset $S = \mathrm{e}^X$ is modeled as the exponential of a Markov diffusion process X that satisfies the SDEs

$$\begin{aligned}\mathrm{d}X_t &= -\frac{1}{2}\sigma^2(X_t,Y_t)\ \mathrm{d}t + \sigma(X_t,Y_t)\ \mathrm{d}W_t^{\mathbb{Q}},\\ \mathrm{d}Y_t &= f(X_t,Y_t)\ \mathrm{d}t + \beta(X_t,Y_t)\ \mathrm{d}B_t^{\mathbb{Q}},\\ \mathrm{d}\langle W^{\mathbb{Q}},B^{\mathbb{Q}}\rangle_t &= \rho\ \mathrm{d}t,\end{aligned}$$

where $W^{\mathbb{Q}}$ and $B^{\mathbb{Q}}$ are correlated Brownian motions. This is as in the model discussed in Section 7.1.2.4 except we remove the explicit t-dependence in the coefficients for simplicity. Note that the drift of X is $-\frac{1}{2}\sigma^2$, which ensures that $S = \mathrm{e}^X$ is a $\mathbb{Q}$-martingale and so the stock price is arbitrage free.

The generator $\mathcal{A}$ of X is given by

$$\mathcal{A} = a(x,y)(\partial_{xx}^2 - \partial_x) + f(x,y)\partial_y + b(x,y)\partial_{yy}^2 + c(x,y)\partial_{xy}^2, \tag{7.21}$$

where we have defined

$$a(x,y) := \frac{1}{2}\sigma^2(x,y), \qquad b(x,y) := \frac{1}{2}\beta^2(x,y), \qquad c(x,y) := \rho\sigma(x,y)\beta(x,y).$$

For general coefficients (σ, β, f), there is no closed-form (or even semiclosed-form) expression of Γ, the fundamental solution corresponding to $\mathcal{A}$. Thus, we seek a decomposition of $\mathcal{A} = \sum_{n=0}^{\infty} \mathcal{A}_n$ for which the order-zero operator $\mathcal{A}_0$ admits a closed-form fundamental solution Γ_0.

The LPP approach is to expand the coefficients of $\mathcal{A}$ in polynomial basis functions where the zeroth-order terms in the expansion are constant. Specifically,

$$\chi(x,y) = \sum_{n=0}^{\infty} \chi_n(x,y), \qquad \chi \in \{a,b,c,f\},$$

where χ_0 is a constant and $\chi_n(x,y)$ depends polynomially on x and y for every $n \geq 1$. For example, Taylor series:

$$\chi_n(x,y) = \sum_{k=0}^{n} \mathcal{X}_{k,n-k}(x-\bar{x})^k(y-\bar{y})^{n-k}, \qquad \mathcal{X}_{k,n-k} = \frac{1}{k!(n-k)!}\partial_x^k \partial_y^{n-k}\chi(\bar{x},\bar{y}), \tag{7.22}$$

or Hermite polynomials:

$$\chi_n(x,y) = \sum_{k=0}^{n} \mathcal{X}_{k,n-k} H_{k,n-k}(x,y), \qquad \mathcal{X}_{k,n-k} = \langle H_{k,n-k}, \chi\rangle\langle H_{k,n-k}, \chi\rangle.$$

In the Taylor series example, $(\bar{x},\bar{y})$ is a fixed point in $\mathbb{R}^2$. In the Hermite polynomial example, the brackets $\langle\cdot,\cdot\rangle$ indicate an L^2 inner product with a Gaussian weighting. The Hermite polynomials $(H_{n,m})$ form a complete basis in this space and (properly weighted) are orthonormal $\langle H_{n,m}, H_{i,j}\rangle = \delta_{n,i}\delta_{m,j}$.

Upon expanding the coefficients of $\mathcal{A}$ in polynomial basis functions, one can formally write the operator $\mathcal{A}$ as

$$\mathcal{A} = \sum_{n=0}^{\infty} \mathcal{A}_n, \quad \text{with} \quad \mathcal{A}_n = \sum_{k=0}^{n} \mathcal{A}_{k,n-k}, \tag{7.23}$$

where

$$\mathcal{A}_{k,n-k} = a_{k,n-k}(x,y)(\partial_{xx}^2 - \partial_x) + f_{k,n-k}(x,y)\partial_y + b_{k,n-k}(x,y)\partial_{yy}^2 + c_{k,n-k}(x,y)\partial_{xy}^2.$$

Note that the operator $\mathcal{A}$ in (7.23) is of the form (7.9) if one sets $\varepsilon = 1$.

Moreover, the order-zero operator

$$\mathcal{A}_0 = \mathcal{A}_{0,0} = a_{0,0}(\partial_{xx}^2 - \partial_x) + f_{0,0}\partial_y + b_{0,0}\partial_{yy}^2 + c_{0,0}\partial_{xy}^2.$$

has a fundamental solution Γ_0, which is a Gaussian density:

$$\Gamma_0(t,x,y;T,\xi,\omega) = \frac{1}{2\pi\sqrt{|\mathbf{C}|}} \exp\left(-\frac{1}{2}(\boldsymbol{\eta}-\mathbf{m})^{\mathrm{T}}\mathbf{C}^{-1}(\boldsymbol{\eta}-\mathbf{m})\right), \qquad \boldsymbol{\eta} = \begin{pmatrix}\xi\\ \omega\end{pmatrix},$$

with covariance matrix $\mathbf{C}$ and mean vector $\mathbf{m}$ given by

$$\mathbf{C} = (T-t)\begin{pmatrix}2a_{0,0} & c_{0,0}\\ c_{0,0} & 2b_{0,0}\end{pmatrix}, \qquad \mathbf{m} = \begin{pmatrix}x-(T-t)a_{0,0}\\ y+(T-t)f_{0,0}\end{pmatrix}.$$

Since Γ_0 is available in closed form, one can use (7.11) and (7.12) to find u_0 and the sequence of functions $(u_n)_{n\geq 1}$, respectively. After a bit of algebra, one can show that

$$u_n(t,x,y) = \mathcal{L}_n u_0(t,x,y), \tag{7.24}$$

where the operator $\mathcal{L}_n$ is of the form

$$\mathcal{L}_n = \sum_{k,m} \eta^{(n)}_{k,m}(t,x,y)\partial_y^k \partial_x^m (\partial_x^2 - \partial_x).$$

The precise form of the coefficients $(\eta^{(n)}_{k,m})$ will depend on the choice of polynomial basis function. If the coefficients of $\mathcal{A}$ are smooth and bounded, then the small-time-to-maturity accuracy of the price approximation is

$$\sup_{x,y} |u(t,x,y) - \bar{u}_n(t,x,y)| = \mathcal{O}((T-t)^{(n+3)/2}), \qquad \bar{u}_n(t,x,y) := \sum_{k=0}^{n} u_k(t,x,y).$$

The proof can be found in Lorig *et al.* (2015a).

The LPP price expansion also leads to closed-form expressions for implied volatility. Indeed, for European call options, one can easily show that $u_0 = u^{\mathrm{BS}}(\sqrt{2a_{0,0}})$. Therefore, the series expansion for u is of the form (7.13), and hence (I_0, I_1, I_2, I_3) can be computed using (7.15)–(7.16). Moreover, the model-dependent terms $(u_n/\partial_\sigma u^{\mathrm{BS}})$ appearing in (7.15)–(7.16) can be computed explicitly with no numerical integration. This is due to the fact that u_n can be written in the form (7.24). For details, we refer the reader to Lorig *et al.* (2015b).

Example 7.1 (Heston Model) *Consider the Heston (1993) model, under which the risk-neutral dynamics of X are given by*

$$\begin{aligned} dX_t &= -\frac{1}{2}e^{Y_t}dt + e^{Y_t/2}dW_t^{\mathbb{Q}},\\ dY_t &= \left((\kappa\theta - \frac{1}{2}\delta^2)e^{-Y_t} - \kappa\right)dt + \delta\, e^{-Y_t/2}dB_t^{\mathbb{Q}},\\ d\langle W^{\mathbb{Q}}, B^{\mathbb{Q}}\rangle_t &= \rho\, dt. \end{aligned}$$

The generator of (X,Y) is given by

$$\mathcal{A} = \frac{1}{2}e^y(\partial_x^2 - \partial_x) + \left((\kappa\theta - \frac{1}{2}\delta^2)e^{-y} - \kappa\right)\partial_y + \frac{1}{2}\delta^2 e^{-y}\partial_y^2 + \rho\,\delta\partial_x\partial_y.$$

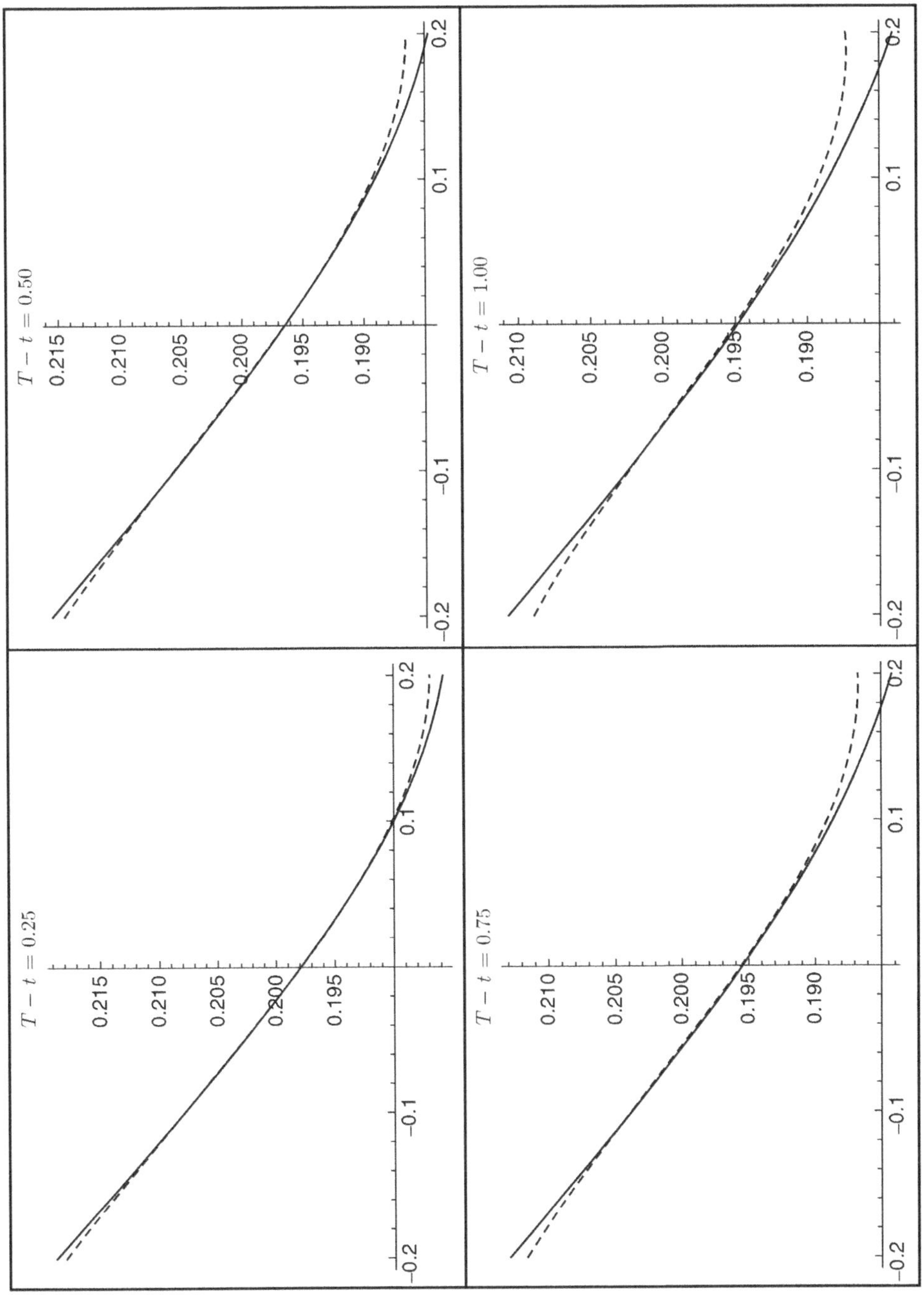

Figure 7.2 Exact (solid) and approximate (dashed) implied volatilities in the Heston model. The horizontal axis is log-moneyness $(k - x)$. Parameters: $\kappa = 1.15$, $\theta = 0.04$, $\delta = 0.2$, $\rho = -0.40$ $x = 0.0$, $y = \log \theta$.

Thus, using (7.21), we identify

$$a(x,y) = \frac{1}{2}e^{y}, \quad b(x,y) = \frac{1}{2}\delta^2 e^{-y}, \quad c(x,y) = \rho\delta, \quad f(x,y) = \left((\kappa\theta - \frac{1}{2}\delta^2)e^{-y} - \kappa \right).$$

We fix a time to maturity τ *and* log-*strike* k. *Assuming a Taylor series expansion (7.22) of the coefficients of* $\mathcal{A}$ *with* $(\bar{x}, \bar{y}) = (x, y)$, *the time* t *levels of* (X, Y), *one computes*

$$\begin{aligned}
I_0 &= e^{y/2},\\
I_1 &= \frac{1}{8}e^{-y/2}\tau(-\delta^2 + 2(-e^y + \theta)\kappa + e^y\delta\rho) + \frac{1}{4}e^{-y/2}\delta\rho(k - x),\\
I_2 &= \frac{-e^{-3y/2}}{128}\tau^2(\delta^2 - 2\theta\kappa)^2 + \frac{e^{y/2}}{96}\tau^2(5\kappa^2 - 5\delta\kappa\rho + \delta^2(-1 + 2\rho^2))\\
&\quad + \frac{e^{-y/2}}{192}\tau(-4\tau\theta\kappa^2 - \tau\delta^3\rho + 2\tau\delta\theta\kappa\rho + 2\delta^2(8 + \tau\kappa + \rho^2))\\
&\quad + \frac{1}{96}e^{-3y/2}\tau\delta\rho(5\delta^2 + 2(e^y - 5\theta)\kappa - e^y\delta\rho)(k - x) + \frac{1}{48}e^{-3y/2}\delta^2(2 - 5\rho^2)(k - x)^2.
\end{aligned}$$

The expression for I_3 *is also explicit, but omitted for brevity. In Figure 7.2, we plot the approximate implied volatility* $(I_0 + I_1 + I_2 + I_3)$ *as well as the exact implied volatility* I, *which can be computed using the pricing formula given in Heston (1993) and then inverting the Black–Scholes formula numerically.*

7.2.5.2 Lévy-type models

In this section, we explore how the LPP method can be applied to compute approximate option prices in a one-dimensional Lévy-type setting. Specifically, we consider an asset $S = \mathrm{e}^X$, where X is a scalar Lévy-type Markov process. Under some integrability conditions on the size and intensity of jumps, every scalar Markov process on $\mathbb{R}$ can be expressed as the solution of a Lévy–Itô SDE of the form:

$$\mathrm{d}X_t = \mu(X_t)\mathrm{d}t + \sqrt{2a(X_t)}\mathrm{d}W_t^{\mathbb{Q}} + \int z\mathrm{d}\widetilde{N}_t(X_{t-}, \mathrm{d}z),$$

where $W^{\mathbb{Q}}$ is a standard Brownian motion; and $\widetilde{N}$ is a state-dependent compensated Poisson random measure

$$\mathrm{d}\widetilde{N}_t(x, \mathrm{d}z) = \mathrm{d}N_t(x, \mathrm{d}z) - \nu(x, \mathrm{d}z)\mathrm{d}t.$$

Note that jumps are now described by a *Lévy kernel* $\nu(x, \mathrm{d}z)$, which is a Lévy measure for every $x \in \mathbb{R}$. In order for S to be a martingale, the drift μ must be given by

$$\mu(x) = -a(x) - \int_{\mathbb{R}} \nu(x, \mathrm{d}z)(\mathrm{e}^z - 1 - z).$$

The generator $\mathcal{A}$ of X is

$$\mathcal{A} = \mu(x)\partial_x + a(x)\partial_{xx}^2 + \int \nu(x, \mathrm{d}z)(\theta_z - 1 - z\partial_x).$$

For general drift $\mu(x)$ variance $a(x)$ and Lévy kernel $\nu(x, dz)$, there is no closed-form (or even semiclosed-form) expression for the fundamental solution Γ corresponding to $\mathcal{A}$. Thus, we seek a decomposition of generator $\mathcal{A} = \sum_n \mathcal{A}_n$ for which $\mathcal{A}_0$ has a fundamental solution Γ_0 available in semiclosed form.

Once again, the LPP approach of expanding the coefficients and Levy kernel of $\mathcal{A}$ in polynomial basis functions will lead to the desired form for the expansion of $\mathcal{A}$. Specifically, let

$$\mu(x) = \sum_{n=0}^{\infty} \mu_n(x), \qquad a(x) = \sum_{n=0}^{\infty} a_n(x), \qquad \nu(x, dz) = \sum_{n=0}^{\infty} \nu_n(x, dz),$$

where $\mu_0(x) = \mu_0$; $a_0(x) = a_0$; and $\nu_0(x, dz) = \nu_0(dz)$; and higher order terms $\mu_n(x)$, $a_n(x)$, and $\nu_n(x, dz)$ depend polynomially on x. For example,

$$\begin{array}{ll} \text{Taylor series:} & a_n(x) = \frac{1}{n!}\partial_x^n a(\bar{x})\cdot(x-\bar{x})^n, \\ \text{Hermite polynomial:} & a_n(x) = \langle H_n, a\rangle\cdot H_n(x), \end{array}$$

and similarly for μ and ν. Upon expanding the coefficients of $\mathcal{A}$ in polynomial basis functions, one can formally write the operator $\mathcal{A}$ as $\mathcal{A} = \sum_{n=0}^{\infty} \mathcal{A}_n$, where $\mathcal{A}_0$ is given by

$$\mathcal{A}_0 = \mu_0 \partial_x + a_0 \partial_{xx}^2 + \int \nu_0(dz)(\theta_z - 1 - z\partial_x), \tag{7.25}$$

and each $\mathcal{A}_n$ for $n \geq 0$ is of the form

$$\mathcal{A}_n = \mu_n(x)\partial_x + a_n(x)\partial_{xx}^2 + \int \nu_n(x, dz)(\theta_z - 1 - z\partial_x).$$

Comparing (7.25) with (7.19), we see that $\mathcal{A}_0$ is the generator of a Lévy process. Thus, using (7.20), the fundamental solution Γ_0 corresponding to $\mathcal{A}_0$ can be written as a Fourier integral

$$\Gamma_0(t, x; T, y) = \frac{1}{2\pi} \int d\xi \; e^{i\xi(x-y)+(T-t)\Phi_0(\xi)},$$

where

$$\Phi_0(\xi) = i\mu_0\xi - a_0\xi^2 + \int \nu_0(dz)(e^{i\xi z} - 1 - i\xi z).$$

Since Γ_0 is available in closed form, one can use (7.11) and (7.12) to find u_0 and u_n, respectively. Explicit computations are carried out in Lorig *et al.* (2015c).

In this case, it is convenient to express $u_n(t, x)$ as an (inverse) Fourier transform of $\hat{u}_n(t, x)$. Defining

$$\begin{array}{ll} \text{Fourier transform:} & \hat{u}_n(t, \xi) = \mathcal{F}[u_n(t, \cdot)](\xi) := \int dx \; e^{-i\xi x} u_n(t, x), \\ \text{Inverse transform:} & u_n(t, x) = \mathcal{F}^{-1}[\hat{u}_n(t, \cdot)](x) := \frac{1}{2\pi} \int dx \; e^{i\xi x} \hat{u}_n(t, \xi), \end{array}$$

we have

$$\begin{aligned} \hat{u}_n(t, \xi) = & \sum_{k=1}^{n} \int_t^T dt_1 \; e^{(t_1-t)\Phi_0(\xi)} \int_{\mathbb{R}_0^d} \nu_k(i\partial_\xi, dz)(e^{i\xi z} - 1 - i\xi z)\hat{u}_{n-k}(t_1, \xi) \\ & + \sum_{k=1}^{n} \int_t^T dt_1 \; e^{(t_1-t)\Phi_0(\xi)} (\mu_k(i\partial_\xi)(i\xi)\hat{u}_{n-k}(t_1, \xi) + a_k(i\partial_\xi)(i\xi)^2 \hat{u}_{n-k}(t_1, \xi)), \end{aligned}$$

where $\hat{u}_0(t,\xi)$ is given by

$$\hat{u}_0(t,\xi) = \mathrm{e}^{\mathrm{i}\xi x+(T-t)\Phi_0(\xi)}\hat{\phi}(\xi).$$

7.2.6 Small "Vol of Vol" Expansion

Lewis (2000) considers a stochastic volatility model of the form

$$\begin{aligned} \mathrm{d}X_t &= -\frac{1}{2}Y_t\,\mathrm{d}t + \sqrt{Y_t}\,\mathrm{d}W_t^{\mathbb{Q}}, \\ \mathrm{d}Y_t &= \alpha(Y_t)\,\mathrm{d}t + \varepsilon\beta(Y_t)\,\mathrm{d}B_t^{\mathbb{Q}}, \\ \mathrm{d}\langle W^{\mathbb{Q}}, B^{\mathbb{Q}}\rangle_t &= \rho\,\mathrm{d}t. \end{aligned}$$

The parameter ε is referred to as the *volatility of volatility*, or *vol of vol* for short. The generator $\mathcal{A}$ of (X,Y) is given by

$$\mathcal{A} = \mathcal{A}_0 + \varepsilon\mathcal{A}_1 + \varepsilon^2\mathcal{A}_2,$$

where

$$\mathcal{A}_0 = \frac{y}{2}(\partial_x^2 - \partial_x) + \alpha(y)\partial_y, \qquad \mathcal{A}_1 = \rho\sqrt{y}\beta(y)\partial_x\partial_y, \qquad \mathcal{A}_2 = \frac{1}{2}\beta^2(y)\partial_y^2.$$

Thus, $\mathcal{A}$ is of the form (7.9). Moreover, the solution of

$$(\partial_t + \mathcal{A}_0)\Gamma_0 = 0, \qquad \Gamma_0(T,x,y;T,\xi) = \delta(x-\xi),$$

is given by a Gaussian

$$\Gamma_0(t,x,y;T,\xi,\zeta) = \frac{1}{\sqrt{2\pi\sigma^2(t,T)}}\exp\left(-\frac{(\xi - x - \sigma^2(t,T)/2)^2}{2\sigma^2(t,T)}\right)\delta(\zeta - \eta(t)),$$

where $\sigma^2(t,T) = \int_t^T \eta(s)\,\mathrm{d}s$, and $\eta(s)$ is the solution of the following ODE:

$$\tfrac{\mathrm{d}\eta(s)}{\mathrm{d}s} = \alpha(\eta(s)), \qquad \eta(t) = y.$$

Since Γ_0 is available in closed form up to finding η, one can use (7.11) and (7.12) to find u_0 and the sequence of functions (u_n), respectively. In particular, for European calls, one finds

$$u_0 = u^{\mathrm{BS}}(\bar{\sigma}(t,T)), \qquad \bar{\sigma}(t,T) = \sqrt{\tfrac{\sigma^2(t,T)}{T-t}}$$

As the order-zero price is equal to the Black–Scholes price, computed with volatility $\bar{\sigma}(t,T)$, one can apply the implied volatility asymptotics of Section 7.2.3 to find approximate implied volatilities. Explicit expressions for approximate implied vols are given in Lewis (2000) for the case $\alpha(y) = \kappa(\theta - y)$ and $\beta(y) = y^\gamma$.

7.2.7 Separation of Timescales Approach

In Fouque *et al.* (2000, 2011), the authors develop analytic price approximations for interest rate, credit, and equity derivatives. The authors' approach (henceforth referred to as FPSS) exploits the separation of timescales that is observed in volatility time-series data. More specifically, the authors consider a class of *multiscale* diffusion models in which the volatility of an underlying is driven by two factors, Y and Z, operating on fast and slow timescales, respectively. Lorig and Lozano-Carbassé (2015) extend the FPSS method to models with jumps. Here, for simplicity, we consider an exponential Lévy-type model with a single slowly varying factor, which drives both the volatility and jump-intensity.

Consider a model for a stock $S = e^X$, where X is modeled under the risk-neutral pricing measure as the solution of the following Lévy–Itô SDE

$$\begin{aligned} dX_t &= \mu(Z_t)\,dt + \sigma(Z_t)\,dW_t^{\mathbb{Q}} + \int_{\mathbb{R}} s\,d\tilde{N}_t(Z_{t-}, ds), \\ dZ_t &= (\varepsilon^2 c(Z_t) - \varepsilon\,\Gamma(Z_t)\,g(Z_t))\,dt + \varepsilon\,g(Z_t)\,dB_t^{\mathbb{Q}}, \qquad \text{(risk-neutral measure } \mathbb{Q}) \\ d\langle W^{\mathbb{Q}}, B^{\mathbb{Q}}\rangle_t &= \rho\,dt. \end{aligned}$$

where $W^{\mathbb{Q}}$ and $B^{\mathbb{Q}}$ are correlated Brownian motions; and $\widetilde{N}$ is a compensated state-dependent Poisson random measure with which we associate a Lévy kernel $\zeta(z)\nu(ds)$. The drift $\mu(z)$ is fixed by the Lévy kernel and volatility so that $S = e^X$ is a martingale:

$$\mu(z) = -\frac{1}{2}\sigma^2(z) - \zeta(z)\int_{\mathbb{R}} \nu(ds)(e^s - 1 - s).$$

Note that the volatility σ and Lévy kernel $\zeta\nu$ are driven by Z, which is *slowly varying* in the following sense. Under the physical measure $\mathbb{P}$, the dynamics of Z are given by

$$dZ_t = \varepsilon^2 c(Z_t)\,dt + \varepsilon\,g(Z_t)\,d\tilde{B}_t, \qquad \text{(physical measure } \mathbb{P})$$

where $\tilde{B}_t = B_t^{\mathbb{Q}} - \int_0^t \Gamma(Z_s)\,ds$ is a standard $\mathbb{P}$-Brownian motion (Γ is known as the *market price of risk* associated with $\tilde{B}$). The infinitesimal generator of Z under the physical measure

$$\mathcal{A}_Z = \varepsilon^2\left(\frac{1}{2}g^2(z)\partial_z^2 + c(z)\partial_z\right)$$

is scaled by ε^2, which is assumed to be a small parameter: $\varepsilon^2 << 1$. Thus, Z fluctuates over an intrinsic timescale $1/\varepsilon^2$, which is large.

Due to the separation of timescales, the (X, Z) process has a generator that naturally factors into three terms of different powers of ε, just as in (7.9):

$$\mathcal{A} = \mathcal{A}_0 + \varepsilon\mathcal{A}_1 + \varepsilon^2\mathcal{A}_2,$$

where $\mathcal{A}_0$, $\mathcal{A}_1$, and $\mathcal{A}_2$ are given by

$$\begin{aligned} \mathcal{A}_0 &= \mu(z)\partial_x + \frac{1}{2}\sigma^2(z)\partial_x^2 + \zeta(z)\int_{\mathbb{R}} \nu(ds)(\theta_s - 1 - s\partial_x), \qquad (7.26) \\ \mathcal{A}_1 &= -\Gamma(z)g(z)\partial_z + \rho g(z)\sigma(z)\partial_x\partial_z, \\ \mathcal{A}_2 &= \frac{1}{2}g^2(z)\partial_z^2 + c(z)\partial_z. \end{aligned}$$

Here, the shift operator θ_s acts on the x variable. Comparing equations (7.19) and (7.26), we observe that, for fixed z, the operator $\mathcal{A}_0$ above is the generator of a Lévy process. As the operator $\mathcal{A}_0$ in (7.26) acts only on the variable x, the variable z serves only as a parameter. Thus, we have

$$\Gamma_0(t,x,z;T,y) = \frac{1}{2\pi}\int \mathrm{d}\xi\ \mathrm{e}^{\mathrm{i}\xi(x-y)+(T-t)\Phi_0(\xi,z)},$$

where

$$\Phi_0(\xi,z) := \mathrm{i}\mu(z)\xi - \frac{1}{2}\sigma^2(z)\xi^2 + \zeta(z)\int \nu(\mathrm{d}s)(\mathrm{e}^{\mathrm{i}\xi s} - 1 - \mathrm{i}\xi s).$$

Because $\mathcal{A}$ is of the form (7.9) (set $\mathcal{A}_n = 0$ for $n \geq 3$), and since the fundamental solution Γ_0 corresponding to $\mathcal{A}_0$ is known, we can use (7.11) and (7.12) to write u_0 and u_1 explicitly. For an option with payoff $\varphi(X_T)$, we have

$$u_0(t,x,z) = \int \mathrm{d}x\ \varphi(y)\frac{1}{2\pi}\int \mathrm{d}\xi\ \mathrm{e}^{\mathrm{i}\xi(x-y)+(T-t)\Phi_0(\xi,z)} = \frac{1}{2\pi}\int \mathrm{d}\xi\ \mathrm{e}^{\mathrm{i}\xi(x-y)+(T-t)\Phi_0(\xi,z)}\hat{\varphi}(\xi),$$

where $\hat{\varphi}$ is the Fourier transform of ϕ.

A similar computation yields the following expression for u_1:

$$u_1(t,x,z) = \frac{(T-t)/2}{2\pi}\int_{\mathbb{R}} d\xi\ \mathrm{e}^{\mathrm{i}\xi x + t\Phi_0(\xi,z)}\hat{\varphi}(\xi)M_\xi(z),$$

where

$$\begin{aligned} M_\xi(z) = {} & V_1(z)(-\mathrm{i}\xi^3 + \xi^2) + U_1(z)\left(\xi^2\int_{\mathbb{R}} \nu(\mathrm{d}s)(\mathrm{e}^s - 1 - s) + \mathrm{i}\xi\int_{\mathbb{R}} \nu(ds)(\mathrm{e}^{i\xi s} - 1 - \mathrm{i}\xi s)\right) \\ & + V_0(z)(-\xi^2 - i\xi) + U_0(z)\left(-i\xi\int_{\mathbb{R}} \nu(\mathrm{d}s)(\mathrm{e}^s - 1 - s) + \int_{\mathbb{R}} \nu(ds)(\mathrm{e}^{\mathrm{i}\xi s} - 1 - \mathrm{i}\xi s)\right), \end{aligned}$$

and

$$V_1(z) = \frac{1}{2}g(z)\rho\sigma(z)\partial_z\sigma^2(z), \qquad V_0(z) = -\frac{1}{2}g(z)\Gamma(z)\partial_z\sigma^2(z),$$

$$U_1(z) = g(z)\rho\sigma(z)\partial_z\zeta(z), \qquad U_0(z) = -g(z)\Gamma(z)\partial_z\zeta(z).$$

If the coefficients of $\mathcal{A}$ and the payoff function are smooth and bounded, then one can establish the following accuracy for the first-order price approximation

$$|u(t,x,z) - (u_0(t,x,z) + \varepsilon u_1(t,x,z))| = \mathcal{O}(\varepsilon^2),$$

which holds pointwise. Note that, to compute the approximate price of an option $u_0 + \varepsilon u_1$, one does not need full knowledge of $(g,\sigma,\Gamma,\zeta,\rho,z,\varepsilon)$. Rather, at order ε, the information contained in these five functions and two variables is entirely captured by four *group parameters* $(\varepsilon U_1(z), \varepsilon U_0(z), \varepsilon V_1(z), \varepsilon V_0(z))$. The values of these four parameters can be obtained by calibrating to observed call and put prices, as described in detail in Lorig and Lozano-Carbassé (2015).

7.2.8 *Comparison of the Expansion Schemes*

The approximation methods presented in Sections 7.2.5, 7.2.6, and 7.2.7 exploit different small parameters in $\mathcal{A}$ and therefore work best (i.e., provide the most accurate approximations) in different regimes. Rigorous accuracy results must be obtained on a case-by-case basis. However, it is generally true that the accuracy of $\bar{u}_n := \sum_{k=0}^{n} u_k$ depends on how well $\bar{\mathcal{A}}_n := \sum_{k=0}^{n} \mathcal{A}_k$ approximates $\mathcal{A}$.

For example, the Taylor series expansion described in Section 7.2.5 works best if the coefficients of $\mathcal{A}$ are slowly varying. In this case, the coefficients can be well approximated by a Taylor series of low order. Thus, a highly accurate approximation of $\mathcal{A}$ can be obtained with only a few terms.

The approximation considered in Section 7.2.6 works best when the diffusion coefficient of the volatility-driving process Y is small in comparison to the drift coefficient of Y and the drift and diffusion coefficients of X.

Finally, the approximation considered in Section 7.2.7 is when the drift coefficient of the volatility-driving process Z is small in comparison to the diffusion coefficient of Z, which in turn is small in comparison to the drift and diffusion coefficients of X. As mentioned in this chapter, this means that the intrinsic timescale of Z must be slow in comparison to the intrinsic timescale of X.

7.3 Merton Problem with Stochastic Volatility: Model Coefficient Polynomial Expansions

A landmark pair of papers on optimal investment strategy by Robert Merton analyzed the problem of how an investor should optimally allocate his wealth between a riskless bond and some risky assets, in order to maximize his expected utility of wealth. This problem and its variation are now referred to as the *Merton problem*.

In the original papers, each of the risky assets follows geometric Brownian motion with constant volatility. This modeling assumption is convenient from the standpoint of analytic tractability but is not realistic in practice, as it does not allow for stochastic volatility. Because of this, there has been much interest in analyzing how an investor's optimal investment strategy changes in the presence of stochastic volatility. Here, we focus on asymptotic methods for analyzing the stochastic control problem associated with portfolio optimization.

Analysis with multiscale stochastic volatility models described in Section 7.2.7 was presented in the case of simple power utilities in Fouque *et al.* (2000 Section 10.1), and expansions for a hedging problem in Jonsson and Sircar (2002a, b) in the dual optimization problem, both for *fast mean-reverting* stochastic volatility. In Fouque *et al.* (2013), expansions are constructed directly in the primal problem under both fast and slow volatility fluctuations. Indifference pricing approximations with *exponential utility* and fast volatility were studied in Sircar and Zariphopoulou (2005). In this section, we present some new approximations for the Merton problem with stochastic volatility.

7.3.1 Models and Dynamic Programming Equation

The polynomial expansion techniques outlined in Section 7.2.5 can be extended to find explicit asymptotic solutions to the Merton problem in a general local-stochastic volatility (LSV) setting, that is, $\sigma_t = \sigma(t, S_t, Y_t)$. To fix ideas, however, we consider the simpler stochastic volatility (SV) setting, in which a risky asset S, under the physical probability measure $\mathbb{P}$, is the solution of the following SDE:

$$dS_t = \mu(Y_t)S_t \, dt + \sigma(Y_t)S_t \, dB_t^S, \qquad dY_t = c(Y_t) \, dt + \beta(Y_t) \, dB_t^Y, \qquad d\langle B^S, B^Y \rangle_t = \rho dt,$$

where B^S and B^Y are standard Brownian motions with correlation ρ. Let W denote the wealth process of an investor who holds π_t units worth of currency in S at time t, and has $(W_t - \pi_t)$ units of currency in a bond. For simplicity, we assume the risk-free rate of interest is zero. As such, the wealth process W satisfies

$$dW_t = \frac{\pi_t}{S_t} \, dS_t = \pi_t \mu(Y_t) \, dt + \pi_t \sigma(Y_t) \, dB_t^S.$$

Observe that S does not appear in the dynamics of the wealth process W.

An investor chooses π_t to maximize his expected utility of wealth at a time T in the future, where utility is measured by a smooth, increasing, and strictly concave function $U : \mathbb{R}_+ \to \mathbb{R}$, and the objective to maximize is $\mathbb{E} \; U(W_T)$. Increasing describes a preference for more wealth than less, whereas concavity captures risk aversion, with more concave being more risk averse. The analysis is illustrated with power utility functions in Section 7.3.3.

We define the investor's *value function* u by

$$u(t, y, w) := \sup_{\pi \in \Pi} \mathbb{E}[U(W_T) | Y_t = y, W_t = w],$$

where Π is the set of *admissible strategies*:

$$\Pi := \left\{ \pi \text{ adapted: } \mathbb{E} \int_0^T \pi_t^2 \sigma^2(Y_t) \, dt < \infty \text{ and } W_t \geq 0 \text{ a.s.} \right\},$$

where adapted means adapted to the filtration generated by (B^S, B^Y).

Assuming that $u \in C^{1,2}([0, T], \mathbb{R}, \mathbb{R}_+)$, the value function solves the Hamilton–Jacobi–Bellman partial differential equation (HJB-PDE) problem:

$$(\partial_t + \mathcal{A}^Y)u + \max_{\pi \in \mathbb{R}} \mathcal{A}^\pi u = 0, \qquad u(T, y, w) = U(w), \tag{7.27}$$

where $(\mathcal{A}^Y + \mathcal{A}^\pi)$ is the generator of (Y, W), assuming a Markov investment strategy $\pi_t = \pi(t, Y_t, W_t)$. Specifically, the operators $\mathcal{A}^Y$ and $\mathcal{A}^\pi$ are given by

$$\mathcal{A}^Y = c(y)\partial_y + \frac{1}{2}\beta^2(y)\partial_y^2,$$

$$\mathcal{A}^\pi = \pi(t, y, w)\mu(y)\partial_w + \frac{1}{2}\pi^2(t, y, w)\sigma^2(y)\partial_w^2 + \pi(t, y, w)\rho\sigma(y)\beta(y)\partial_y\partial_w.$$

The optimal strategy π^* is given by

$$\pi^* = \arg\max_{\pi \in \mathbb{R}} \mathcal{A}^\pi u = -\frac{\mu(\partial_w u) + \rho\beta\sigma(\partial_y \partial_w u)}{\sigma^2(\partial_w^2 u)}, \tag{7.28}$$

where, for simplicity (and from now on), we have omitted the arguments (t, y, w).

Inserting the optimal strategy π^* into the HJB-PDE (7.27) yields

$$(\partial_t + \mathcal{A}^Y)u + \mathcal{N}(u) = 0, \tag{7.29}$$

where $\mathcal{N}(u)$ is a nonlinear term:

$$\mathcal{N}(u) = -\frac{1}{2}\lambda^2 \frac{(\partial_w u)^2}{\partial_w^2 u} - \rho\beta\lambda \frac{(\partial_w u)(\partial_y \partial_w u)}{\partial_w^2 u} - \frac{1}{2}\rho^2\beta^2 \frac{(\partial_y \partial_w u)^2}{\partial_w^2 u}.$$

Here, we have introduced the *Sharpe ratio* $\lambda(y) := \mu(y)/\sigma(y)$.

7.3.2 Asymptotic Approximation

For general $\{\beta, c, \lambda\}$, there is no closed-form solution of (7.29). Hence, we seek an asymptotic approximation for u. To this end, using equation (7.22) from Section 7.2.5.1 as a guide, we expand the coefficients in (7.29) in a Taylor series about an arbitrary point $\bar{y}$. Specifically, for any function $\chi : \mathbb{R} \to \mathbb{R}$, we may formally write

$$\chi(y) = \sum_{n=0}^{\infty} \varepsilon^n \chi_n(y), \qquad \chi_n(y) := \frac{1}{n!}\partial_y^n \chi(\bar{y})\cdot(y - \bar{y})^n, \qquad \varepsilon = 1, \tag{7.30}$$

where we have once again introduced ε for purposes of accounting. We also expand the function u as a power series in ε

$$u = \sum_{n=0}^{\infty} \varepsilon^n u_n, \qquad \varepsilon = 1. \tag{7.31}$$

Now, for each group of coefficients appearing in (7.29), we insert an expansion of the form (7.30), and we define

$$\mathcal{A}_n := c_n \partial_y + (\tfrac{1}{2}\beta^2)_n \partial_y^2, \qquad n \in \{0\} \cup \mathbb{N}. \tag{7.32}$$

We also insert into (7.29) our expansion (7.31) for u.

Next, collecting terms of like powers of ε, we obtain at lowest order

$$(\partial_t + \mathcal{A}_0)u_0 - (\frac{1}{2}\lambda^2)_0 \frac{(\partial_w u_0)^2}{\partial_w^2 u_0} - (\rho\beta\lambda)_0 \frac{(\partial_w u_0)(\partial_y \partial_w u_0)}{\partial_w^2 u_0} - (\frac{1}{2}\rho^2\beta^2)_0 \frac{(\partial_y \partial_w u_0)^2}{\partial_w^2 u_0} = 0,$$

with $u_0(T, y, w) = U(w)$. We can look for a solution $u_0 = u(t, w)$ that is independent of y, and then we have

$$\partial_t u_0 - (\frac{1}{2}\lambda^2)_0 \frac{(\partial_w u_0)^2}{\partial_w^2 u_0} = 0, \qquad u_0(T, w) = U(w). \tag{7.33}$$

We observe that (7.33) is the same *nonlinear* PDE problem that arises when one considers an underlying that has a constant drift $\mu_0 = \mu(\bar{y})$, diffusion coefficient $\sigma_0 = \sigma(\bar{y})$, and Sharpe ratio $\lambda_0 = \lambda(\bar{y}) = \mu(\bar{y})/\sigma(\bar{y})$.

It is convenient to define the *risk-tolerance* function

$$R_0 := \frac{-\partial_w u_0}{\partial_w^2 u_0},$$

and the operators

$$\mathcal{D}_k = R_0^k \partial_w^k, \quad k = 1, 2, \cdots .$$

We now proceed to the order $\mathcal{O}(\varepsilon)$ terms. Using $u_0 = u_0(t, w)$ and (7.33), we obtain

$$(\partial_t + \mathcal{A}_0)u_1 + (\frac{1}{2}\lambda_0^2)\mathcal{D}_2 u_1 + \lambda_0^2 \mathcal{D}_1 u_1 + (\rho\beta\lambda)_0 \mathcal{D}_1 \partial_y u_1 = -(\frac{1}{2}\lambda^2)_1 \mathcal{D}_1 u_0, \tag{7.34}$$

$$u_1(T, y, w) = 0, \tag{7.35}$$

which is a *linear* PDE problem for u_1.

We can rewrite equations (7.34)–(7.35) more compactly as

$$(\partial_t + \mathcal{A}_0 + \mathcal{B}_0(t))u_1 + H_1 = 0, \quad u_1(T, y, w) = 0, \tag{7.36}$$

where the linear operator $\mathcal{B}(t)$ and the source term H_1 are given by

$$\mathcal{B}_0(t) = \tfrac{1}{2}\lambda_0^2 \mathcal{D}_2 + \lambda_0^2 \mathcal{D}_1 + (\rho\beta\lambda)_0 \mathcal{D}_1 \partial_y, \quad H_1 = (\tfrac{1}{2}\lambda^2)_1 R_0 \partial_w u_0.$$

Observe that (7.36) is a linear PDE for u_1.

The following change of variables (see Fouque *et al.*, 2013) will be useful for solving the PDE problem (7.36). Define

$$u_1(t, y, w) = q_1(t, y, z(t, w)), \quad z(t, w) = -\log \partial_w u_0(t, w) + \tfrac{1}{2}\lambda_0^2(T - t). \tag{7.37}$$

Inserting (7.37) into (7.36), we find that q_1 satisfies

$$0 = (\partial_t + \mathcal{A}_0 + \mathcal{C}_0)q_1 + Q_1, \quad q_1(T, y, z) = 0, \tag{7.38}$$

where the operator $\mathcal{C}_0$ is given by

$$\mathcal{C}_0 = \frac{1}{2}\lambda_0^2 \partial_z^2 + (\rho\beta\lambda)_0 \partial_y \partial_z, \tag{7.39}$$

and the function Q_1 satisfies $H_1(t, y, w) = Q_1(t, y, z(t, w))$.

Now, from (7.32) and (7.39), we observe that the operator $(\mathcal{A}_0 + \mathcal{C}_0)$ is the infinitesimal generator of a diffusion in $\mathbb{R}^2$ whose drift vector and covariance matrix are constant. The *semigroup* $\mathcal{P}_0(t, t')$ generated by $(\mathcal{A}_0 + \mathcal{C}_0)$ is given by

$$\mathcal{P}_0(t, T)G(y, z) := \int_{\mathbb{R}^2} \mathrm{d}\eta \, \mathrm{d}\zeta \; \Gamma_0(t, y, z; T, \eta, \zeta) G(\eta, \zeta),$$

where Γ_0, the *fundamental solution* corresponding to $(\partial_t + \mathcal{A}_0 + \mathcal{C}_0)$, is a Gaussian kernel:

$$\Gamma_0(t, y, z; T, \eta, \zeta) = \frac{1}{\sqrt{(2\pi)^3 |\mathbf{C}|}} \exp\left(-\frac{1}{2}\mathbf{m}^{\mathrm{T}}\mathbf{C}^{-1}\mathbf{m}\right),$$

with covariance matrix $\mathbf{C}$ and vector $\mathbf{m}$ given by

$$\mathbf{C} = (T - t)\begin{pmatrix} (\beta^2)_0 & (\rho\beta\lambda)_0 \\ (\rho\beta\lambda)_0 & (\lambda^2)_0 \end{pmatrix}, \qquad \mathbf{m} = \begin{pmatrix} \eta - y - (T - t)c_0 \\ \zeta - z \end{pmatrix}.$$

By Duhamel's principle, the unique classical solution to (7.38) is given by

$$q_1(t) = \int_t^T \mathrm{d}s \; \mathcal{P}_0(t, s) Q_1(s),$$

In the case of a general utility function, (7.33) is easily solved numerically, for instance by solving the fast diffusion (or Black's) equation for the risk tolerance function R_0 (see Fouque *et al.*, 2013). Then, u_1 can also be computed numerically using the formulas above. In the case of power utility, there are explicit formulas, as given in Section 7.3.3.

Having obtained an approximation for the value function $u \approx u_0 + u_1$, we now seek an expansion for the optimal control $\pi^* \approx \pi_0^* + \pi_1^*$. Inserting the expansion (7.30) of the coefficients and the expansion (7.31) for u into (7.28), and collecting terms of like powers of ε, we obtain

$$\mathcal{O}(1): \qquad \pi_0 = -\frac{\mu_0(\partial_w u_0)}{(\sigma^2)_0(\partial_w^2 u_0)}, \tag{7.40}$$

$$\mathcal{O}(\varepsilon): \qquad \pi_1 = -\pi_0\frac{(\sigma^2)_1}{(\sigma^2)_0} - \pi_0\frac{(\partial_w^2 u_1)}{(\partial_w^2 u_0)} - \mu_1\frac{(\partial_w u_0)}{(\sigma^2)_0(\partial_w^2 u_0)}$$
$$- \mu_0\frac{(\partial_w u_1)}{(\sigma^2)_0(\partial_w^2 u_0)} - (\rho\beta\sigma)_0\frac{(\partial_y\partial_w u_1)}{(\sigma^2)_0(\partial_w^2 u_0)}. \tag{7.41}$$

Higher order terms for both the value function u and the optimal control π^* can be obtained in the same manner as u_1 and π_1. Analysis of the asymptotic formulas for different utility functions and stochastic volatility models is presented in more detail in Lorig and Sircar (2015).

7.3.3 *Power Utility*

Finally, we consider a utility function U from the constant relative risk aversion (CRRA), or power family:

$$\text{CRRA utility:} \qquad U(w) := \frac{w^{1-\gamma}}{1-\gamma}, w > 0, \qquad \gamma > 0, \gamma \neq 1,$$

where γ is called the risk aversion coefficient. Here, all the quantities above can be computed explicitly.

The explicit solution u_0 to (7.33) is

$$u_0(t,w) = U(w)\exp\left(\frac{1}{2}\lambda_0^2\left(\frac{1-\gamma}{\gamma}\right)(T-t)\right).$$

The risk-tolerance function is $R_0 = \frac{w}{\gamma}$, and the transformation in (7.37) is then

$$z(t,w) = \gamma w + (T-t)\left(\frac{2\gamma-1}{\gamma}\right)\frac{1}{2}\lambda_0^2.$$

An explicit computation reveals that u_1 is given by

$$u_1(t,y,w) = q_1(t,y,z(t,w))$$
$$= \frac{1-\gamma}{\gamma}u_0(t,w)\left(\frac{1}{2}\lambda^2(\bar{y})\right)'\left((T-t)(y-\bar{y}) + \frac{1}{2}(T-t)^2\left(c_0 + \frac{1-\gamma}{\gamma}\rho\beta_0\lambda_0\right)\right).$$

For the specific case $\bar{y} = y$, the above expression simplifies to

$$u_1(t,y,w) = \frac{1-\gamma}{\gamma} u_0(t,w) \left(\frac{1}{2}\lambda^2(y) \right)' \left(\frac{1}{2}(T-t)^2 \left(c_0 + \frac{1-\gamma}{\gamma} \rho \beta_0 \lambda_0 \right) \right).$$

Using these explicit representations of u_0 and u_1, the expressions (7.40) and (7.41) for the optimal stockholding approximations become

$$\pi_0^* = \frac{\mu_0}{\gamma \sigma_0^2},$$

$$\pi_1^*(t,y) = (y-\bar{y}) \left(\frac{\mu'(\bar{y})}{\gamma \sigma_0^2} - \frac{\mu_0}{\gamma \sigma_0^4} (\sigma^2(\bar{y}))' \right) + \frac{(1-\gamma)(T-t)}{\gamma \sigma_0} \left(\rho \beta_0 \frac{1}{\gamma} \left(\frac{1}{2}\lambda^2(\bar{y}) \right)' \right).$$

For the specific case $\bar{y} = y$, the formula for π_1^* simplifies to

$$\pi_1^*(t,y) = \frac{(1-\gamma)(T-t)}{\gamma \sigma_0} \rho \beta_0 \frac{1}{\gamma} \left(\frac{1}{2}\lambda^2(y) \right)'.$$

7.4 Conclusions

Asymptotic methods can be used to analyze and simplify pricing and portfolio optimization problems, and we have presented some examples and methodologies. A key insight is to perturb problems with stochastic volatility around their constant volatility counterparts to obtain the principle effect of volatility uncertainty.

These approaches reduce the dimension of the effective problems that have to be solved, and often lead to explicit formulas that can be analyzed for intuition. In the context of portfolio problems accounting for stochastic volatility, recent progress has been made in cases where there are transaction costs (Bichuch and Sircar, 2014, Kallsen and Muhle-Karbe, 2013) or stochastic risk aversion that varies with market conditions (Dong and Sircar, 2014), and under more complex local-stochastic volatility models (Lorig and Sircar, 2015).

Acknowledgements

RL's work is partially supported by NSF grant DMS-0739195. RS's work is partially supported by NSF grant DMS-1211906.

References

Aït-Sahalia, Y. and Jacod, J. (2014). *High-frequency financial econometrics*. Princeton: Princeton University Press.

Bakshi, G., Cao, C. and Chen, Z. (1997). Empirical performance of alternative option pricing models. *Journal of Finance*, 52 (5), 2003–2049.

Berestycki, H., Busca, J. and Florent, I. (2004). Computing the implied volatility in stochastic volatility models. *Communications on Pure and Applied Mathematics*, 57 (10), 1352–1373.

Bichuch, M. and Sircar, R. (2014). Optimal investment with transaction costs and stochastic volatility. Submitted.

Black, F. and Scholes, M. (1972). The valuation of option contracts and a test of market efficiency. *Journal of Finance*, 27, 399–417.

Black, F. and Scholes, M. (1973). The pricing of options and corporate liabilities. *The Journal of Political Economy*, 81(3), 637–654.

Dong, Y. and Sircar, R. (2014). Time-inconsistent portfolio investment problems. In *Stochastic analysis and applications 2014 – in honour of Terry Lyons* (ed. Crisan, D. Hambly, B. and Zariphopoulou T.). Berlin: Springer.

Dupire, B. (1994). Pricing with a smile. *Risk*, 7 (1), 18–20.

Engle, R. (1982). Autoregressive conditional heteroscedasticity with estimates of the variance of United Kingdom inflation. *Econometrica*, 50 (4), 987–1008.

Fouque, J.-P., Papanicolaou, G. and Sircar, R. (2000). *Derivatives in financial markets with stochastic volatility*. Cambridge: Cambridge University Press.

Fouque, J.-P., Papanicolaou, G., Sircar, R. and Solna, K. (2011). *Multiscale stochastic volatility for equity, interest rate, and credit derivatives*. Cambridge: Cambridge University Press.

Fouque, J.-P., Sircar, R. and Zariphopoulou, T. (2013). Portfolio optimization & stochastic volatility asymptotics. *Mathematical Finance*, forthcoming.

Gulisashvili, A. (2012). *Analytically tractable stochastic stock price models*. New York: Springer.

Heston, S. (1993). A closed-form solution for options with stochastic volatility with applications to bond and currency options. *Review of Financial Studies*, 6 (2), 327–343.

Hull, J. and White, A. (1987). The pricing of options on assets with stochastic volatilities. *Journal of Finance*, 42 (2), 281–300.

Jonsson, M. and Sircar, R. (2002a). Optimal investment problems and volatility homogenization approximations. In *Modern methods in scientific computing and applications* (ed. Bourlioux, A., Gander, M. and Sabidussi, G.), NATO Science Series II vol. 75. New York: Kluwer, pp. 255–281.

Jonsson, M. and Sircar, R. (2002b). Partial hedging in a stochastic volatility environment. *Mathematical Finance*, 12 (4), 375–409.

Kallsen, J. and Muhle-Karbe, J. (2013). The general structure of optimal investment and consumption with small transaction costs. Submitted.

Lee, R.W. (2004). The moment formula for implied volatility at extreme strikes. *Mathematical Finance*, 14 (3), 469–480.

Lewis, A. (2000). *Option valuation under stochastic volatility*. Newport Beach, CA: Finance Press.

Lorig, M. and Lozano-Carbassé, O. (2015). Multiscale exponential Lévy models. *Quantitative Finance*, 15 (1), 91–100.

Lorig, M., Pagliarani, S. and Pascucci, A. (2014). A Taylor series approach to pricing and implied vol for LSV models. *Journal of Risk*, 17, 1–17.

Lorig, M., Pagliarani, S. and Pascucci, A. (2015a). Analytical expansions for parabolic equations. *SIAM Journal on Applied Mathematics*, 75, 468–491.

Lorig, M., Pagliarani, S. and Pascucci, A. (2015b). Explicit implied volatilities for multifactor local-stochastic volatility models. *Mathematical Finance*, forthcoming.

Lorig, M., Pagliarani, S. and Pascucci, A. (2015c). A family of density expansions for Lévy-type processes with default. *Annals of Applied Probability*, 25 (1), 235–267.

Lorig, M., Pagliarani, S. and Pascucci, A. (2015d). Pricing approximations and error estimates for local Lévy-type models with default. *Computers and Mathematics with Applications*, forthcoming.

Lorig, M. and Sircar, R. (2015). Portfolio optimization under local-stochastic volatility: coefficient Taylor series approximations & implied Sharpe ratio. Submitted.

Merton, R. (1976). Option pricing when underlying stock returns are discontinuous. *Journal of Financial Economics*, 3 (1), 125–144.

Merton, R. (1992). *Continuous-time finance*. Oxford: Blackwell.

Merton, R.C. (1969). Lifetime portfolio selection under uncertainty: the continuous-time case. *Review of Economics and Statistics*, 51 (3), 247–257.

Merton, R.C. (1971). Optimum consumption and portfolio rules in a continuous-time model. *Journal of Economic Theory*, 3 (1–2), 373–413.

Nelson, D. (1990). ARCH models as diffusion approximations. *Journal of Econometrics*, 45 (1–2), 7–38.

Nelson, D. (1991). Conditional heteroskedasticity in asset returns: a new approach. *Econometrica*, 59 (2), 347–370.

Scott, L. (1987). Option pricing when the variance changes randomly: theory, estimation, and an application. *Journal of Financial and Quantitative Analysis*, 22 (4), 419–438.

Sircar, R. and T. Zariphopoulou (2005). Bounds and asymptotic approximations for utility prices when volatility is random. *SIAM Journal of Control and Optimization*, 43 (4), 1328–1353.

Tehranchi, M.R. (2009). Asymptotics of implied volatility far from maturity. *Journal of Applied Probability*, 629–650.

Wiggins, J. (1987). Option values under stochastic volatility. *Journal of Financial Economics*, 19 (2), 351–372.

8

Statistical Measures of Dependence for Financial Data

David S. Matteson, Nicholas A. James and William B. Nicholson
Cornell University, USA

8.1 Introduction

The analysis of financial and econometric data is typified by non-Gaussian multivariate observations that exhibit complex dependencies: heavy-tailed and skewed marginal distributions are commonly encountered; serial dependence, such as autocorrelation and conditional heteroscedasticity, appear in time-ordered sequences; and nonlinear, higher-order, and tail dependence are widespread. Illustrations of serial dependence, nonnormality, and nonlinear dependence are shown in Figure 8.1.

When data are assumed to be jointly Gaussian, all dependence is linear, and therefore only pairwise among the variables. In this setting, Pearson's product-moment correlation coefficient uniquely characterizes the sign and strength of any such dependence.

Definition 8.1 *For random variables X and Y with joint density* f_{XY}, **Pearson's correlation coefficient** *is defined as*

$$\rho_P(X,Y) = E\left[\frac{(X-\mu_X)}{\sqrt{\sigma_X^2}}\frac{(Y-\mu_Y)}{\sqrt{\sigma_Y^2}}\right] = \int\int \frac{(x-\mu_X)}{\sqrt{\sigma_X^2}}\frac{(y-\mu_Y)}{\sqrt{\sigma_Y^2}} f_{XY}(x,y)\,dx\,dy,$$

where $\mu_X = E(X) = \int\int x f_{XY}(x,y)\,dx\,dy$, *and* $\sigma_X^2 = E[(X-\mu_X)^2] = \int\int (x-\mu_X)^2 f_{XY}(x,y)\,dx\,dy$, *are the mean and variance of X, respectively, and* μ_Y *and* σ_Y^2 *are defined similarly.*

This conventional measure of pairwise linear association is well-defined provided σ_X^2 and σ_Y^2 are positive and finite, in which case $\rho_P(X,Y) \in [-1,+1]$. A value of -1 or $+1$ indicates

Financial Signal Processing and Machine Learning, First Edition.
Edited by Ali N. Akansu, Sanjeev R. Kulkarni and Dmitry Malioutov.
© 2016 John Wiley & Sons, Ltd. Published 2016 by John Wiley & Sons, Ltd.

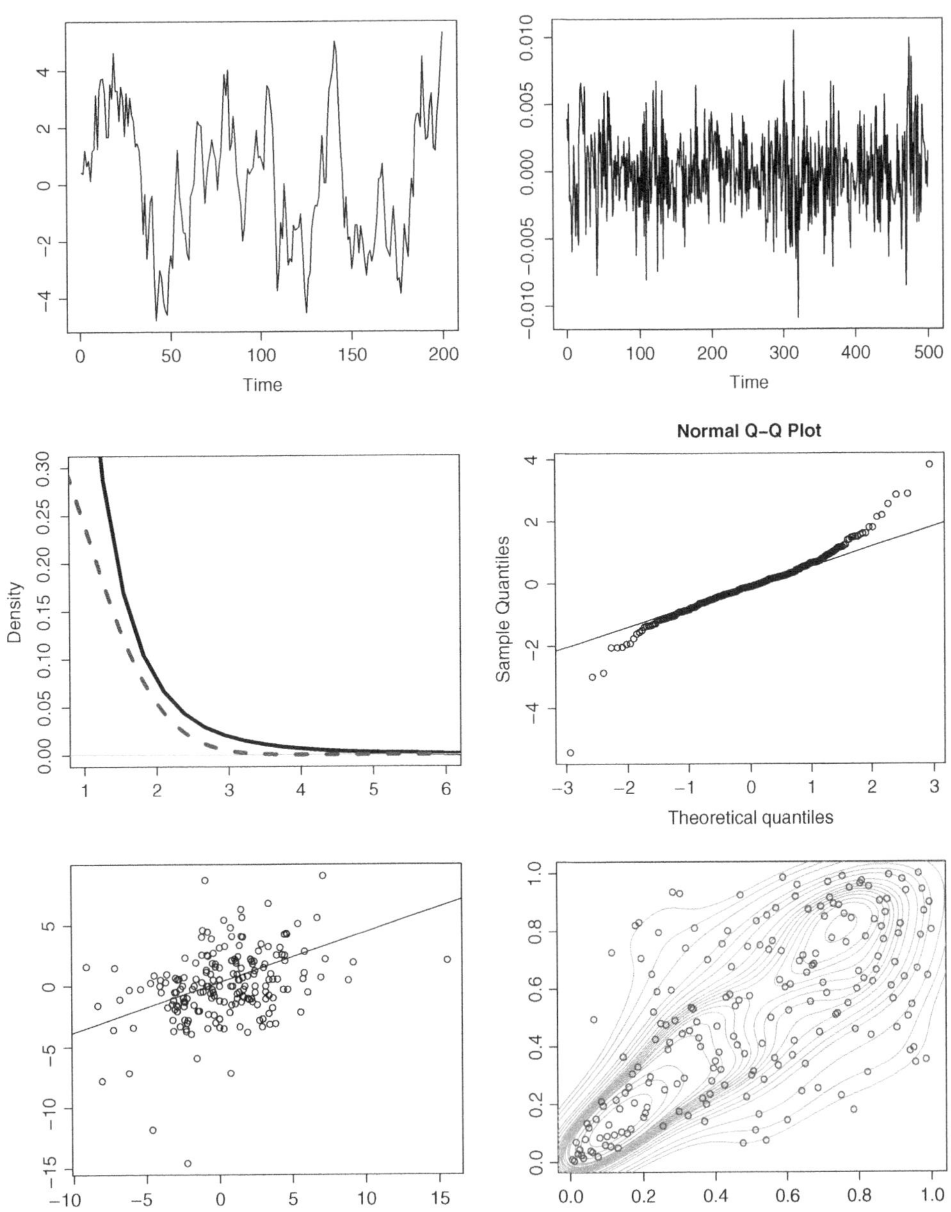

Figure 8.1 Top left: Strong and persistent positive autocorrelation, that is, persistence in local level; top right: moderate volatility clustering, that is, i.e., persistence in local variation. Middle left: Right tail density estimates of Gaussian versus heavy- or thick-tailed data; middle right: sample quantiles of heavy-tailed data versus the corresponding quantiles of the Gaussian distribution. Bottom left: Linear regression line fit to non-Gaussian data; right: corresponding estimated density contours of the normalized sample ranks, which show a positive association that is stronger in the lower left quadrant compared to the upper right.

perfect negative or positive linear dependence, respectively, and in either case X and Y have an exact linear relationship. Negative and positive values indicate negative and positive linear associations, respectively, while a value of 0 indicates no linear dependence.

A sample estimator $\hat{\rho}_P$ is typically defined by replacing expectations with empirical expectations in the above definition. For a paired random sample of n observations $(\boldsymbol{X}_{1:n}, \boldsymbol{Y}_{1:n}) = \{(x_i, y_i)\}_{i=1}^n$, define

$$\hat{\rho}_P(\boldsymbol{X}_{1:n}, \boldsymbol{Y}_{1:n}) = \frac{1}{n-1} \sum_{i=1}^{n} \frac{(x_i - \hat{\mu}_X)}{\sqrt{\hat{\sigma}_X^2}} \frac{(y_i - \hat{\mu}_Y)}{\sqrt{\hat{\sigma}_Y^2}} \; : \; \hat{\mu}_X = \frac{1}{n} \sum_{i=1}^{n} x_i; \; \hat{\sigma}_X^2 = \frac{1}{n-1} \sum_{i=1}^{n} (x_i - \hat{\mu}_X)^2,$$

where $\hat{\mu}_Y$ and $\hat{\sigma}_Y^2$ are defined similarly. For jointly Gaussian variables (X, Y), zero correlation is equivalent to independence. For an independent and identically distributed (i.i.d.) sample from the bivariate normal distribution, inference regarding ρ_P can be conducted using the following asymptotic approximation:

$$\sqrt{n}(\hat{\rho}_P - \rho_P) \xrightarrow{\mathcal{D}} \mathcal{N}[0, (1 - \rho_P^2)^2],$$

in which $\xrightarrow{\mathcal{D}}$ denotes convergence in distribution. Under the null hypothesis of zero correlation, this expression simplifies to $\sqrt{n}\hat{\rho}_P \xrightarrow{\mathcal{D}} \mathcal{N}(0, 1)$. More generally, for an i.i.d sample from an arbitrary distribution with finite fourth moments, $E(X^4)$ and $E(Y^4)$, a variance-stabilizing transformation may be applied (Fisher's transformation) to obtain the alternative asymptotic approximation (cf. Ferguson, 1996)

$$\frac{\sqrt{n}}{2} \left(\log \frac{1 + \hat{\rho}_P}{1 - \hat{\rho}_P} - \log \frac{1 + \rho_P}{1 - \rho_P} \right) \xrightarrow{\mathcal{D}} \mathcal{N}(0, 1).$$

Although the previous approximation is quite general, assuming finite fourth moments may be unreasonable in numerous financial applications where extreme events are common. Furthermore, Pearson's correlation coefficient measures the strength of linear relationships only. There are many situations in which correlations are zero but a strong nonlinear relationship exists, such that variables are highly dependent. In Section 8.2, we discuss several robust measures of correlation and pairwise association, and illustrate their application in measuring serial dependence in time-ordered data. In Section 8.3, we consider multivariate extensions and Granger causality, and introduce measures of mutual independence. Finally, in Section 8.4 we explore copulas and their financial applications.

8.2 Robust Measures of Correlation and Autocorrelation

Financial data are often time-ordered, and intertemporal dependence is commonplace. The autocovariance and autocorrelation functions are extensions of covariance and Pearson's correlation to a time-dependent setting, respectively.

Definition 8.2 *For a univariate ordered sequence of random variables $\{X_t\}$, the* **autocovariance** γ *and* **autocorrelation** ρ *at indices q and r (Shumway and Stoffer, 2011) are defined as*

$$\gamma(X_q, X_r) = \gamma(q, r) = E[(X_q - \mu_q)(X_r - \mu_r)]$$

and

$$\rho_P(X_q, X_r) = \rho_P(q, r) = \frac{\gamma(q, r)}{\sqrt{\gamma(q, q)}\sqrt{\gamma(r, r)}},$$

respectively, in which $\mu_q = E(X_q)$ and $\mu_r = E(X_r)$. The above quantities are well-defined provided $E(X_t^2)$ is finite for all t; however, estimating these quantities from an observed sequence $\boldsymbol{X}_{1:n} = \{x_t\}_{t=1}^n$ requires either multiple i.i.d. realizations of the entire sequence (uncommon in finance), or some additional assumptions. The first basic assumption is that the observations are equally spaced, and t denotes their discrete-time index. We will refer to such sequences generically as *time-series*. In finance, this assumption may only hold approximately. For example, a sequence of daily market closing asset prices may only be available for weekdays, with additional gaps on holidays, or intraday asset transaction prices may be reported every hundredth of a second, but there may be no transaction at many of these times. In either case, the consecutive observations are commonly regarded as equally spaced, for simplicity.

The next basic assumption is some form of distributional invariance over time, such as stationarity.

Definition 8.3 *A univariate sequence of random variables* $\{X_t\}$ *is* **weakly** *(or* **covariance***)* **stationary** *if and only if*

$$E(X_t) = E(X_{t-h}) = \mu \quad \text{and} \quad \gamma(t, t-h) = \gamma(|h|) = \gamma_h \quad \forall t, h, \text{ and} \quad \gamma_0 < \infty.$$

This implies that the means and variances are finite and constant, and the autocovariance is constant with respect to t, and only depends on the relative time lag h between observations.

For any k-tuple of indices $t_1, t_2, \ldots, t_k$*, let* $F_{t_1,t_2,\ldots,t_k}(\cdot)$ *denote the joint distribution function of* $(X_{t_1}, X_{t_2}, \ldots, X_{t_k})$*. Then, the sequence is* **strictly stationary** *if and only if*

$$F_{t_1,t_2,\ldots,t_k}(\cdot) = F_{t_1-h,t_2-h,\ldots,t_k-h}(\cdot) \quad \forall h, k, \text{ and } \forall t_1, t_2, \ldots, t_k.$$

This implies that the joint distributions of all k-tuples are invariant to a common time shift h such that their relative time lags remain constant.

Strict stationary implies weak stationarity provided the variance is also finite.

Now, under the weak stationarity assumption, the parameters $\gamma_h = \gamma(h)$ for $h = 0, 1, 2, \ldots$ denote the autocovariance function of $\{X_t\}$ with respect to the time lag h, and the corresponding autocorrelation function (ACF) is defined as $\rho_P(h) = \gamma_h/\gamma_0$. Under the weak stationarity assumption, the joint distribution of the random variables $(X_1, \ldots, X_n)$ has mean vector $\boldsymbol{\mu}_X = \mathbf{1}\mu$, where $\mathbf{1}$ denotes a length n vector of ones, and a symmetric Toeplitz covariance matrix Σ_X, with $[\Sigma_X]_{i,j} = \gamma(|i-j|)$. Furthermore, both Σ_X and the corresponding correlation matrix Ω_X are positive definite for any stationary sequence. For an observed stationary time-series $\boldsymbol{X}_{1:n} = \{x_t\}_{t=1}^n$, the mean is estimated as before ($\hat{\mu}_X$), while autocovariances and autocorrelations are commonly estimated as $\hat{\gamma}(\boldsymbol{X}_{1:n}; h) = \hat{\gamma}(h) = \frac{1}{n}\sum_{t=h+1}^n (x_t - \hat{\mu}_X)(x_{t-h} - \hat{\mu}_X)$ and $\hat{\rho}_P(\boldsymbol{X}_{1:n}; h) = \hat{\rho}_P(h) = \hat{\gamma}(h)/\hat{\gamma}(0)$, respectively. Using the scaling $\frac{1}{n}$ as opposed to $\frac{1}{n-h}$ assures that the corresponding estimated covariance and correlation matrices $[\hat{\Sigma}_X]_{i,j} = \hat{\gamma}(|i-j|)$ and $[\hat{\Omega}_X]_{i,j} = \hat{\rho}_P(|i-j|)$ are both positive definite (McLeod and Jimenéz, 1984).

8.2.1 Transformations and Rank-Based Methods

Pearson's correlation and the autocorrelation function are commonly interpreted under an implicit joint normality assumption on the data. The alternative measures discussed in this subsection offer robustness to outlying and extreme observations and consider nonlinear dependencies, all of which are common in financial data.

8.2.1.1 Huber-type Correlations

Transformations may be applied to define a robust correlation measure between pairs of random variables (X, Y). For example, let μ_R and σ_R denote robust location and scale parameters, such as the trimmed mean and trimmed standard deviation. And let ψ denote a bounded monotone function, such as $\psi(x;k) = xI_{|x|\leq k} + \text{sgn}(x)kI_{|x|>k}$ (cf. Huber, 1981), where k is some positive constant and I_A is the indicator function of an event A. Then, a robust covariance and correlation may be defined as

$$\gamma_R(X,Y) = \sigma_R(X)\sigma_R(Y)E\left[\psi\left(\frac{X-\mu_R(X)}{\sigma_R(X)}\right)\psi\left(\frac{Y-\mu_R(Y)}{\sigma_R(Y)}\right)\right]$$

and

$$\rho_R(X,Y) = \frac{\gamma_R(X,Y)}{\sqrt{\gamma_R(X,X)}\sqrt{\gamma_R(Y,Y)}},$$

respectively (cf. Maronna *et al.*, 2006). Although these measures are robust to outlying and extreme observations, they depend on the choice of transformation ψ. They provide an intuitive measure of association, however, for an arbitrary joint distribution on (X, Y), $\rho_R(X,Y) \neq \rho_P(X,Y)$, in general. Sample versions are obtained by replacing the expectation, μ_R and σ_R, by their sample estimates in the above expressions. Asymptotic sampling distributions can be derived, but the setting is more complicated than above. Finally, robust pairwise covariances and correlations such as these may be used to define corresponding covariance and correlation matrices for random vectors, but the result is not positive definite (or affine equivariant), in general.

8.2.1.2 Kendall's Tau

Rank-based methods measure all monotonic relationships and are resistant to outliers. Kendall's tau (Kendall, 1938) is a nonparametric measure of dependence, for which the sample estimate considers the pairwise agreement between two ranked lists.

Definition 8.4 *For random variables X and Y,* **Kendall's tau** *is defined as*

$$\begin{aligned}\tau(X,Y) &= P[(X-X^*)(Y-Y^*)>0] - P[(X-X^*)(Y-Y^*)<0]\\ &= E\{sgn[(X-X^*)(Y-Y^*)]\}, \qquad (8.1)\end{aligned}$$

in which (X^*, Y^*) *denotes an i.i.d. copy of* (X, Y).

A sample estimator $\hat{\tau}$ considers the pairwise agreement between two ranked lists. For a paired random sample of n observations $(\boldsymbol{X}_{1:n}, \boldsymbol{Y}_{1:n}) = \{(x_i, y_i)\}_{i=1}^{n}$, the sample analog of (8.1) is defined as

$$\hat{\tau}(\boldsymbol{X}_{1:n}, \boldsymbol{Y}_{1:n}) = \binom{n}{2}^{-1} \sum_{j=2}^{n} \sum_{i=1}^{j-1} \mathrm{sgn}(x_j - x_i)\mathrm{sgn}(y_j - y_i),$$

and, in the absence of ties, is also equal to

$$\frac{\text{number of concordant pairs} - \text{number of discordant pairs}}{\text{number of concordant pairs} + \text{number of discordant pairs}},$$

in which a pair of observations $(x_j, y_j), (x_i, y_i)$ are concordant if $\mathrm{sgn}(x_j - x_i) = \mathrm{sgn}(y_j - y_i)$, and discordant otherwise. These definitions may also be extended to define analogous autocorrelation functions.

Definition 8.5 *The* **ACF based on Kendall's tau** *(Bingham and Schmidt, 2006) for a stationary sequence of random variables $\{X_t\}$ is defined as*

$$\begin{aligned} \tau(X_t; h) = \tau_h &= P[(X_t - X_t^*)(X_{t-h} - X_{t-h}^*) > 0] - P[(X_t - X_t^*)(X_{t-h} - X_{t-h}^*) < 0] \\ &= E\{sgn[(X_t - X_t^*)(X_{t-h} - X_{t-h}^*)]\}, \end{aligned} \tag{8.2}$$

in which (X_t^*, X_{t-h}^*) denotes an i.d.d. copy of (X_t, X_{t-h}).

For an observed time-series $\boldsymbol{X}_{1:n} = \{x_t\}_{t=1}^{n}$, the sample analog of (8.2) is defined as

$$\hat{\tau}(\boldsymbol{X}_{1:n}; h) = \hat{\tau}_h = \binom{n-h}{2}^{-1} \sum_{t=2}^{n-h} \sum_{i=1}^{t-1} \mathrm{sgn}(x_t - x_i)\mathrm{sgn}(x_{t+h} - x_{i+h}),$$

which is not assured to have a corresponding estimated correlation matrix that is positive definite.

In certain applications, Kendall's tau has a distinct advantage over Pearson's correlation in that it is invariant to increasing monotonic transformations of one or both variables. For example, $\tau(X, Y) = \tau(X, \log\ Y) = \tau(\log\ X, \log\ Y)$, for positive random variables, and similarly for $\hat{\tau}$. Bingham and Schmidt (2006) consider an example based on a time-series of squared asset returns, which exhibits abnormally high dependence at a specific lag. This is exaggerated in a Pearson ACF, as it only measures linear dependence. After applying a data standardization, the magnitude of the Pearson ACF at this lag decreases substantially, whereas the ACF based on Kendall's tau is very similar under both specifications.

Figure 8.2 offers an example of a scenario in which the autocorrelation based on Kendall's tau differs substantially from that produced from a Pearson ACF. It depicts the Bank of America (BOA) stock price and the estimated ACF of the squared daily stock returns from June 2008 to January 2009, a very volatile period that included the failure of Lehman Brothers, Washington Mutual, and several other financial institutions. The ACF based on Kendall's tau detects persistent serial dependence across several lags, while the Pearson ACF detects dependence only at the first and 10th lags. This discrepancy suggests that the underlying serial dependence is highly nonlinear and non-Gaussian in nature.

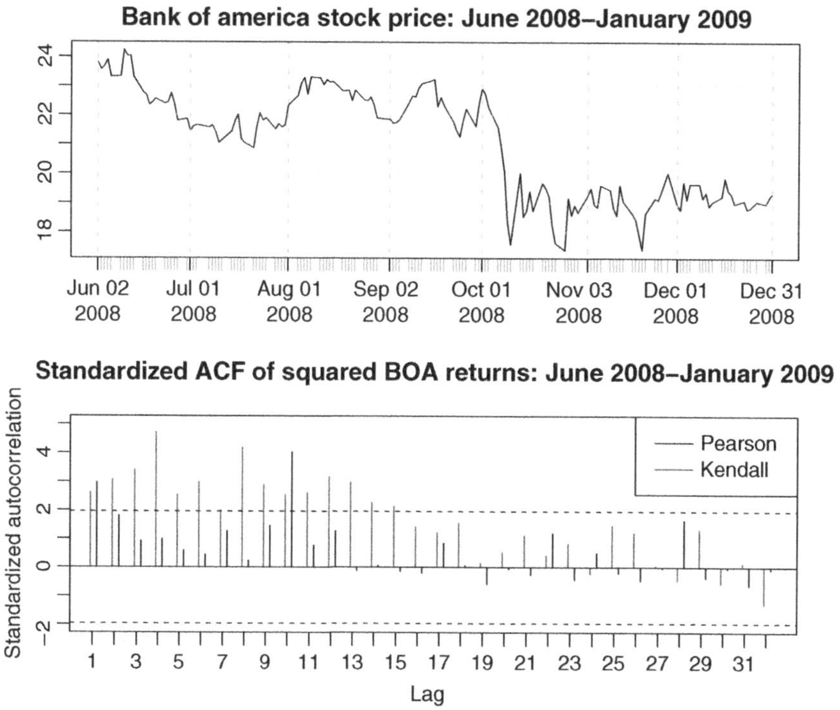

Figure 8.2 Bank of America (BOA) daily closing stock price. Bottom: Standardized (Fisher's transformation) ACF based on Kendall's tau and Pearson's correlation coefficient for the squared daily stock returns.

8.2.1.3 Spearman's Rho

One of the shortcomings of Kendall's tau, as discussed in Helsel and Hirsch (1992), is that since it is purely based on sign, the magnitude of concordance or discordance is not taken into account. Spearman's rho (Spearman, 1904), which computes Pearson's correlation on the normalized ranks in the sample case, does explicitly take magnitude into account. Though Kendall's tau and Spearman's rho are asymptotically equivalent when testing for independence, Spearman's rho can be preferable in certain scenarios. For example, Xu *et al.* (2013) conduct a simulation study with bivariate contaminated normal data and conclude that Spearman's rho is preferable to Kendall's tau in small sample sizes when the population correlation is moderate.

Definition 8.6 *The* **ACF based on Spearman's rho** *(Hollander and Wolfe, 1999) for a stationary sequence of random variables* $\{X_t\}$ *with common marginal distribution* F *is defined as*

$$\rho_S(X_t; h) = \rho_S(h) = \rho_P[F(X_t), F(X_{t-h})],$$

that is, the Pearson correlation of $F(X_t)$ *and* $F(X_{t-h})$.

For an observed time-series $X_{1:n} = \{x_t\}_{t=1}^n$, a sample version may be defined as

$$\hat{\rho}_S(X_{1:n}; h) = \hat{\rho}_S(h) = \frac{12}{n(n^2-1)} \sum_{t=1+h}^{n} \left(r_t - \frac{n+1}{2}\right)\left(r_{t-h} - \frac{n+1}{2}\right),$$

in which $r_t = rank\{x_t : x_t \in X_{1:n}\}$.

8.2.2 *Inference*

8.2.2.1 Kendall's Tau

We can test the hypothesis of zero correlation, that is,

$$H_0 : \tau_h = 0 \quad vs. \quad H_1 : \tau_h \neq 0,$$

with the decision rule: Reject H_0 if $|\hat{\tau}_h| \leq \kappa_{\alpha/2}$, in which $\kappa_{\alpha/2}$ can be computed from a table of critical values. For larger sample sizes, a normal approximation may be utilized on a standardized version of $\hat{\tau}_h$

$$\hat{\tau}_h^* = \frac{\hat{\tau}_h - E_0(\hat{\tau}_h)}{\sqrt{Var_0(\hat{\tau}_h)}},$$

where $E_0(\hat{\tau}_h)$ and $Var_0(\hat{\tau}_h)$ represent the respective mean and variance of $\hat{\tau}_h$ under the null hypothesis of independence. Under serial independence, Ferguson *et al.* (2000) derive

$$E_0(\hat{\tau}_h) = \begin{cases} \frac{(3n-3h-1)(n-h)}{12} & \text{if } 1 \leq h \leq \frac{n}{2} \\ \frac{(n-h)(n-h-1)}{4} & \text{if } \frac{n}{2} \leq h < n-1. \end{cases}$$

The variance is derived in the case where $h = 1$ as

$$Var_0(\hat{\tau}_h) = \frac{10n^3 - 37n^2 + 27n + 74}{360}.$$

A formula for a general lag h (assuming $n \geq 4h$) is constructed by Šiman (2012) as:

$$Var_0(\hat{\tau}_h) = \frac{2(10n^3 + (7-30h)n^2 + (30h^2 + 46h - 49)n + (-10h^3 - 29h^2 + 114h))}{45(n-h)^2(n-h-1)^2}.$$

It was additionally shown in Ferguson *et al.* (2000) that $\hat{\tau}_h$ is asymptotically standard normal and independent across lags h.

8.2.2.2 Spearman's Rho

When n is large, we consider Fisher's transformation of Spearman's rho

$$\hat{z}_h = \frac{1}{2} \log \left(\frac{1 + \hat{\rho}_S(h)}{1 - \hat{\rho}_S(h)}\right)$$

to test the hypothesis

$$H_0 : \rho_S(h) = 0 \qquad vs. \qquad H_1 : \rho_S(h) \neq 0.$$

The corresponding test statistic is $\hat{z}_h^* = \hat{z}_h / \sqrt{Var(\hat{z}_h)}$. Per Anderson (1954), the standard error of z is approximated as $\sqrt{Var(\hat{z}_h)} = \sqrt{n-3}$, and $\hat{z}_h^*$ can be compared to the standard normal distribution.

Example 8.1 *To illustrate a setting in which rank-based measures are more informative than Pearson's correlation coefficient, we consider data simulated from a cubic moving average model (adapted from Ashley and Patterson, 2001). Let*

$$X_t = a_t + 0.02a_{t-1}^3, \; with \qquad a_t \overset{\text{i.i.d.}}{\sim} F(0, \sigma_a^2).$$

Clearly, there is serial dependence at lag $h = 1$, but it is nonlinear in nature.

We consider three distributions F for the mean zero i.i.d. innovations a_t: standard normal; Student's T distribution with 5 degrees of freedom (hence, heavy tails); and a stable Pareto distribution with shape parameter $\alpha = 1.93$, which has characteristic function $E(e^{i\theta X}) = \exp(-|\theta|^\alpha)$. The choice of $\alpha = 1.93$ is believed by Fama (1965) to adequately capture the leptokurtic nature of US stock returns. From its construction, it is clear that the second moment of this distribution does not exist. One simulation from each process is shown in Figure 8.3. At each iteration,

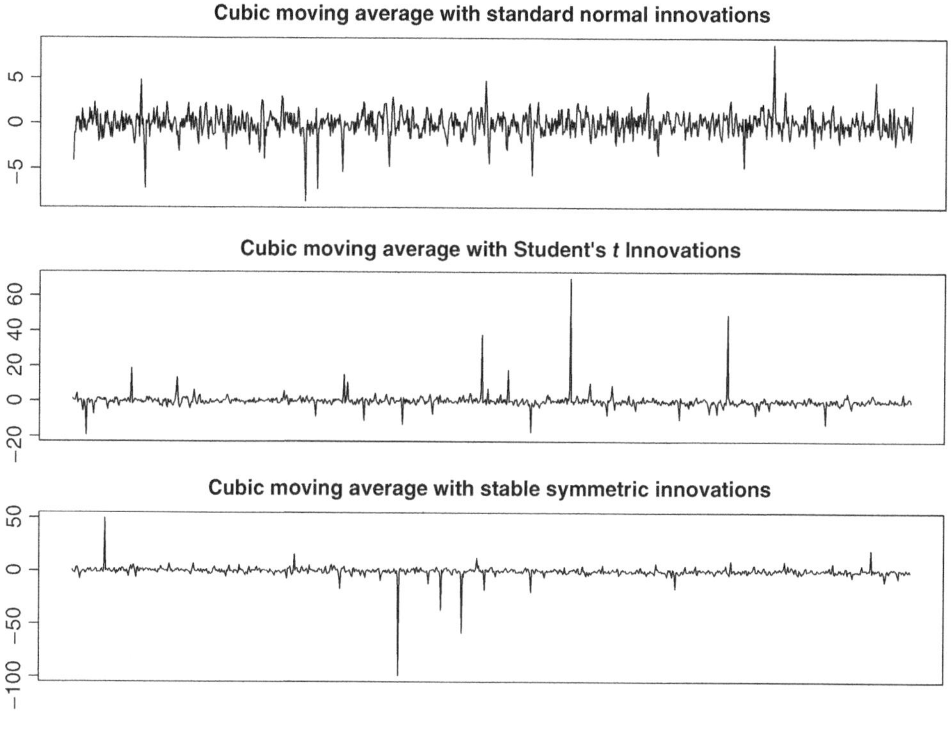

Figure 8.3 Realized time-series simulated from each of the three process models discussed in Example 8.1.

Table 8.1 Percentage of tests failing to reject H_0 of no lag-1 correlation

Error distribution	Pearson	Kendall	Spearman
Standard normal	0.00	0	0
Student's T	0.13	0	0
Stable Pareto	0.53	0	0

1000 observations were drawn from each model and the lag-1 correlations were tested under the hypotheses stated in this chapter. This process was repeated 5000 times, and the results are summarized in Table 8.1.

These simulations provide evidence to support the claim that Pearson's correlation coefficient is robust to nonlinearity, but performs poorly in the presence of leptokurtic distributions and outliers. Hence, one should be judicious in measuring and testing autocorrelation, especially when working with data exhibiting these characteristics.

8.2.3 Misspecification Testing

8.2.3.1 Ljung–Box Test

Assuming a time-series $\boldsymbol{X}_{1:n}$ is i.i.d. with finite variance, it has no autocorrelation at any lag, that is, $\rho_P(h) = 0$ for all $h \geq 1$. Additionally, $\sqrt{n}\hat{\rho}_P(h)$ is asymptotically $N(0, 1)$, and $\hat{\rho}_P(1), \ldots, \hat{\rho}_P(m)$ are asymptotically independent. Thus, we can perform a joint test with the hypotheses

$$H_0 : \rho_P(1) = \rho_P(2) = \ldots = \rho_P(m) = 0 \quad vs. \quad H_1 : \rho_P(h) \neq 0 \text{ for some } h = 1, \ldots, m.$$

This hypothesis is often tested on the residuals of a fitted time-series model. For example, consider the residuals $\hat{\boldsymbol{a}}_{1:n}$ from an autoregressive moving average [ARMA(p,q)] model. Either the Ljung–Box statistic $Q^*(m)$ (Ljung and Box, 1978) or the Box–Pierce statistic $Q(m)$ (Box and Pierce, 1970), which are asymptotically equivalent, can be used:

$$Q^*(\hat{\boldsymbol{a}}_{1:n}; m) = n(n+2)\sum_{h=1}^{m} \hat{\rho}(\hat{\boldsymbol{a}}_{1:n}; h)^2/(n-h), Q(\hat{\boldsymbol{a}}_{1:n}; m) = n\sum_{h=1}^{m} \hat{\rho}(\hat{\boldsymbol{a}}_{1:n}; h)^2. \tag{8.3}$$

The null hypothesis is rejected at significance level α if the statistic exceeds the $1 - \alpha$ quantile of the χ^2_{m-p-q} distribution. Per Lai and Xing (2008), simulation studies suggest that the Ljung–Box statistic is better approximated by a χ^2_{m-p-q} than the Box–Pierce statistic, hence it is preferred for small to moderate-sized samples. Analogous joint tests are formed for both $\tau(h)$ and $\rho_S(h)$ by leveraging the asymptotic normality and independence across lags by applying similar transformations.

8.2.3.2 Conditional Heteroscedasticity

Traditionally, the Ljung–Box test is performed on the residuals of a linear or conditional mean for a given series. However, in financial applications, conditional variance (also known as *volatility*), as opposed to conditional mean, is often a primary parameter of interest. Per Tsay (2010), in many financial applications, a daily asset return series $\boldsymbol{X}_{1:n} = \{x_t\}_{t=1}^n$ generally is either serially uncorrelated or has minor correlations that can be accounted for with a low-order ARMA model. However, substantial autocorrelation may exist in the squared return series $\{x_t^2\}$, and the volatility $\{\sigma_t^2\}$ sequence is the focus of modeling. The autoregressive conditional heteroscedasticity (ARCH) model and its generalization are popular models for forecasting volatility.

Definition 8.7 *Let $\{a_t\}$ denote the innovations from a return series centered by its conditional mean (e.g., ARMA) equation. Then, the* **autoregressive conditional heteroscedasticity model** *of order r* (ARCH$_r$ *model) is defined as*

$$a_t = \sigma_t \varepsilon_t, \quad \varepsilon_t \overset{iid}{\sim} F(0,1),$$
$$\sigma_t^2 = \omega + \alpha_1 a_{t-1}^2 + \ldots + \alpha_r a_{t-r}^2,$$

in which $\{\varepsilon_t\}$ represents a sequence of i.i.d. standardized innovations (Engle, 1982).

Normal random variables are commonly used for F, but other asymmetric and heavy-tailed distributions, such as a standardized skewed Student's T, can be substituted. The volatility equation intercept ω is positive, and the coefficients $\alpha_1, \ldots, \alpha_r$ are constrained to be nonnegative and sum to less than one to ensure positivity of $\{\sigma_t^2\}$ and stationarity of $\{(a_t, \sigma_t^2)\}$. The ARCH model was generalized to allow for a more parsimonious representation when r is large.

Definition 8.8 *The* **generalized ARCH model** (GARCH$_{r,s}$) *for $\{a_t\}$ is defined as above, but with*

$$\sigma_t^2 = \omega + \sum_{i=1}^{r} \alpha_i a_{t-i}^2 + \sum_{j=1}^{s} \beta_j \sigma_{t-j}^2,$$

with $\omega > 0$, $\alpha_i, \beta_j \geq 0$ for positivity, and $\sum_{i=1}^r \alpha_i + \sum_{j=1}^s \beta_j < 1$ for stationarity (Bollerslev, 1986).

8.2.3.3 Parametric Tests for ARCH Effects

Observations from an ARCH or GARCH process are dependent. In particular, the sequence of squared observations $\boldsymbol{a}_{1:n}^2 = \{a_t^2\}_{t=1}^n$ have autocorrelation, which is referred to as ARCH effects. Perhaps the most straightforward test for ARCH effects is the McLeod–Li test, which simply computes the Ljung–Box statistic (8.3) on the squared innovations a_t^2 (or on squared residuals $\hat{a}_t^2$), in place of the innovations a_t. It was shown in McLeod and Li (1983) that under the null hypothesis of independent observations, for fixed m, the scaled correlations $\sqrt{n}[\hat{\rho}_P(\boldsymbol{a}_{1:n}^2; 1), \ldots, \hat{\rho}_P(\boldsymbol{a}_{1:n}^2; m)]$ are asymptotically normally distributed with mean zero and identity covariance matrix. The McLeod–Li statistic is then computed as

$$Q_{MC}^*(\boldsymbol{a}_{1:n}^2; m) = n(n+2) \sum_{h=1}^{m} [\hat{\rho}_P(\boldsymbol{a}_{1:n}^2; h)]^2/(n-h),$$

which is asymptotically χ_m^2 distributed under H_0, and no adjustment is needed when $\boldsymbol{a}_{1:n}^2$ is replaced by $\hat{\boldsymbol{a}}_{1:n}^2$.

A popular alternative to test for ARCH effects is the Lagrange multiplier test of Engle (1982). Per Tsay (2010), the test is equivalent to using a standard F-statistic to conduct the joint test that all $\alpha_i = 0$ in the regression equation

$$a_t^2 = \omega + \alpha_1 a_{t-1}^2 + \ldots + \alpha_m a_{t-m}^2 + e_t, \tag{8.4}$$

where m is a prespecified lag order. The hypotheses are

$$H_0 : \alpha_1 = \ldots = \alpha_m = 0 \quad vs. \quad H_1 : \alpha_h \neq 0 \text{ for some } h = 1, \ldots, m.$$

To calculate the F-statistic, first define $\text{SSR}_0 = \sum_{t=m+1}^{n}(a_t^2 - \hat{\omega}_0)$, where $\hat{\omega}_0 = \frac{1}{n}\sum_{t=m+1}^{n} a_t^2$, and $\text{SSR}_1 = \sum_{t=m+1}^{n} \hat{e}_t^2$, where $\{\hat{e}_t\}$ represents the residuals from (8.4). The F-statistic is then defined as

$$F(\boldsymbol{a}_{1:n}^2; m) = \frac{(\text{SSR}_0 - \text{SSR}_1)/m}{\text{SSR}_1/(n - 2m - 1)}, \tag{8.5}$$

which is also asymptotically χ_m^2 distributed under H_0 (Luukkonen *et al.*, 1988).

Another test developed by Tsay (1986) specifically looks for quadratic serial dependence by incorporating cross-product terms. Let $\boldsymbol{M}_{t-1} = \text{vech}[(1, x_{t-1}, \ldots, x_{t-p})(1, x_{t-1}, \ldots, x_{t-p})']$. Then, consider the regression

$$x_t = (1, x_{t-1}, \ldots, x_{t-p})' \boldsymbol{\phi} + \boldsymbol{M}_{t-1}\boldsymbol{\alpha} + a_t,$$

where $\boldsymbol{\phi}$ is a $(p+1) \times 1$ coefficients vector and $\boldsymbol{\alpha}$ is a $p(p-1)/2 \times 1$ coefficients vector. If the AR(p) model is adequate, then $\boldsymbol{\alpha}$ should be zero. This can be tested using a standard F-statistic similar to (8.5), but with $[p(p+1)/2, n - p - p(p+1)/2 - 1]$ degrees of freedom.

8.2.3.4 Nonparametric Tests for the ARCH Effect

The BDS test (Brock *et al.*, 1996) originated as a test for the detection of nonrandom chaotic dynamics, but has gained traction as a test for nonlinear serial dependence in financial time-series. The BDS test is based on a correlation integral. To construct a correlation integral for a series $\boldsymbol{X}_{1:n} = \{x_t\}_{t=1}^{n}$, define a sequence of m-vectors as $\boldsymbol{x}_t^m = (x_t, x_{t-1}, \ldots, x_{t-m+1})'$, in which m is the embedding dimension. The correlation integral measures the fraction of pairs of m-vectors that are close to each other (given some range parameter $\epsilon > 0$). As described in Patterson and Ashley (2000), it counts the number of m-vectors that lie within a hypercube of size ϵ of each other.

Definition 8.9 An empirical correlation integral *of order m and range $\epsilon > 0$ is*

$$C_{m,\epsilon}(\boldsymbol{X}_{1:n}) = \frac{2}{n_m(n_m - 1)} \sum_{m \le s < t \le n} I(x_t^m, x_s^m; \epsilon), \tag{8.6}$$

in which $n_m = n - m + 1$, and $I(x_t^m, x_s^m; \epsilon)$ is one if $|x_{t-i} - x_{s-i}| < \epsilon$ for $i = 0, \ldots, m-1$, and zero otherwise.

Under the null hypothesis of serial independence, the correlation integral factorizes as $C_{m,\epsilon} = (C_{1,\epsilon})^m$. Correlation integrals are U-statistics, so asymptotic theory can readily be applied. The

BDS test statistic is defined in Zivot and Wang (2007) as

$$V_{m,\epsilon} = \sqrt{n}\frac{C_{m,\epsilon} - (C_{1,\epsilon})^m}{s_{m,\epsilon}},$$

where $s_{m,\epsilon}$ is an estimate of the asymptotic standard deviation of $C_{m,\epsilon} - (C_{1,\epsilon})^m$, as derived in Brock *et al.* (1996).

Under mild regularity conditions, the BDS statistic converges in distribution to a standard normal distribution, though convergence can be very slow. As stated in Patterson and Ashley (2000), convergence for large m requires an extremely large sample size. The null hypothesis of i.i.d. data is rejected if $|V_{m,\epsilon}| > Z_{\alpha/2}$. Per Diks (2009), the range parameter ϵ is typically set to 0.5 to 1.5 times the sample standard deviation of the observed series. In practice, the test is constructed for several values of ϵ, which are then compared against one another as a stability check. As an alternative to relying on asymptotic normality, Genest *et al.* (2007) developed a rank-based extension of the BDS test whose finite sample p-values can be constructed by simulation.

The BDS test also functions as a model misspecification test. Brock *et al.* (1996) shows that the asymptotic distribution of the test statistic for residuals is the same as that of true innovations. Similar to the aforementioned parametric tests, it is often performed on the residuals of a fitted model in order to test for any remaining dependence. However, as pointed out by Diks (2009), though this holds for autoregressive models, it does not hold for models in the ARCH family. While it is not free of nuisance parameters for GARCH models, Caporale *et al.* (2005) found the BDS test to perform very well on the logged squared residuals of a GARCH(1,1) model.

8.2.3.5 Comparative Performance of Misspecification Tests

Ashley and Patterson (2001) conducted a simulation study of the BDS, Mcleod–Li, Tsay, Lagrange multiplier, and several other tests over a wide variety of data-generating processes. They found that the BDS test has the highest power against all alternatives, rendering it very useful as a so-called nonlinearity screening test, which can detect any residual nonlinear serial dependence. However, it is not informative as to the type of nonlinear structure that may be present in the observations. The Tsay test was also found to perform well in the detection of self-excited threshold autoregression models.

8.3 Multivariate Extensions

The vector autoregression (VAR) is the multivariate analog of the autoregressive model. It allows simultaneous and dynamic linear dependence across multiple components.

Definition 8.10 *A pth order* **vector autoregression** *is defined as*

$$\mathbf{y}_t = \boldsymbol{\phi} + \boldsymbol{\Phi}_1\mathbf{y}_{t-1} + \ldots + \boldsymbol{\Phi}_p\mathbf{y}_{t-p} + \mathbf{a}_t, \tag{8.7}$$

in which $\boldsymbol{\phi} \in \mathbb{R}^k$, $\mathbf{y}_t, \mathbf{a}_t \in \mathbb{R}^k$ *for all t; each* $\boldsymbol{\Phi}_i$ *is a* $k \times k$ *coefficient matrix; and* $\mathbf{a}_t$ *is a serially uncorrelated weakly stationary sequence (white noise) with mean vector zero and nonsingular* $k \times k$ *covariance matrix* $\boldsymbol{\Sigma}_a$.

8.3.1 Multivariate Volatility

The vector autoregression (8.7) operates under the assumption that the innovations are white noise, but it allows for nonlinear serial dependence. Such dependence is common in financial data, particularly multivariate ARCH effects. In this case, we may decompose the linear and nonlinear dynamics as

$$\mathbf{y}_t = \boldsymbol{\mu}_t + \mathbf{a}_t, \quad \mathbf{a}_t = \boldsymbol{\Sigma}_t^{1/2}\boldsymbol{\epsilon}_t, \quad \boldsymbol{\epsilon}_t \overset{iid}{\sim} F(\mathbf{0}, \boldsymbol{I}),$$

in which $\boldsymbol{\mu}_t$ denotes the conditional mean given the past observations $E(\mathbf{y}_t|\mathbf{y}_{t-1}, \mathbf{y}_{t-2}, \ldots)$ (possibly estimated with a VAR); and $\boldsymbol{\epsilon}_t$ represents an i.i.d sequence with mean zero and identity covariance. As with the univariate ARCH case, we want to consider processes in which the conditional covariance matrix $\boldsymbol{\Sigma}_t$ is dependent across time, where $\boldsymbol{\Sigma}_t = Cov(\mathbf{y}_t|\mathbf{y}_{t-1}, \mathbf{y}_{t-2}, \ldots) = Cov(\mathbf{a}_t|\mathbf{y}_{t-1}, \mathbf{y}_{t-2}, \ldots)$. As outlined in Tsay (2014), there are many challenges in multivariate volatility modeling. The dimension of $\boldsymbol{\Sigma}_t$ increases quadratically with k, and restrictions are needed to ensure $\boldsymbol{\Sigma}_t$ is positive definite for all t.

One way to model multivariate volatility is to extend the conventional GARCH models to a multivariate setting. A simple approach is to model a half-vectorization (vech) of the diagonal and lower triangle elements of $\boldsymbol{\Sigma}_t$.

Definition 8.11 *The* **VECH GARCH model** *(Bollerslev* et al.*, 1992) for a k-dimensional series has volatility defined by the recursion*

$$vech(\boldsymbol{\Sigma}_t) = vech(\boldsymbol{\Omega}) + \sum_{i=1}^{r} \boldsymbol{A}_i\ vech(\boldsymbol{\epsilon}_{t-i}\boldsymbol{\epsilon}_{t-i}^{'}) + \sum_{j=1}^{s} \boldsymbol{B}_j\ vech(\boldsymbol{\Sigma}_{t-j}),$$

in which $\boldsymbol{\Omega}$ is a $k \times k$ symmetric, positive definite matrix; and the coefficient matrices $\boldsymbol{A}_i$ and $\boldsymbol{B}_j$ each have dimension $k(k+1)/2 \times k(k+1)/2$.

As stated in Jondeau *et al.* (2007), this model requires estimating on the order of k^4 parameters, and it is difficult to incorporate restrictions that guarantee the conditional covariances $\boldsymbol{\Sigma}_t$ will be positive definite.

One popular model that addresses some of these concerns is the Baba, Engle, Kraft, and Kroner (BEKK) representation.

Definition 8.12 *The* **BEKK GARCH model** *(Engle and Kroner, 1995) for a k-dimensional series has volatility defined by the recursion*

$$\boldsymbol{\Sigma}_t = \tilde{\boldsymbol{\Omega}} + \sum_{i=1}^{r} \tilde{\boldsymbol{A}}_i^{'}\boldsymbol{\epsilon}_{t-i}\boldsymbol{\epsilon}_{t-i}^{'}\tilde{\boldsymbol{A}}_i + \sum_{j=1}^{s} \tilde{\boldsymbol{B}}_j^{'}\boldsymbol{\Sigma}_{t-j}\tilde{\boldsymbol{B}}_j,$$

in which $\tilde{\boldsymbol{\Omega}}$ is a $k \times k$ symmetric, positive definite matrix; and the coefficient matrices $\tilde{\boldsymbol{A}}_i$ and $\tilde{\boldsymbol{B}}_j$ each have dimension $k \times k$.

The major advantage of this model is that $\boldsymbol{\Sigma}_t$ is positive definite as long as $\tilde{\boldsymbol{\Omega}}$ is positive definite and the sequence is initialized at positive definite values. Estimation includes $k(k+1)/2 + (r+s)k^2$ parameters. In certain cases, to reduce the computational burden, the matrices

$\tilde{A}_i$ and $\tilde{B}_j$ are assumed to be diagonal. However, the BEKK model does not scale well, even in the diagonal case; the number of parameters grows quadratically with the number of variables. For additional model structures, the interested reader can consult Bauwens *et al.* (2006), which conducts a comprehensive survey of multivariate volatility models.

8.3.2 Multivariate Misspecification Testing

Many univariate misspecification tests have multivariate extensions. Referencing Equation (8.7), the lagged cross-covariance matrices of $\{\mathbf{a}_t\}$ can be estimated as

$$\boldsymbol{C}_h = \frac{1}{n}\sum_{t=h+1}^{n} \mathbf{a}_t \mathbf{a}_{t-h}'.$$

Let $\boldsymbol{D}$ denote a diagonal matrix with $[\boldsymbol{D}]_{ii} = [\boldsymbol{C}_0]_{ii}^{1/2}$, then the lagged cross-correlation matrices of $\{\mathbf{a}_t\}$ can be estimated as

$$\boldsymbol{R}_h = \boldsymbol{D}^{-1}\boldsymbol{C}_h\boldsymbol{D}^{-1}.$$

The Ljung–Box test was extended to a multivariate setting by Baillie and Bollerslev (1990), to test the hypotheses

$$H_0 : \boldsymbol{R}_1 = \cdots = \boldsymbol{R}_m = \mathbf{0} \quad vs. \quad H_1 : \mathbf{R}_h \neq \mathbf{0}, \text{ for some } h = 1, \ldots, m.$$

The test statistic recommended by Lütkepohl (2007) is

$$\bar{Q}(\{\mathbf{a}_t\}; m) = n^2 \sum_{h=1}^{m} \text{trace}(\boldsymbol{C}_h' \boldsymbol{C}_0^{-1} \boldsymbol{C}_h \boldsymbol{C}_0^{-1})/(n-h),$$

which asymptotically follows a $\chi^2_{mk^2}$ distribution under H_0. For residuals $\{\hat{\mathbf{a}}_t\}$ from an estimated $VAR_k(p)$ model, the $\boldsymbol{C}_h$ are defined analogously, as is $\bar{Q}(\{\hat{\mathbf{a}}_t\}; m)$, which instead has a $\chi^2_{(m-p)k^2}$ asymptotic distribution under H_0. Furthermore, analogous hypotheses and tests can be conducted for the multivariate version of the ARCH effect based on $\{\mathbf{a}_t^2\}$, $\{\hat{\mathbf{a}}_t^2\}$, or $\{\hat{\epsilon}_t^2\}$, where $\hat{\epsilon}_t = \hat{\boldsymbol{\Sigma}}_t^{-1/2}\hat{\mathbf{a}}_t$, for an estimated volatility sequence $\{\hat{\boldsymbol{\Sigma}}_t\}$.

8.3.3 Granger Causality

In financial applications, it is widely believed (cf. Gallant *et al.*, 1992) that the joint dynamics of stock prices and trading volume can be more informative as to the underlying state of the stock market than stock prices alone. Many studies have explicitly tested for a causal link between stock prices and trading volume. The notion of Granger causality, developed by Granger (1969), can be used to determine whether one time-series helps to predict another. As a simple example, consider a bivariate time-series $\{(x_t, y_t)\}_{t=1}^n$ specified by the equations

$$x_t = \nu_1 + \phi x_{t-1} + \gamma y_{t-1} + a_{1t},$$
$$y_t = \nu_2 + \beta y_{t-1} + \delta x_{t-1} + a_{2t}$$

where a_{1t}, a_{2t}iid$\sim N(0, \sigma^2)$. The series y_t is said to "Granger-cause" x_t if $\gamma \neq 0$.

More generally, given a prespecified lag length p, consider the marginal model for $\{x_t\}$ defined by

$$x_t = \nu + \phi_1 x_{t-1} + \phi_2 x_{t-2} + \cdots + \phi_p x_{t-p} + \gamma_1 y_{t-1} + \gamma_2 y_{t-2} + \cdots + \gamma_p y_{t-p} + a_t \tag{8.8}$$

To test for Granger causality, we may test the hypotheses

$$H_0 : \gamma_1 = \cdots = \gamma_p = 0 \quad vs. \quad H_1 : \gamma_h \neq 0 \text{ for some } \ h = 1, \ldots, p.$$

As a test statistic, one can use the sum of squared residuals from (8.8)

$$RSS_{\text{Full}} = \sum_{t=1}^{n} \hat{a}_t^2,$$

which can then be compared with the sum of squared residuals from a univariate autoregression of $\{x_t\}$, for example,

$$x_t = \nu + \phi_1 x_{t-1} + \phi_2 x_{t-2} + \cdots + \phi_p x_{t-p} + u_t, \quad RSS_{\text{Null}} = \sum_{i=1}^{n} \hat{u}_i^2.$$

Then a F-statistic is constructed as

$$F(\{(x_t, y_t)\}; p) = \frac{(RSS_{\text{Null}} - RSS_{\text{Full}})/p}{RSS_{\text{Full}}/(n - 2p - 1)} \tag{8.9}$$

which is asymptotically distributed as $F(p, 2p - 1)$ under H_0. There are several additional methods to test for Granger causality, but simulation studies by Geweke *et al.* (1983) suggest that (8.9) achieves the best performance. For an overview of additional tests, the interested reader is referred to Chapter 11 of Hamilton (1994).

8.3.4 Nonlinear Granger Causality

The aforementioned test for Granger causality will only elucidate whether $\{y_t\}$ can help to predict $\{x_t\}$ linearly. Potential nonlinear relationships will remain undetected. Consider the following example, procured from Baek and Brock (1992),

$$x_t = \beta y_{t-q} x_{t-p} + a_t$$

There exists an obvious but nonlinear relationship that will not be evident in Granger causality testing. In a similar fashion to the BDS test, Baek and Brock (1992) develop a test for nonlinear Granger causality based on the correlation integral (8.6).

Definition 8.13 (Nonlinear Granger Causality) *$\{y_t\}$ does not nonlinearly Granger-cause $\{x_t\}$ if:*

$$Pr(|x_t - x_s| < \epsilon_1 \,|\, |x_{t,p} - x_{s,p}| < \epsilon_1, |y_{t,q} - y_{s,q}| < \epsilon_2) = Pr(|x_t - x_s| < \epsilon_1 \,|\, |x_{t,p} - x_{s,p}| < \epsilon_1),$$

in which $z_{t,m} = (z_{t-1}, \ldots, z_{t-m})$. In words: "Given ϵ_1 and ϵ_2, q lags of Y do not incrementally help to predict next period's value given p lags of X."

The notion of nonlinear Granger causality has been extensively developed to analyze relationships in finance. In particular, Hiemstra and Jones (1994) examine the relationship between aggregate stock prices and trading volume.

They expand upon Baek and Brock (1992) and give a detailed description of a hypothesis test (henceforth, the H-J test).

Under the null hypothesis of the H-J test, it is assumed that $\{X_t\}$ does not nonlinearly Granger cause $\{Y_t\}$. This is determined by testing conditional independence using finite lags ℓ_x and ℓ_y:

$$Y_{t+1}|X_t^{\ell_x}, Y_t^{\ell_y} \sim Y_{t+1}|Y_t^{\ell_y},$$

where $X_t^{\ell_x} = (X_{t-\ell_x+1}, \ldots, X_t)$ and, $Y_t^{\ell_y} = (Y_{t-\ell_y+1}, \ldots, Y_t)$. Assuming the bivariate time-series $\{X_t, Y_t\}$ is strictly stationary, the H-J test can be interpreted as a statement about the invariant distribution of the $(\ell_x + \ell_y + 1)$ dimensional random vector $W_t = (X_t^{\ell_x}, Y_t^{\ell_y}, Z_t)$, where

$Z_t = Y_{t+1}$. For notational ease, we will drop the time index and refer to this vector as $W = (X, Y, Z)$. Under the null hypothesis, the joint probability density function $f_{X,Y,Z}(x, y, z)$ and its marginals must satisfy:

$$\frac{f_{X,Y,Z}(x,y,z)}{f_{X,Y}(x,y)} = \frac{f_{Y,Z}(y,z)}{f_Y(y)} \iff \frac{f_{X,Y,Z}(x,y,z)}{f_Y(y)} = \frac{f_{X,Y}(x,y)}{f_Y(y)} \frac{f_{Y,Z}(y,z)}{f_Y(y)}, \tag{8.10}$$

for every (x, y, z) in the support of (X, Y, Z). Correlation integrals are used to measure the discrepancies between the left and right sides of (8.10). Given $\epsilon > 0$, the correlation integral for a multivariate random vector is the probability of finding two independent realizations of the vector V: V_1 and V_2, at a distance less than or equal to ϵ, that is,

$$C_v(\epsilon) = P(\|V_1 - V_2\| \le \epsilon).$$

This leads to the test case:

$$\frac{C_{X,Y,Z}(\epsilon)}{C_{X,Y}(\epsilon)} = \frac{C_{Y,Z}(\epsilon)}{C_Y(\epsilon)}. \tag{8.11}$$

It is then estimated via the sample analog:

$$C_{W,n}(\epsilon) = \frac{2}{n(n-1)} \sum \sum_{i<j} I(\|W_i - W_j\| \le \epsilon). \tag{8.12}$$

However, as shown in Diks and Panchenko (2006), in certain situations this test tends to reject too often under the null hypothesis of no Granger causality. Diks and Panchenko (2006) show that the test statistic used in Hiemstra and Jones (1994) is biased and converges to a nonzero limit, while the variance decreases to zero, which will generate significant values for the test statistic as the length of the series increases. An important instance in which this occurs i between two series with independent ARCH effects.

Instead, Diks and Panchenko (2006) construct a new test statistic that measures dependence between X and Z given $Y = y_i$ locally for each y_i. Define the local density estimator as:

$$\hat{f}_W(W_i) = \frac{(2\epsilon)^{d_W}}{n-1} \sum_{j,j\neq i} I_{ij}^W,$$

where $I_{ij}^W = I(\|W_i - W_j\| < \epsilon)$. The test statistic can then be expressed as:

$$T_n(\epsilon_n) = \frac{(n-1)}{n(n-2)} \sum_i (\hat{f}_{X,Y,Z}(X_i, Y_i, Z_i)\hat{f}_Y(Y_i) - \hat{f}_{X,Y}(X_i, Y_i)\hat{f}_{X,Y}(X_i, Y_i)\hat{f}_{Y,Z}(Y_i, Z_i)).$$

The test is shown to be consistent for $d_x = d_y = d_z = 1$, and bandwidth chosen as:

$$\epsilon_n = Cn^{-\beta},$$

for any $C > 0$ and $\beta \in (\frac{1}{4}, \frac{1}{3})$. Details of derivation for the optimal choice of C are discussed in Diks and Panchenko (2006). The authors then show that under these conditions, the test is asymptotically normally distributed:

$$\sqrt{n}\frac{T_n(\epsilon_n) - q}{S_n} \xrightarrow{d} N(0, 1)$$

where $q = E(f_{X,Y,Z}(X, Y, Z)f_Y(Y) - f_{X,Y}(X, Y)f_{Y,Z}(Y, Z))$; and S_n is an autocorrelation robust estimate of the asymptotic variance. They repeat the empirical study of Hiemstra and Jones (1994) and find weaker evidence of nonlinear Granger causality between S&P 500 returns and volume.

8.4 Copulas

Suppose that we have two random variables X and Y, which we use to create the random vector $\mathbf{Z} = (X, Y)'$. Then the joint distribution of $\mathbf{Z}$ can be decomposed into two parts, one that only depends on the marginal distributions $F_X(x)$ and $F_Y(y)$, and another that is only associated with the dependence between X and Y. The latter is referred to as the copula. More formally, we have the following definition.

Definition 8.14 (Sklar's Theorem (Sklar, 1959)) *Let X and Y be continuous random variables with marginal distributions F_X and F_Y. Furthermore, let U and V be uniform random variables such that $U = F_X(X)$ and $V = F_Y(Y)$. We then define the copula C of (X, Y) as the joint distribution of (U, V).*

One way to model the dependence between two random variables is through the use of a copula. As can be seen from Definition 8.14, a copula is a multivariate distribution for which all marginal distributions are uniform. Thus, copulas are a natural tool for modeling dependence between two random variables since we can ignore the effects of their marginal distributions. For instance, we can express Kendall's tau and Spearman's rho coefficients in terms of the copula.

Theorem 8.1 *Let X and Y be continuous random variables with copula C. Then, if we let $\tau(X, Y)$ and $\rho_S(X, Y)$ represent their corresponding Kendall's tau and Spearman's rho coefficients, we have the following equations:*

$$\tau(X, Y) = 4 \int\int_{[0,1]^2} C(u, v)\, dC(u, v) - 1,$$

$$\rho_S(X, Y) = 12 \int\int_{[0,1]^2} uv\, dC(u, v) - 3 = 12 \int_0^1 \int_0^1 C(u, v)\, du\, dv - 3.$$

Proof. The proof for these equalities can be found in Embrechts *et al.* (2003).

Below are some useful notes regarding copulas.

- $C(0, v) = C(u, 0) = 0.$
- $C(1, v) = v$ and $C(u, 1) = u.$
- For all $0 \leq u_1 < u_2 \leq 1$ and $0 \leq v_1 < v_2 \leq 1$, we have that

$$C(u_2, v_2) - C(u_1, v_2) - C(u_2, v_1) + C(u_1, v_1) > 0,$$

  which implies that C is increasing in both variables.
- Copulas satisfy the following Lipschitz condition:

$$|C(u_2, v_2) - C(u_1, v_1)| \leq |u_2 - u_1| + |v_2 - v_1|.$$

8.4.1 Fitting Copula Models

There are numerous methods for fitting a copula model to data, including parametric, semi-parametric, and nonparametric (Choroś *et al.*, 2010; Kim *et al.*, 2007). Suppose we wish to fit bivariate data $\{(x_i, y_i)\}_{i=1}^n$ to a copula C. In many cases, it is more natural to work on the density scale and with the corresponding copula density. The copula density c is obtained through differentiating the copula C, yielding the following relation between the joint density $f(x, y)$ and marginal densities f_X and f_Y:

$$f(x, y) = c[F_X(x), F_Y(y)] \cdot f_X(x) \cdot f_Y(y).$$

First, suppose that we believe the marginal distributions and copula to belong to known families that can be indexed by parameters θ_1, θ_2, and Ω respectively. Since we are selecting a model from a known family, we can select the best model through maximum likelihood. For this, we select $(\hat{\theta}_1, \hat{\theta}_2, \hat{\Omega})$ so as to maximize

$$L(\theta_1, \theta_2, \Omega) = \sum_{i=1}^{n} \log[c(F_X(x_i|\theta_1), F_Y(y_i|\theta_2)|\Omega)] + \sum_{i=1}^{n} \log(f_X(x_i|\theta_1)) + \sum_{i=1}^{n} \log(f_Y(y_i|\theta_2)).$$

The typical maximum likelihood approach would attempt to estimate θ_1, θ_2, and Ω all at once. However, the inference function for margins (IFM) method of Joe (1997) estimates the parameters in sequence. First, the parameters for the marginal distributions are estimated to obtain $\hat{\theta}_1$ and $\hat{\theta}_2$. Then, these estimates are used to estimate Ω, which maximizes

$$L(\Omega) = \sum_{i=1}^{n} \log\left[c\left(F_X(x_i|\hat{\theta}_1), F_Y(y_i|\hat{\theta}_2)|\Omega\right)\right].$$

The IFM method has a significant computational advantage to maximum likelihood estimation, and Joe (1997) shows that it is almost as efficient as maximum likelihood estimation in most cases.

If, however, the model is misspecified because of an incorrect marginal model choice, then the maximum likelihood estimates may lose their efficiency, and may in fact not be consistent (Kim *et al.*, 2007). For this reason, semiparametric approaches have also been suggested for fitting copula models. In this setting, we usually have enough data to make accurate inferences about the marginal distributions, but not enough to easily model the intervariable dependence. Thus, we will assume that the copula belongs to a family that can be indexed by parameter Ω, while no specification about the marginals is made. The approach taken by Genest *et al.* (1995) fits the marginal distributions by using the empirical distribution function to get estimates $\hat{F}_X$ and $\hat{F}_Y$. Then, the fitted copula is selected so as to have parameter $\tilde{\Omega}$, which maximizes

$$L(\Omega) = \sum_{i=1}^{n} \log \left[c \left(\hat{F}_X(x_i), \hat{F}_Y(y_i) | \Omega \right) \right].$$

In Genest *et al.* (1995), it is shown that the estimate $\tilde{\Omega}$ is consistent and asymptotically normal. Furthermore, approaches to estimate the variance of the $\tilde{\Omega}$ estimator are also presented.

Finally, there are nonparametric approaches that can be used to fit copula models. Unlike the marginals, the copula is a hidden model, and thus selecting an appropriate parametric model is difficult. Thus, employing nonparametric methods is a natural choice. In this setting, many approaches fit the copula itself instead of the density because of increased rates of convergence. The natural starting point is to work with the empirical copula process:

$$\hat{C}_n(u, v) = \frac{1}{n} \sum_{i=1}^{n} 1 \left\{ x_i \leq \hat{F}_X^{-1}(u), y_i \leq \hat{F}_Y^{-1}(v) \right\}.$$

For the case where C is the independence copula, Deheuvels (1979) shows consistency and asymptotic normality for the empirical copula; whereas Fermanian *et al.* (2004) show consistency for a more general class of copulas, as well as consistency when using kernel functions to approximate the copula. However, when using kernel approaches, care must be taken when dealing with the boundary, so as to reduce the effect of boundary bias associated with kernels. This issue has been addressed through the use of locally linear kernels in Chen and Huang (2007).

8.4.2 *Parametric Copulas*

The simplest copula to work with is the product copula $C(u, v) = uv$. This corresponds to the random variables X and Y being independent, and thus is also commonly referred to as the independence copula. Other popular copulas include the Gaussian, t, and Archimedean copulas. The Gaussian copula and some Archimedean copulas include the independence copula as a special case. Additionally, the t copula contains the independence copula as a limiting case.

If the vector $(X, Y)'$ has a bivariate normal distribution, then all of the dependence between X and Y is captured by the correlation matrix Ω. Thus, the copulas for bivariate normal random vectors can be parameterized by their correlation matrix, $C(\cdot\ |\Omega)$. The copulas that are created in such a manner are called Gaussian copulas. If the random vector $(X, Y)'$ has a Gaussian copula, then it is said to have a meta-Gaussian distribution; this is because it is not required that the univariate marginal distributions be normal. For a given correlation matrix, Ω, the Gaussian copula's density, $c(u, v|\Omega)$, is given by

$$\frac{1}{\sqrt{\det(\Omega)}} \exp \left\{ -\frac{1}{2} \begin{pmatrix} \Phi^{-1}(u) \\ \Phi^{-1}(v) \end{pmatrix}' (\Omega^{-1} - I_2) \begin{pmatrix} \Phi^{-1}(u) \\ \Phi^{-1}(v) \end{pmatrix} \right\},$$

where Φ^{-1} is the standard normal quantile function and I_2 is the two-dimensional identity matrix. Fitting this copula to a given dataset can be accomplished by using any of the techniques of Section 8.4.1. In this case, we can also directly estimate the correlation matrix by using Kendall's tau or Spearman's rho. Using Kendall's tau, we have the following estimate for the correlation matrix:

$$\Omega_{i,j} = \sin\left\{\frac{\pi}{2}\rho_\tau(Z_i, Z_j)\right\} \tag{8.13}$$

where $\mathbf{Z} = (Z_1, Z_2)' = (X, Y)'$. Once this calculation has been carried out, we can either directly use this estimate of Ω, or use the estimate as a starting value for maximum likelihood or pseudo-maximum likelihood estimation. A similar relation exists for Spearman's rho and can be found in Ruppert (2015).

In a similar fashion, we can define a copula that is based upon the bivariate t-distribution. This distribution is completely described by its correlation matrix Ω along with a shape parameter ν. The parameter ν affects both the marginal distributions as well as the dependence, and must thus be included in the copula parameterization, $C(\cdot|\nu, \Omega)$. A random vector that has a t-copula is said to have a t-meta distribution.

Regardless of which approach we choose to employ to fit the model, there is a relatively simple way to obtain an estimate of the correlation matrix, and it closely resembles that used for the Gaussian copula. Using Equation 8.13, we obtain a matrix that we refer to as $\Omega^{(1)}$. There is a chance that the matrix $\Omega^{(1)}$ is not positive definite because it has nonpositive eigenvalues. In this case, we have that $\Omega^{(1)} = O^T D^{(1)} O$, where $D^{(1)}$ is a diagonal matrix of eigenvalue and O is an orthogonal matrix of the corresponding eigenvectors. To obtain positive eigenvalues, we transform $D^{(1)}$ to $D^{(2)}$ by $D^{(2)}_{i,i} = \max(\epsilon, D^{(1)}_{i,i})$ for some small positive constant ϵ. Then $\Omega^{(2)} = O^T D^{(2)} O$ is a positive definite matrix; however, its diagonal entries may not be equal to one. To fix this, we multiply the ith row and column by $(\Omega^{(2)}_{i,i})^{-1/2}$. After performing these steps, we are left with a true correlation matrix Ω. We can then estimate the shape parameter by holding Ω constant and maximizing the following equation:

$$L(\Omega, \nu) = \sum_{i=1}^{n} \log\left[c\left(\hat{F}_X(x_i), \hat{F}_Y(y_i)|\Omega, \nu\right)\right].$$

The final class of copulas we will discuss are Archimedean copulas. This class of copula provides modeling characteristics that are not available with either the Gaussian or t-copulas.

Definition 8.15 *A copula C is said to be Archimedean if it has the following representation:*

$$C(u, v) = \phi^{-1}[\phi(u) + \phi(v)].$$

where $\phi : [0, 1] \mapsto [0, \infty]$ *is a function, called the generator, with the following properties;*

1. ϕ *is a decreasing convex function.*
2. $\phi(0) = \infty$ *and* $\phi(1) = 0$.

Depending on the choice of generator, we are able to model a wide array of dependencies. For instance, the Frank copula can be obtained by selecting

$$\phi(t|\theta) = -\log\left\{\frac{e^{-\theta t} - 1}{e^{-\theta} - 1}\right\}$$

Table 8.2 A table of common Archimedean copulas

Copula	$\phi(t\|\theta)$	$C(u,v)$	Range
Clayton	$\frac{1}{\theta}(t^{-\theta}-1)$	$\max(\{u^{-\theta}+v^{-\theta}\}^{-1/\theta},0)$	$\theta \in [-1,\infty)\backslash\{0\}$
Ali–Mikhail–Haq	$\log\left(\frac{1-\theta(1-t)}{t}\right)$	$\frac{uv}{1-\theta(1-u)(1-v)}$	$\theta \in [-1,1)$
Joe	$-\log(1-(1-t)^{\theta})$	$1-\left((1-u)^{\theta}+(1-v)^{\theta}-(1-u)^{\theta}(1-v)^{\theta}\right)^{1/\theta}$	$\theta \in [1,\infty)$

with $\theta \in \mathbb{R}$. The choice of generator gives the following copula:

$$C(u,v|\theta) = -\frac{1}{\theta}\log\left\{1+\frac{(e^{-\theta u}-1)(e^{-\theta v}-1)}{e^{-\theta}-1}\right\}.$$

By noting that $\lim_{\theta\to 0}\phi(t|\theta) = -\log(t)$, we have that as θ approaches 0, the Frank copula converges to the independence copula; whereas as $\theta \to -\infty$, we obtain the copula for the random vector $(U, 1-U)'$, $C(u,v) = \max(0, u+v-1)$, which is the copula for perfect negative dependence. Similarly, as $\theta \to \infty$, we obtain the copula for the random vector $(U,U)'$, which has perfect positive dependence, $C(u,v) = \min(u,v)$.

Now let us consider the generator $\phi(t|\theta) = [-\log(t)]^{\theta}$ for $\theta \geq 1$. This gives the Gumbel copula

$$C(u,v|\theta) = \exp\ \{-([\log(u)]^{\theta}+[\log(v)]^{\theta})^{1/\theta}\}.$$

Selecting $\theta = 1$ results in the independence copula. Also, note that as $\theta \to \infty$, the Gumbel copula converges to the copula for perfect positive dependence. However, unlike the Frank copula, the Gumbel copula is unable to model negative dependence.

Table 8.2 contains some additional examples of Archimedean copulas along with their generators.

8.4.3 *Extending beyond Two Random Variables*

So far, we have only seen how to model the dependence between two random variables. Naturally, we would also like to model dependence between $m \geq 3$ random variables. In this case, our copula would be a function $C : [0,1]^m \mapsto [0,1]$. However, some difficulties arise when trying to extend the tools for creating bivariate copulas to multivariate settings. For instance, Genest *et al.* (1995) show that if C is a bivariate copula and $F_m(x_1,\dots,x_m)$, $G_n(y_1,\dots,y_n)$ are multivariate marginal distributions, then the only copula that satisfies $H(x_1,\dots,x_m,y_1,\dots,y_n) = C(F_m(x), G_n(y))$ is the independence copula.

Because of these difficulties, we will only be considering ways to extend Archimedean copulas, as these extensions still remain relatively easy to analyze. More specifically, we will consider the exchangeable Archimedean copula (EAC), nested Archimedean copula (NAC), and pair copula construction (PCC).

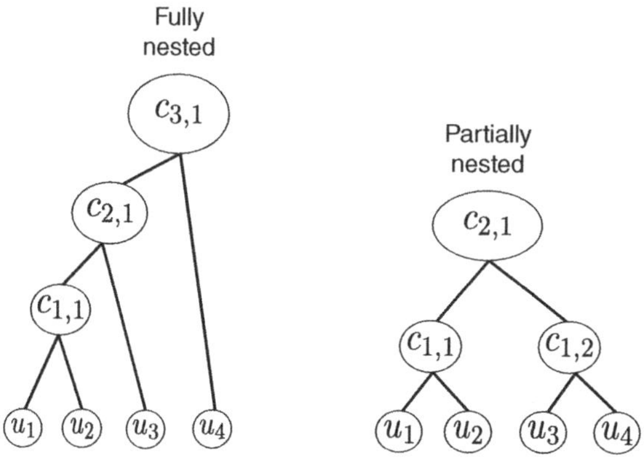

Figure 8.4 Tree representation of the fully nested (left) and partially nested (right) Archimedean copula construction. Leaf nodes represent uniform random variables, while the internal and root nodes represent copulas. Edges indicate which variables or copulas are used in the creation of a new copula.

To start, we consider the simplest extension, EACs. As with the bivariate Archimedean copula, these are tied to a generating function ϕ. In this case, we have that

$$C(u_1, \dots, u_m) = \phi^{-1}\{\phi(u_1) + \cdots + \phi(u_m)\}.$$

This extension, however, comes at the cost of a greater restriction on the generator ϕ. This restriction is d-monotonic; see McNeil and Nešlehová (2009) for more information.

NACs provide more flexibility than EACs because they allow for the combination of different types of bivariate copulas to obtain higher order copulas. The bivariate copulas are combined hierarchically and thus allow for a convenient graphical representation. Figure 8.4 shows a graphical representation for fully and partially nested Archimedean copulas in the case of four variables. As can be seen from Figure 8.4, NACs only allow one to model $m - 1$ copulas. Thus, all other possible interactions are predefined by the hierarchical structure. For instance, in the partially nested structure, both (u_1, u_3) and (u_1, u_4) are modeled by $c_{2,1}(u_1, u_2, u_3, u_4)$. Fitting a general NAC can be done with maximum likelihood, but in general this must be done recursively and becomes extremely computationally intensive as the number of variables increases. Similarly, sampling from such copulas is also a difficult task (Aas and Berg, 2009), but in the case of fully and partially nested models Hofert (2010) presents algorithms that are efficient for sampling any number of variables.

We finally consider PCCs, which like NACs are created by a hierarchical process. The most popular types of PCCs are canonical vines and drawable vines, commonly referred to as C-vines and D-vines, respectively. For both C- and D-vines, there exist analytic expressions for the densities; see Czado (2010). Thus, both models can be fit by maximum likelihood; however, as with NACs, this must be done in a recursive fashion and can thus be computationally expensive. Unlike NACs, drawing samples from PCCs is much simpler (Aas and Berg, 2009). Furthermore, using PCCs, we are able to model all $m(m - 1)/2$ copulas, which is not possible when using NAC extensions. Figure 8.5 shows the graphical representations of C- and D-vines on four variables.

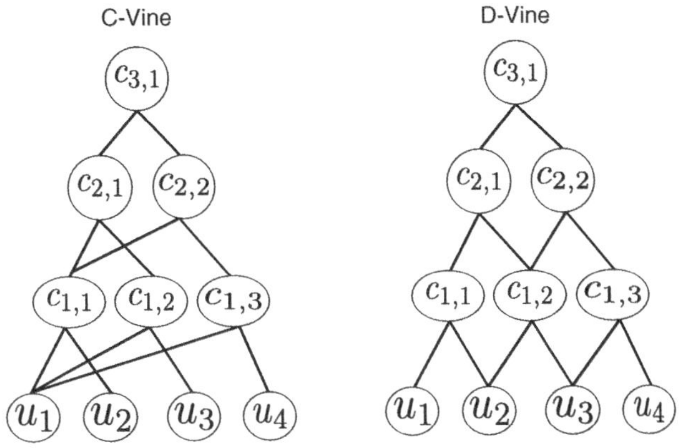

Figure 8.5 Graphical representation of the C-vine (left) and D-vine (right) Archimedean copula construction. Leaf nodes labeled u_i represent uniform random variables, whereas nodes labeled $c_{i,j}$ represent the jth copula at the ith level. Edges indicate which variables or copulas are used in the creation of a new copula.

8.4.4 Software

There are a number of software packages to choose from when working with copulas. Here, we mention a few for working with copulas in R. All of these packages are designed to perform statistical inference of various copula models. They also all have methods for performing model estimation and random sampling. The **copula** package (Hofert *et al.*, 2014) works with elliptical, Archimedean, and a few other copula families. **VineCopula** (Schepsmeier *et al.*, 2012) is designed to perform analysis of vine copulas, while **CDVine** (Brechmann and Schepsmeier, 2013) is specialized for C- and D-vines.

8.5 Types of Dependence

In Section 8.4, we showed that copulas can be used to model the dependence between random variables. We now turn our attention to the dependence structure itself, and when appropriate make connections to copulas. We first describe different types of dependence, and then provide theoretical background.

When dealing with only two random variables, the concept of positive and negative dependence is more straightforward. Suppose that we have random variables X and Y, then positive dependence means that an increase/decrease in the value of X is likely to accompany an increase/decrease in the value of Y. However, if X and Y are negatively dependent, then an increase/decrease in X is likely to be accompanied by a decrease/increase in Y.

8.5.1 Positive and Negative Dependence

We now examine ways to extend the notion of positive and negative dependence to more than two random variables. In this section, we will use the following notation; if $x, y \in \mathbb{R}^d$, we say that $\mathbf{x} > \mathbf{y}$ if $x_i > y_i$ for all $i = 1, 2, \ldots, d$.

In the bivariate case, the random vector $X = (X_1, \ldots, X_d)$ is positively dependent if all of its components tend to move in the same direction. However, when $d > 2$, there are many different interpretations.

Definition 8.16 *A random vector $\boldsymbol{X} = (X_1, \ldots, X_d)$ is positive upper orthant dependent if for all $\mathbf{x} \in \mathbb{R}^d$*

$$P(\boldsymbol{X} > \mathbf{x}) \geq \prod_{i=1}^{d} P(X_i > x_i). \tag{8.14}$$

Similarly, if

$$P(\boldsymbol{X} \leq \mathbf{x}) \geq \prod_{i=1}^{d} P(X_i \leq x_i), \tag{8.15}$$

we say that $\boldsymbol{X}$ is positive lower orthant dependent.

If we generalize the concept of an orthant, we obtain another form of positive dependence.

Definition 8.17 *A set $\mathcal{U}$ is called an upper set if $\mathbf{x} \in \mathcal{U}$ and $\mathbf{y} > \mathbf{x}$ implies that $\mathbf{y} \in \mathcal{U}$. The complement of an upper set is called a lower set.*

It should be noted that since the lower orthant is not the complement of the upper orthant, it is not a lower set. Then, the concept of positive upper/lower set dependence can be easily expressed using inequalities similar to those in Equations 8.14 and 8.15.

Definition 8.18 *A random vector $\boldsymbol{X} = (X_1, \ldots, X_d)$ is said to be positive upper set dependent, or positive lower set dependent, if*

$$P(\boldsymbol{X} \in \bigcap \mathcal{U}_k) \geq \prod_k P(\boldsymbol{X} \in \mathcal{U}_k)$$

and

$$P(\boldsymbol{X} \in \bigcap \mathcal{L}_k) \geq \prod_k P(\boldsymbol{X} \in \mathcal{L}_k)$$

respectively, where $\mathcal{U}_k$ are upper sets and $\mathcal{L}_k$ are lower sets.

Unlike positive dependence, the concept of negative dependence is not as easily extended to multiple random variables. For instance, it is impossible for more than two random variables to have perfect negative dependence. To see this, suppose that X_1 has perfect negative dependence with X_2 and X_3; this would imply that X_2 and X_3 are positively dependent.

One way to obtain a type of negative dependence is by reversing the inequalities in Equations 8.14 and 8.15. This results in what are called negative upper orthant and negative lower orthant dependence, respectively. However, conditions under which these inequalities hold are difficult to show when $d > 2$. One such condition, based upon the behavior of the multinomial distribution, is given by Block *et al.* (1982).

The versions of positive and negative dependence for multiple random variables presented here are elementary. There are notions of positive/negative association that are related to the covariance structure, as well as other less intuitive concepts such as setwise dependence. Further details about these and other types of dependence can be found in the book by Drouet Mari and Kotz (2001).

8.5.2 Tail Dependence

One type of dependence that we will discuss in more detail has major applications in finance. Tail dependence describes how likely it is that a set of random variables will simultaneously take on extreme values. For instance, when working with a portfolio, tail dependence provides information on how likely it is to observe large, simultaneous losses from a portfolio's assets.

Suppose that we have random variables X and Y with distributions F_X and F_Y, respectively, and let $p \in (0, 1)$. In the most basic case, upper tail dependence is concerned with the following quantity:

$$u(p) = P(X > F_X^{-1}(p) | Y > F_Y^{-1}(p)) = \frac{1 - 2p + C(p, p)}{1 - p} \tag{8.16}$$

where $C(\cdot, \cdot)$ is the copula for X and Y. Similarly, lower tail dependence is concerned with the quantity

$$l(p) = P\left(X \le F_X^{-1}(p) | Y \le F_Y^{-1}(p)\right) = \frac{C(p, p)}{p}. \tag{8.17}$$

Let us assume that the variables X and Y measure the amount lost from the components of a two-asset portfolio. We then say that X and Y are asymptotically independent if $u^* = \lim_{p \to 1} u(p)$ exists and is equal to zero. Otherwise, we say that X and Y are asymptotically dependent. Poon *et al.* (2004) provide a nonparametric method for estimating the value of u^*, as well as a way to measure upper tail dependence even if X and Y are asymptotically independent. They also demonstrate the impact of knowing whether variables are asymptotically dependent or independent on the conclusions drawn from statistical analyses.

We now turn our attention to the type of dependence that is under consideration in Equation 8.17. As with Equation 8.16, we will say that X and Y are asymptotically independent if $l^* = \lim_{p \to 0} l(p)$ exists and is equal to zero, and asymptotically dependent if the limit exists and is nonzero. As mentioned by Schmid and Schmidt (2007), one issue with this method is that it only examines values along the diagonal of the copula. Thus, the quantity

$$\frac{C(p, p^2)}{p^2}$$

might have a different limit as $p \to 0$ than l^*. For this reason, Schmid and Schmidt (2007) propose the following measure of tail dependence, which is closely related to Spearman's rho:

$$\rho_L = \lim_{p \to 0} \frac{3}{p^3} \int_0^p \int_0^p C(u, v)\, du\, dv. \tag{8.18}$$

The measures discussed so far have only been for bivariate processes; however, in financial cases, multivariate approaches are typically more applicable. Extensions to higher dimensions are rather simple for Equations 8.16 and 8.17, requiring only a change in the set of variables over which we condition and those for which we wish to find the probability of observing an extreme event. Schmid and Schmidt (2007) provide a general multivariate version of Equation 8.18.

Financial applications are typically multivariate and consider a portfolio of assets rather than individual assets. Such portfolios may exhibit complex dependencies and evolutions. The methods described in this chapter allow robust analysis of non-Gaussian and time-series data, and provide intuitive and tractable tools for multivariate data. They also make relatively weak assumptions about the underlying dependence structure to maintain broad applicability for financial data analysis.

References

Aas, K. and Berg, D. (2009). Models for construction of multivariate dependence: a comparison study. *The European Journal of Finance*, 15(7–8), 639–659.

Anderson, T.W. (1954). *An introduction to multivariate statistical analysis*. New York: John Wiley & Sons.

Ashley, R.A. and Patterson, D.M. (2001). Nonlinear model specification/diagnostics: insights from a battery of nonlinearity tests. Economics Department Working Paper E99-05. Blacksburg, VA: Virginia Tech.

Baek, E. and Brock, W. (1992). A general test for nonlinear granger causality: Bivariate model. Working Paper. Ames and Madison: Iowa State University and University of Wisconsin at Madison.

Baillie, R.T. and Bollerslev, T. (1990). A multivariate generalized arch approach to modeling risk premia in forward foreign exchange rate markets. *Journal of International Money and Finance*, 9(3), 309–324.

Bauwens, L., Laurent, S. and Rombouts, J.V. (2006). Multivariate GARCH models: a survey. *Journal of Applied Econometrics*, 21(1), 79–109.

Bingham, N.H. and Schmidt R (2006). Interplay between distributional and temporal dependence: an empirical study with high-frequency asset returns. In *From stochastic calculus to mathematical finance*. Berlin: Springer, pp. 69–90.

Block, H.W., Savits, T.H., Shaked, M., *et al.* (1982). Some concepts of negative dependence. *The Annals of Probability*, 10(3), 765–772.

Bollerslev, T. (1986). Generalized autoregressive conditional heteroskedasticity. *Journal of Econometrics*, 31(3), 307–327.

Bollerslev, T., Chou, R.Y. and Kroner, K.F. (1992). Arch modeling in finance. *Journal of Econometrics*, 52, 5–59.

Box, G.E. and Pierce, D.A. (1970). Distribution of residual autocorrelations in autoregressive-integrated moving average time series models. *Journal of the American Statistical Association*, 65(332), 1509–1526.

Brechmann, E.C. and Schepsmeier, U. (2013). Modeling dependence with c- and d-vine copulas: The R package CDVine. *Journal of Statistical Software*, 52(3), 1–27.

Brock, W., Scheinkman, J.A., Dechert, W.D. and LeBaron, B. (1996). A test for independence based on the correlation dimension. *Econometric Reviews*, 15(3), 197–235.

Caporale, G.M., Ntantamis, C., Pantelidis, T. and Pittis, N. (2005). The BDS test as a test for the adequacy of a GARCH (1, 1) specification: a Monte Carlo study. *Journal of Financial Econometrics*, 3(2), 282–309.

Chen, S.X. and Huang, T.M. (2007). Nonparametric estimation of copula functions for dependence modelling. *Canadian Journal of Statistics*, 35(2), 265–282.

Chorós, B., Ibragimov, R. and Permiakova, E. (2010). Copula estimation. In *Copula theory and its applications* (ed. Jaworski P, Durante F, Hrdle WK and Rychlik T). Lecture Notes in Statistics. Berlin: Springer pp. 77–91.

Czado, C. (2010). Pair-copula constructions of multivariate copulas. In *Copula theory and its applications* (ed. Jaworski P, Durante F, Hrdle WK and Rychlik T). Lecture Notes in Statistics. Berlin: Springer pp. 93–109.

Deheuvels, P. (1979). La fonction de dépendance empirique et ses propriétés: un test non paramétrique d'indépendance. *Academie Royale de Belgique, Bulletin de la Classe des Sciences*, 65(6), 274–292.

Diks, C. (2009). Nonparametric tests for independence. In *Encyclopedia of complexity and systems science*. New York: Springer, pp. 6252–6271.

Diks, C. and Panchenko, V. (2006). A new statistic and practical guidelines for nonparametric Granger causality testing. *Journal of Economic Dynamics and Control*, 30(9), 1647–1669.

Drouet Mari, D. and Kotz, S. (2001). *Correlation and dependence*, vol. 518. Singapore: World Scientific.

Embrechts, P., Lindskog, F. and McNeil, A. (2003). Modelling dependence with copulas and applications to risk management. *Handbook of Heavy Tailed Distributions in Finance*, 8(1), 329–384.

Engle, R.F. (1982). Autoregressive conditional heteroscedasticity with estimates of the variance of United Kingdom inflation. *Econometrica*, 50, 987–1007.

Engle, R.F. and Kroner, K.F. (1995). Multivariate simultaneous generalized arch. *Econometric Theory*, 11(01), 122–150.

Fama, E.F. (1965). The behavior of stock-market prices. *Journal of Business*, 38, 34–105.

Ferguson, T.S., Genest, C. and Hallin, M. (2000). Kendall's tau for serial dependence. *Canadian Journal of Statistics*, 28(3), 587–604.

Fermanian, J.D., Radulovic, D., Wegkamp, M., *et al.* (2004). Weak convergence of empirical copula processes. *Bernoulli*, 10(5), 847–860.

Gallant, A.R., Rossi, P.E. and Tauchen, G. (1992). Stock prices and volume. *Review of Financial Studies*, 5(2), 199–242.

Genest, C., Ghoudi, K. and Rémillard, B. (2007). Rank-based extensions of the Brock, Dechert, and Scheinkman test. *Journal of the American Statistical Association*, 102, 1363–1376.

Genest, C., Quesada Molina, J. and Rodríguez Lallena, J. (1995). De l'impossibilité de construire des lois à marges multidimensionnelles données à partir de copules. *Comptes rendus de l'Académie des sciences, Série 1, Mathématique*, 320(6), 723–726.

Geweke, J., Meese, R. and Dent, W. (1983). Comparing alternative tests of causality in temporal systems: analytic results and experimental evidence. *Journal of Econometrics*, 21(2), 161–194.

Granger, C.W. (1969). Investigating causal relations by econometric models and cross-spectral methods. *Econometrica*, 37, 424–438.

Hamilton, J.D. (1994). *Time series analysis*, vol. 2. Princeton, NJ: Princeton University Press.

Helsel, D.R. and Hirsch, R.M. (1992). *Statistical methods in water resources*, vol. 49. Amsterdam: Elsevier.

Hiemstra, C. and Jones, J.D. (1994). Testing for linear and nonlinear granger causality in the stock price-volume relation. *The Journal of Finance*, 49(5), 1639–1664.

Hofert, M. (2010). Construction and sampling of nested Archimedean copulas. In *Copula theory and its applications* (ed. Jaworski P, Durante F, Hrdle WK and Rychlik T). Lecture Notes in Statistics. Berlin: Springer pp. 147–160.

Hofert, M., Kojadinovic, I., Maechler, M. and Yan, J. (2014). *Copula: multivariate dependence with copulas*. R package version 0.999-8.

Hollander, M. and Wolfe, D.A. (1999). *Nonparametric statistical methods*. New York: John Wiley & Sons.

Joe, H. (1997). *Multivariate models and multivariate dependence concepts*, vol. 73. Boca Raton, FL: CRC Press.

Jondeau, E., Poon, S.H. and Rockinger, M. (2007). *Financial modeling under non-Gaussian distributions*. Berlin: Springer.

Kendall, M.G. (1938). A new measure of rank correlation. *Biometrika*, 30(1–2): 81–93.

Kim, G., Silvapulle, M.J. and Silvapulle, P. (2007). Comparison of semiparametric and parametric methods for estimating copulas. *Computational Statistics & Data Analysis*, 51(6), 2836–2850.

Lai, T.L. and Xing, H. (2008). *Statistical models and methods for financial markets*. New York: Springer.

Ljung, G.M. and Box, G.E. (1978). On a measure of lack of fit in time series models. *Biometrika*, 65(2), 297–303.

Lütkepohl, H. (2007). *New introduction to multiple time series analysis*. New York: Springer.

Luukkonen, R., Saikkonen, P. and Teräsvirta, T. (1988). Testing linearity against smooth transition autoregressive models. *Biometrika*, 75(3), 491–499.

Maronna, R.A., Martin, D.R. and Yohai, V.J. (2006). *Robust statistics: theory and methods*. Hoboken, NJ: John Wiley & Sons.

McLeod, A.I. and Jimenéz, C. (1984). Nonnegative definiteness of the sample autocovariance function. *The American Statistician*, 38(4), 297–298.

McLeod, A.I. and Li, W.K. (1983). Diagnostic checking arma time series models using squared-residual autocorrelations. *Journal of Time Series Analysis*, 4(4), 269–273.

McNeil, A.J. and Nešlehovà, J. (2009). Multivariate Archimedean copulas, d-monotone functions and 1-norm symmetric distributions. *The Annals of Statistics*, 37, 3059–3097.

Patterson, D.M. and Ashley, R.A. (2000). *A nonlinear time series workshop: a toolkit for detecting and identifying nonlinear serial dependence*, vol. 2. New York: Springer.

Poon, S.H., Rockinger, M. and Tawn, J. (2004). Extreme value dependence in financial markets: diagnostics, models, and financial implications. *Review of Financial Studies*, 17(2), 581–610.

Ruppert, D., and Matteson, D.S. (2015). *Statistics and data analysis for financial engineering*.

Schepsmeier, U., Stoeber, J., Brechmann, E. and Gräler, B. (2012). *Vinecopula: statistical inference of vine copulas*. R package version.

Schmid, F. and Schmidt, R. (2007). Multivariate conditional versions of Spearman's rho and related measures of tail dependence. *Journal of Multivariate Analysis*, 98(6), 1123–1140.

Shumway, R.H., and Stoffer, D.S. (2011). *Time Series Analysis and Its Applications: With R Examples*. Springer Texts in Statistics.

Ŝiman, M. (2012). On Kendall's autocorrelations. *Communications in Statistics: Theory and Methods*, 41(10), 1733–1738.

Sklar, M. (1959). Fonctions de Répartition à n Dimensions et Leurs Marges. *Publications de l'Institut de Statistique de L'Université de Paris*, 8, 229–231.

Spearman, C. (1904). The proof and measurement of association between two things. *American Journal of Psychology*, 15(1), 72–101.

Tsay, R. (2014). *Multivariate time series analysis with R and financial applications*. Berlin: Springer.

Tsay, R.S. (1986). Nonlinearity tests for time series. *Biometrika*, 73(2), 461–466.

Tsay, R.S. (2010). *Analysis of Financial Time Series*. John Wiley & Sons.

Xu, W., Hou, Y., Hung, Y. and Zou, Y. (2013). A comparative analysis of Spearman's rho and Kendall's tau in normal and contaminated normal models. *Signal Processing*, 93(1), 261–276.

Zivot, E. and Wang, J. (2007). *Modeling financial time series with S-PLUSR*, vol. 191. New York: Springer.

9

Correlated Poisson Processes and Their Applications in Financial Modeling

Alexander Kreinin
Risk Analytics, IBM, Canada

9.1 Introduction

Multivariate risk factor models set the foundation of financial risk measurement. The modern financial theory often recommends jump-diffusion models to describe dynamics of the individual risk factors, such as interest rates, foreign exchange rates, stock indices, and volatility surfaces (Kou, 2002; Lipton and Rennie, 2008; Merton, 1976; Musiela and Rutkowsky, 2008), One of the most popular model of jumps, the Poisson model, requires introduction of a codependence structure in the multivariate setting.

The multivariate Gaussian diffusion models are traditionally popular in financial applications (Musiela and Rutkowsky, 2008). In this class of models, the dynamics of the risk factors are described by the Gaussian stochastic processes. The calibration problem in this case can be reduced to the estimation of the drift vector and the diffusion matrix describing the covariance structure of the risk factor space. The only constraint imposed on the covariance matrix is nonnegativity of its eigenvalues.

It is very well known that calibration of the models for equity derivatives pricing can be performed satisfactorily in the class of jump-diffusion processes (Kou, 2002; Kyprianou *et al.* 2005; Merton, 1974). Once the jump processes are introduced, the calibration problem becomes more challenging and, technically, more demanding. In particular, an admissible set of model parameters is described by more sophisticated conditions: not only must an observed covariance matrix have nonnegative eigenvalues but the elements of this matrix should also obey additional inequalities depending on the intensities of the jump processes. The nature

Financial Signal Processing and Machine Learning, First Edition.
Edited by Ali N. Akansu, Sanjeev R. Kulkarni and Dmitry Malioutov.
© 2016 John Wiley & Sons, Ltd. Published 2016 by John Wiley & Sons, Ltd.

of these new constraints and their computation is discussed in Griffiths *et al.*(1979), Whitt (1976), and Duch *et al.* (2014).

The computation of the aditional constraints on the elements of the covariance matrix appeared to be closely related to the analysis of the extreme joint distributions having maximal and minimal correlations. Our approach to the computation of the extreme joint distributions is, in spirit, very close to that developed by Frechet (1951) and Whitt (1976). The only difference is that we propose a pure probabilistic method for the computation of the joint probabilities, based on the Strong Law of Large Numbers and a well-known result on optimal permutations that can be derived from Hardy *et al.* (1952).

Practitioners often set the calibration problem as a matching of the intensities and correlations of the components. One of the difficulties of this problem for multivariate Poisson processes is presented by negative elements of the correlation matrix. We propose a solution to this problem having the following two ingredients:

1. A construction of the joint distribution with extreme correlations at some future point in time, usually at the end of time horizon, T.
2. A backward simulation of Poisson processes in the interval $[0, T]$.

The resulting multivariate Poisson process is a Markov process but not infinitely divisible.[1]

The chapter is organized as follows. In Section 9.2, we discuss some classes of risk factor models with jump processes and their existing and potential applications in different areas of financial risk management. In this section, we discuss some technical obstacles in calibration associated with negative correlations. In Section 9.3, the common shock model and the mixed Poisson model are presented in the bivariate case. The multivariate case of the CSM is discussed in Section 9.4. The first model, traditionally used in the actuarial science, introduces correlations by the simultaneous jumps in the components of the process. This model is too restrictive; it does not solve the problem with negative correlations. The second model introduces a much more general class of processes having time-dependent correlations between the components. This model addresses the negative correlation problems but might have a slow convergence to the stationary distribution, making the calibration problematic.

In Section 9.4, we describe the backward simulation algorithm and discuss the calibration problem for the parameters of the model. We also compare the backward simulation approach with the forward simulation. One of the ingredients of the solution, the computation of the lower and upper bounds for the correlation coefficient, is described in Section 9.5, where the computation of the extreme joint distribution and its support is presented. The computation of the extreme distribution allows one to find the boundaries for the correlation coefficient of the corresponding components of the multivariate Poisson process. We also describe a numerical scheme for the approximation of the joint distribution and computation of the boundaries for the correlation coefficients. These results, leading to the additional necessary conditions on the elements of the correlation matrix, are obtained with the help of a version of the Frechet–Hoeffding theorem[2] presented in Section 9.5.4.

Implementation of the algorithm for the computation of the extreme distributions requires continuity bounds of the approximation of the joint distribution by a distribution with bounded support. These bounds are obtained in Section 9.5.5.

[1] Despite that each component of the process is an infinitely divisible Poisson process.
[2] This version of the theorem is new, to the best of our knowledge.

In Section 9.6, we demonstrate a few patterns of the support of a bivariate Poisson distribution typical for maximal and minimal correlations. These examples are computed using approximation of the joint extreme distribution analyzed in Section 9.5.5. In Section 9.7, we extend the backward simulation (BS) approach to the processes having both Poisson and Wiener components. The idea behind this extension is to bring together, in a common framework, both Poisson and Gaussian processes and to develop a unified simulation algorithm for scenario generation with diffusion and pure jump components. In the Poisson–Wiener case, the time structure of correlations under BS simulation appeared to be linear, as in the case of the multivariate Poisson processes. This feature of the BS simulation put distance between our approach and the traditional forward simulation keeping the correlation constant over time.

The chapter concludes with comments on possible extensions of the BS an method and on the interplay between the probabilistic approach and the infinite-dimensional linear programming optimization problem describing the extreme measures.

9.2 Poisson Processes and Financial Scenarios

In this section, we review some existing and potential applications of the multivariate Poisson processes in financial modeling. The Poisson and the Gaussian risk factor models can be integrated into a general scenario generation framework for portfolio risk measurement. In this framework, the codependence structure of the risk factors is usually described by their correlation matrix.

The correlation matrix of the risk factors contains three submatrices: C_{gg} describing the Gaussian components, C_{pp} describing the Poisson component, and C_{gp} describing the cross-correlations of the risk factors.

Such a complex model for the risk factor space can be used in the integrated market–credit risk portfolio risk framework as well as for pricing of financial derivatives. For operational risk modeling, the matrix C_{pp} should be sufficient.

9.2.1 Integrated Market–Credit Risk Modeling

Monte Carlo methods form the industry standard approach to the computation of risk measures of credit portfolios. The conceptual model of the risk factors dynamics is a combination of the Merton model, developed in Merton (1974), and the conditional independence framework (Vasicek 1987, 2002).

The first dynamic integrated market–credit risk modeling framework was developed in Iscoe *et al.* (1999) for the Gaussian risk factors. Each credit-risky name in the portfolio is characterized by a credit-worthiness index described as a linear combination of the systemic components dependent on market risk factors, macroeconomic indices, and an idiosyncratic component. Calibration of the model is reduced to a series of the default boundary problems that must be solved for each name in the portfolio independently of the other names. This problem is equivalent to the generalized Shiryaev problem (Jaimungal *et al.*, 2013) studied in the Gaussian case. In Jaimungal *et al.* (2013), it was demonstrated that adding a single random jump at time $t = 0$ to the Brownian motion process allows one to cover a very rich class of default time distributions including, in particular, finite mixture of gamma distributions.

Introduction of the general jump processes in the context of credit portfolio modeling leads to a nontrivial calibration problem streaming from heterogeneity of the risk factor space.

Pricing of credit derivatives is also based on the conditional-independence framework (Avellaneda and Zhu, 2001; Hull and White, 2001; Iscoe and Kreinin, 2006, 2007), where randomized, time-dependent default intensities of jumps describe the default process. The number of components of this process is equal to the number of names in the basket credit derivative.

9.2.2 Market Risk and Derivatives Pricing

Geometric Brownian motion has been widely used in the option-pricing framework to model returns of the assets. In Merton (1976), Merton proposed a jump-diffusion model to describe better a phenomenon called volatility smile in the option markets. However, he could not address the leptokurtic feature that the return distribution of assets may have a higher peak and two (asymmetric) heavier tails than those of the normal distribution.

To incorporate both of them, Kou (2002) proposed a double exponential jump-diffusion model. The model is simple enough to produce analytical solutions for a variety of option-pricing problems, including call and put options, interest rate derivatives, and path-dependent options. Detailed accounts of the development in the area of option pricing using jump models can be found in Cont and Tankov (2003, 2006), Boyarchenko and Levendorski (2002), and Carr and Madan (1998). An alternative class of models with non-Gaussian innovations is discussed in Barndorff-Nielsen and Shephard (2001) and Kyprianou *et al.* (2005).

Notice that risk-neutral pricing of basket derivatives with several underlying instruments in the jump-diffusion framework will require a model with several correlated jump processes.

9.2.3 Operational Risk Modeling

Dependent Poisson and compound Poisson processes find many applications in the area of operational risk (OR) modeling (see Aue and Kalkbrener, 2006); Badescu *et al.*, 2013, 2015; Böcker and Klüppelberg, 2010; Chavez-Demoulin *et al.*, 2006; Embrechts and Puccetti, 2006; Panjer, 2006; Peters *et al.*, 2009; Shevchenko, 2011; and references therein). In the OR models, Poisson processes describe random operational events in the business units of a financial organization (Chavez-Demoulin *et al.*, 2006; Lindskog and McNeil, 2001; Nešlehovà *et al.*, 2006; Peters *et al.*, 2009; Shevchenko, 2011). The result of each operational event is a random loss. Thus, the loss process is represented as a multivariate compound Poisson process.

While modeling credit events and credit derivatives pricing in the integrated market–credit risk framework are, probably, the most complicated tasks in financial risk measurement, the isolated operational risk-modeling problem is relatively simple. The operational risk framework is based on the assumption that operational losses of the jth business unit, ($j = 1, 2, \dots, J$), are described by a compound process

$$L_t^{(j)} = \sum_{k=1}^{N_t^{(j)}} \xi_k^{(j)}, \quad j = 1, 2, \dots, J, \tag{9.1}$$

where $N_t^{(j)}$ is the number of operational events occurred by time, t; J is the number of business units in the financial organization; and $\xi_k^{(j)}$ is the loss of the jth unit in the kth event. The problem is how to find distribution of the aggregated losses

$$L_t = \sum_{j=1}^{J} L_t^{(j)}, \quad t > 0,$$

and compute the risk measures of the loss process for the purpose of capital allocation.

Denote the arrival moments of the operational events in the jth unit by $T_k^{(j)}, (k = 1, 2, \dots)$. Then the number of operational events is

$$N_t^{(j)} = \sum_{k=1}^{\infty} \mathbb{1}(T_k^{(j)} \leq t),$$

where $\mathbb{1}(\cdot)$ is the indicator function.

There are two standard models for the processes $N_t^{(j)}$. The first is the classical Poisson model. The second is the negative binomial (NB) model generalizing the classical Poisson model (see Barndorff-Nielsen and Yeo, 1969). In this chapter, we do not discuss the technical details of the NB model, leaving this topic for future publications.

9.2.4 Correlation of Operational Events

Statistical analysis of the operational events indicates the presence of both positive and negative correlations in the multivariate arrival process (Bae, 2012).

In Duch *et al.*, (2014), we considered a fragment of the estimated annual correlation matrix of the operational events, originally studied in Bae, (2012). The estimated correlation matrix $\mathbf{C}_{pp} = \|\rho_{ij}\|$, where $\rho_{ij} = \mathrm{corr}(N_T^{(i)}, N_T^{(j)})$ is the correlation coefficient of the coordinates of the multivariate process, $\mathbf{N}_t$, looks as follows:

$$\mathbf{C}_{pp} = \begin{bmatrix} 1 & 0.14 & 0.29 & 0.32 & 0.15 & 0.16 & 0.03 & 0.05 & -0.06 \\ 0.14 & 1.0 & 0.55 & -0.12 & 0.49 & 0.52 & -0.16 & 0.2 & 0.02 \\ 0.29 & 0.55 & 1.0 & 0.11 & 0.27 & 0.17 & -0.31 & 0.05 & 0.08 \\ 0.32 & -0.12 & 0.11 & 1.0 & -0.12 & 0.23 & 0.19 & -0.18 & -0.11 \\ 0.15 & 0.49 & 0.27 & -0.12 & 1.0 & 0.49 & -0.17 & 0.44 & -0.03 \\ 0.16 & 0.52 & 0.17 & -0.23 & 0.49 & 1.0 & -0.02 & 0.13 & 0.29 \\ 0.03 & -0.16 & -0.31 & 0.19 & -0.17 & -0.02 & 1.0 & 0.32 & 0.5 \\ 0.05 & 0.2 & 0.05 & -0.18 & 0.44 & 0.13 & 0.32 & 1.0 & 0.16 \\ -0.06 & 0.02 & 0.08 & -0.11 & -0.03 & 0.29 & 0.5 & 0.16 & 1.0 \end{bmatrix}$$

One can notice that there are some correlation coefficients that cannot be ignored for the simulation purposes. In particular, $\rho_{32} = 0.55$ and $\rho_{73} = -0.31$ represent the extreme correlations in $\mathbf{C}_{pp}$. Practitioners working in the operational risk area often ignore the correlation structure of the process, $\mathbf{N}_t$, for the sake of simulation simplicity. We believe that such a simplification may result in inaccurate estimation of the operational losses.

9.3 Common Shock Model and Randomization of Intensities

Two possible extensions of the multivariate model with independent components are considered in this section. The first extension is the common shock model (CSM), which has traditional applications in insurance as well as in the area of operational risk modeling.

The second extension is randomization of intensities of the Poisson processes, leading to a more general class of stochastic processes usually called the mixed Poisson processes.

9.3.1 *Common Shock Model*

Correlation of Poisson processes can be controlled by various operations applied to independent processes. One of the most popular operations, often considered in actuarial applications, is a superposition of random processes, the CSM (Lindskog and McNeil, 2001; Powojovsky *et al.*, 2002; Shevchenko, 2011). The idea of this model is described here in the bivariate case. Consider three independent Poisson processes, $\nu_t^{(1)}$, $\nu_t^{(2)}$, and $\nu_t^{(3)}$, and denote their intensities by λ_1, λ_2, and λ_3, respectively. Let us form two new processes using a standard superposition operation of the processes, $N_t^{(1)} = \nu_t^{(1)} \bigoplus \nu_t^{(2)}$ and $N_t^{(2)} = \nu_t^{(2)} \bigoplus \nu_t^{(3)}$, defined as follows. If $\nu_t^{(1)} = \sum_{i=0}^{\infty} \mathbb{1}(t_i^{(1)} \le t)$ and $\nu_t^{(2)} = \sum_{i=0}^{\infty} \mathbb{1}(t_i^{(2)} \le t)$, then the superposition of the processes, $(\nu^{(1)} \bigoplus \nu^{(2)})_t$, is defined by

$$\left(\nu^{(1)} \bigoplus \nu^{(2)}\right)_t := \sum_{j=1}^{2} \sum_{i=0}^{\infty} \mathbb{1}(t_i^{(j)} \le t).$$

Clearly, the processes $N_t^{(1)}$ and $N_t^{(2)}$ are dependent; their correlation coefficient,

$$\rho(N_t^{(1)}, N_t^{(2)}) = \frac{\lambda_2}{\sqrt{(\lambda_1 + \lambda_2)(\lambda_2 + \lambda_3)}},$$

does not depend on time (see Lindskog and McNeil, 2001; Powojovsky *et al.*, 2002) and is always nonnegative.

If all the elements of the correlation matrix, C_{pp}, are positive, the CSM can be used for approximation of the multivariate Poisson processes (Powojovsky *et al.*, 2002). However, if there are negative elements, the CSM has difficulties explaining these correlations and cannot be used unless the estimated negative elements are small.

9.3.2 *Randomization of Intensities*

One promising approach to the analysis and simulation of the jump processes with negative correlations is randomization of intensities. In this case, the resulting processes, usually called mixed Poisson processes, will have a more general probabilistic structure and nonstationary correlations.

Proposition 9.1 (Rolski *et al.*, 1999) The correlation coefficient of the mixed Poisson processes is

$$\begin{aligned}\rho(\nu_t^{(1)},\nu_t^{(2)}) &:= \frac{\mathbb{E}[\nu_t^{(1)}\cdot\nu_t^{(2)}]-\mathbb{E}[\nu_t^{(1)}]\mathbb{E}[\nu_t^{(2)}]}{\sigma(\nu_t^{(1)})\,\sigma(\nu_t^{(2)})}\\ &= \frac{\rho(\lambda_1(X),\lambda_2(X))}{\sqrt{1+\frac{1}{t}\frac{\mathbb{E}[\lambda_1(X)]}{\sigma^2(\lambda_1(X))}}\cdot\sqrt{1+\frac{1}{t}\frac{\mathbb{E}[\lambda_2(X)]}{\sigma^2(\lambda_2(X))}}},\end{aligned} \tag{9.2}$$

Randomization, clearly, increases applicability of the Poisson model. The only drawback of the model is that convergence to the limit value of the correlation coefficient may be very slow.

9.4 Simulation of Poisson Processes

Let us now discuss two approaches to simulation of the Poisson processes. The first method is the forward simulation of the arrival moments. It is, probably, the most natural simulation method for independent Poisson processes. The second method, discussed in this section, is the BS method described in the univariate case in Fox (1996) and in the multivariate case in Duch *et al.*, (2014).

9.4.1 Forward Simulation

Let $\mathbf{N_t} = (N_t^{(1)},\dots,N_t^{(j)}), t\geq 0$, be a J-dimensional Poisson process with independent components. Denote by λ_j the parameter of the jth coordinate of the process, $(j = 1,2,\dots,J)$. Then,

$$\mathbb{E}[N_t^{(j)}] = \lambda_j t.$$

It is very well known that the interarrival times of the jth Poisson process, $\Delta T_k{}^{(j)} := T_k^{(j)} - T_{k-1}^{(j)}$, are mutually independent, for all k and j, and identically distributed, for each j random variables with exponential distribution,

$$\mathbb{P}(\Delta T_k{}^{(j)} \leq t) = 1-\exp(-\lambda_j t), \quad t\geq 0, \quad j=1,2,\dots,J.$$

Probably, the most natural method to generate arrival moments of the Poisson processes is the recursive simulation

$$T_k^{(j)} = T_{k-1}^{(j)} + \Delta T_k{}^{(j)}, \quad k\geq 1, \quad T_0^{(j)} = 0. \tag{9.3}$$

The number of events, $N_t^{(j)}$, in the interval, $[0,t]$, is a stochastic process with independent increments such that

$$\mathbb{P}\left(\bigcap_{j=1}^{J} N_t^{(j)} = k_j\right) = \prod_{j=1}^{J} \exp\,(-\lambda_j t)\cdot\frac{(\lambda_j t)^{k_j}}{k_j!}, \quad j=1,2,\dots,J, \quad k_j\in\mathbb{Z}_+. \tag{9.4}$$

If one is only interested in the number of events in the interval, $[0, t]$, the random vector, $\mathbf{N_t}$, can be sampled directly from the joint distribution (9.4).

It is very well known that if the vectors of the interarrival times are independent, the process $\mathbf{N_t}$ is Markovian in the natural filtration, $\mathfrak{F}_t = \{\sigma(N_\tau)\}_{\tau\leq t}$, generated by the process $\mathbf{N_t}$.

Denote by $P_t(n_1, \dots, n_J) = \mathbb{P}\left(\bigcap_{j=1}^{J} \{N_t^{(j)} = n_j\}\right)$ the probabilities of the states of the Markov process, $\mathbf{N_t}$. Then,

$$\frac{\partial}{\partial t} P_t(n_1, \dots, n_J) = -\Lambda P_t(n_1, \dots, n_J) + \sum_{j=1}^{J} \lambda_j P_t(n_1, \dots, n_j - 1, \dots, n_J), \tag{9.5}$$

where $\Lambda = \sum_{j=1}^{J} \lambda_j$.

Until now, we considered the processes with independent coordinates. In the general case, the following alternative, describing the correlation structure of the (multivariate) Poisson process, can be found in Shreve (2004) and Revus and Yor (1991). The key role in this proposition plays the natural filtration, $\mathfrak{F}_t$, described in Revus and Yor (1991). Recall that in the definition of the multivariate Poisson process given in Revus and Yor (1991) the increments of the components of the process are conditionally independent conditional on the natural filtration, $\mathfrak{F}_t$.

Proposition 9.2 Let $N_t^{(1)}$ and $N_t^{(2)}$ be two Poisson processes measurable with respect to the filtration $\mathfrak{F}_t$. Then, either these processes are independent or they have simultaneous jumps.

Remark 9.1 *If the processes $N_t^{(1)}$ and $N_t^{(2)}$ do not have simultaneous jumps, then the system of equations (9.5) has a product-form solution*

$$P_t(n_1, n_2) = P_t^{(1)}(n_1) \cdot P_t^{(2)}(n_2),$$

where

$$P_t^{(j)}(n_j) = \exp\,(-\lambda_j t) \frac{(\lambda_j t)^{n_j}}{n_j!}, \quad j = 1, 2.$$

The latter is equivalent to independence of $N_t^{(1)}$ and $N_t^{(2)}$. The probabilities $P_t^{(i)}(n)$ satisfy (9.5) with $J = 1$. If the processes $N_t^{(1)}$ and $N_t^{(2)}$ have simultaneous jumps with positive intensity, then the bivariate process has correlated coordinates.[3]

Remark 9.2 *The statement of Proposition 9.2 can be generalized to arbitrary local martingales and semi-martingales; see Revus and Yor (1991). Simulation of correlated processes using joint distribution of the interarrival times is not very convenient: one has to find relations between the correlations of the interarrival times and correlations of the number of events, $N_t^{(j)}$. One commonly used approach by practitioners is to exploit the CSM (Powojovsky* et al., *2002). Let us take M independent Poisson processes, $\eta_t^{(1)}, \eta_t^{(2)}, \dots, \eta_t^{(M)}$, and form the processes $N_t^{(j)}$ obtained by superposition of the subset of the processes $\eta_t^{(m)}$.*

[3] The latter situation is regarded as the CSM.

More precisely, let $T_l^{(j)}$ be a sequence of the arrival times of the jth process, $\eta_t^{(j)}$:

$$\eta_t^{(j)} = \sum_{l=1}^{\infty} \mathbb{1}(T_l^{(j)} \le t), \quad j = 1, 2, \ldots, M.$$

Denote

$$\delta_{jk} = \begin{cases} 1, & \text{if } k\text{th process} \eta_t^{(k)} \text{ contributes to the superposition} \\ 0, & \text{otherwise,} \end{cases}$$

Then the jth process,

$$N_t^{(j)} = \bigoplus_{k=1}^{M} \delta_{jk} \eta_t^{(k)},$$

is

$$N_t^{(j)} := \sum_{k=1}^{M} \delta_{jk} \sum_{l=1}^{\infty} \mathbb{1}(T_l^{(k)} \le t), \quad j = 1, 2, \ldots, J.$$

The intensity, λ_j, of the process $N_t^{(j)}$ satisfies

$$\lambda_j = \sum_{k=1}^{M} \lambda_k^* \delta_{jk}, \tag{9.6}$$

where λ_k^* is intensity of $\eta_t^{(k)}$.

The elements of the correlation matrix are

$$\rho_{ij} = \frac{\lambda_{ij}^c}{\sqrt{\lambda_i \lambda_j}}, \tag{9.7}$$

where

$$\lambda_{ij}^c = \sum_{k=1}^{M} \lambda_k^* \delta_{ik} \delta_{jk}.$$

The calibration problem for the CSM can be formulated as follows. Find M, the matrix of contributions, $\hat{\delta} = \|\delta_{jk}\|$, $(j, k = 1, 2, \ldots, M)$, and a vector of intensities $(\lambda_1^*, \ldots, \lambda_M^*)$ such that Equations (9.6) and (9.7) are satisfied, where $\lambda_1, \ldots, \lambda_J$ are estimated intensities of the multivariate process and ρ_{ij} are estimated correlation coefficients.

Proposition 9.3 The correlation coefficients, ρ_{ij}, satisfy the inequality

$$0 \le \rho_{ij} \le \min\left(\sqrt{\frac{\lambda_i}{\lambda_j}}, \sqrt{\frac{\lambda_j}{\lambda_i}}\right) \tag{9.8}$$

Proof. It immediately follows from (9.7) that the correlation coefficients must be nonnegative and must satisfy the inequality $\lambda_{ij}^c \le \min(\lambda_i, \lambda_j)$, implying the upper bound for the elements of the correlation matrix.

Proposition 9.3 demonstrates that the CSM is too restrictive and may lead to large errors if parameters of the model do not satisfy (9.8).

9.4.2 Backward Simulation

It is well known that the conditional distribution of the arrival moments of a Poisson process in an interval, $[0, T]$, conditional on the number of events in this interval[4] is uniform (Cont and Tankov, 2003). More precisely, let $\mathcal{T} = \{T_1, T_2, \dots, T_n\}$ be a sequence of n independent random variables with a uniform distribution in the interval, $[0, T]$.

Denote by τ_k the kth-order statistic of $\mathcal{T}$, $(k = 1, 2, \dots, n)$:

$$\tau_1 = \min_{1 \le k \le n} T_k, \tau_2 = \min_{1 \le k \le n} \{T_k : T_k > \tau_1\}, \dots, \tau_n = \max_{1 \le k \le n} T_k.$$

Theorem 9.1 (Cont and Tankov, 2003; Rolski *et al.*, 1999; Nawrotzki, 1962) *The conditional distribution of the arrival moments, $\tilde{T}_1 < \tilde{T}_2 < \cdots < \tilde{T}_n$, of a Poisson process, N_t, with finite intensity coincides with the distribution of the order statistics:*

$$\mathbb{P}(\tilde{T}_k \le t \mid N_T = n) = \mathbb{P}(\tau_k \le t), \quad t \le T, \quad k = 1, 2, \dots, n. \tag{9.9}$$

The converse statement, formulated and proved here, is a foundation of the BS method for Poisson processes. Consider a process, N_t, $(0 \le t \le T)$ defined as

$$N_t = \sum_{i=1}^{N_*} \mathbb{1}(T_i \le t),$$

where N_* is a random variable with a Poisson distribution; and the random variables, T_i, are mutually independent and independent of N_*, identically distributed, with the uniform distribution

$$\mathbb{P}(T_i \le t) = tT^{-1}, \quad i = 1, 2, \dots, 0 \le t \le T.$$

Theorem 9.2 *Let N_* have a Poisson distribution with parameter λT. Then, N_t is a Poisson process with intensity λ in the interval* $[0, T]$.

Proof. We give here a combinatorial proof of the statement exploiting the analytical properties of the binomial distribution and binomial thinning. First of all, we notice that $N_T = N_*$. Let us prove that

1. For any interval, $[s, s+t]$, of the length t, $(s + t < T)$, $\mathbb{P}(N_{s+t} - N_s = m) = e^{-\lambda t}\frac{(\lambda t)^m}{m!}$, $m = 0, 1, 2, \dots$.
2. For any m disjoint subintervals $[t_i, t_i + \tau_i] \in [0, T]$, $(i = 1, 2, \dots, m)$, the random variables $N_{t_i+\tau_i} - N_{t_i}$ are mutually independent, $m = 2, 3, \dots$.

Denote $\Delta N_s(t) := N_{s+t} - N_s$. By definition of N_t, we have $N_T = N_*$ and

$$\mathbb{P}(\Delta N_s(t) = k) = \sum_{m=0}^{\infty} \mathbb{P}(\Delta N_s(t) = k \mid N_* = k + m) \cdot \mathbb{P}(N_* = k + m). \tag{9.10}$$

[4] The arrival moments here are not sorted (in ascending order).

Notice that $\Delta N_s(t)$ is the number of events that have occurred in the interval $[s, s+t]$. The conditional distribution of the random variable $\Delta N_s(t)$ is binomial

$$\mathbb{P}(\Delta N_s(t) = k \mid N_* = k+m) = \binom{k+m}{k}\left(\frac{t}{T}\right)^k \cdot \left(1 - \frac{t}{T}\right)^m, \quad m = 0, 1, \ldots$$

and the probability generating function $\hat{p}(z) := \mathbb{E}[z^{N_T}]$ is[5] $\hat{p}(z) = \exp\,(\lambda T(z-1))$. The proof of the first statement of Theorem 9.2 is based on the following.

Lemma 9.1 *Let $\{p_k\}_{k=0}^{\infty}$ be a probability distribution. Denote its generating function by $\hat{p}(z)$:*

$$\hat{p}(z) = \sum_{k=0}^{\infty} p_k z^k, \quad |z| \le 1.$$

Consider a sequence

$$q_k = \sum_{m=0}^{\infty} p_{k+m}\binom{k+m}{k} x^k(1-x)^m, 0 \le x \le 1, k = 0, 1, \ldots. \tag{9.11}$$

The sequence $\{q_k\}_{k\ge 0}$ is a probability distribution with the generating function, $\hat{q}(z)$,

$$\hat{q}(z) = \hat{p}(1 - x + xz). \tag{9.12}$$

Applying Lemma 9.1, with $p_{k+m} = \mathbb{P}(\Delta N_s(t) = k \mid N_* = k+m)$ and satisfying equation (9.10) and $x = t{\cdot}T^{-1}$, we obtain that the generating function, $g(z) := \mathbb{E}[z^{\Delta N_s(t)}] = \exp\,(\lambda t(z-1))$. Therefore, the increments of the process have a Poisson distribution, $\Delta N_s(t) \sim \text{Pois}(\lambda t)$, and this distribution does not depend on s.

To prove the second statement of the theorem, we need the following generalization of Lemma 9.1. Consider a vector $\vec{x} = (x_1, x_2, \ldots, x_m)$, satisfying the conditions

$$x_j \ge 0, \quad j = 1, 2, \ldots, m, \quad \sum_{j=1}^{m} x_j < 1,$$

and denote

$$y = 1 - \sum_{j=1}^{m} x_j.$$

Denote $v_i := N_{t_i+\tau_i} - N_{t_i}, i = 1, 2, \ldots, m$. For an m-dimensional vector, $\vec{k} = (k_1, k_2, \ldots, k_m) \in \mathbb{Z}_+^m$, with nonnegative integer coordinates, $(k_j \ge 0)$, we define the norm of the vector

$$\|\vec{k}\| := \sum_{j=1}^{m} k_j.$$

For any m-dimensional vector, $\vec{x} = (x_1, x_2, \ldots, x_m)$, with nonnegative coordinates, $(x_j \ge 0)$, and $\vec{k} \in \mathbb{Z}_+^m$, we denote

$$\vec{x}^{\,\vec{k}} := \prod_{j=1}^{m} x_j^{k_j}.$$

[5] For the sake of brevity, we will call $\hat{p}(z)$ just generating function, in what follows.

We also introduce a combinatorial coefficient

$$\binom{\vec{k}+l}{\vec{k}} := \prod_{j=1}^{m} \binom{\sum_{i=j}^{m} k_i + l}{k_j}. \tag{9.13}$$

It is not difficult to see that this coefficient can be written in a more symmetric form as a multinomial coefficient

$$\binom{\vec{k}+l}{\vec{k}} = \frac{\left(\sum_{i=1}^{m} k_i + l\right)!}{l! \cdot \prod_{i=1}^{m} k_i!}.$$

Lemma 9.2 *Let $\{p_k\}_{k=0}^{\infty}$ be the probability distribution of a discrete random variable, ξ and let $\hat{p}(z) = \mathbb{E}[z^{\xi}]$ be its generating function. Let $\vec{k} \in \mathbb{Z}_+^m$ and the combinatorial coefficient, $\binom{\vec{k}+l}{\vec{k}}$, be defined by (9.13). Let $\pi : \mathbb{Z}_+^m \to \mathbb{R}$ be defined by*

$$\pi(\vec{k}) = \sum_{l=0}^{\infty} p_{\|\vec{k}\|+l} \binom{\vec{k}+l}{\vec{k}} \cdot \vec{x}^{\,\vec{k}} \cdot y^l, \tag{9.14}$$

and denote by $\hat{\pi}(\vec{z})$ the generating function

$$\hat{\pi}(\vec{z}) := \sum_{\vec{k} \in \mathbb{Z}_+^m} \pi(\vec{k}) \vec{z}^{\,\vec{k}}.$$

Then

$$\hat{\pi}(\vec{z}) = \hat{p}\left(1 - \sum_{j=1}^{m} x_j (1 - z_j)\right). \tag{9.15}$$

Lemma 9.2 is proved in the Appendix. Let us now finish the proof of the second statement.

Denote $p_j = \frac{\tau_j}{T}$, $(j = 1, 2, \ldots, m)$, $\vec{p} = (p_1, \ldots, p_m)$, and $q = 1 - \sum_{j=1}^{m} p_j$. Then we have

$$\mathbb{P}\left(\nu_1 = k_1, \ldots, \nu_m = k_m \mid X_T = \sum_{j=1}^{m} k_j + l\right) = \binom{\vec{k}+l}{\vec{k}} \cdot \vec{p}^{\,\vec{k}} \cdot q^l.$$

Therefore,

$$\mathbb{P}(\nu_1 = k_1, \ldots, \nu_m = k_m) = \sum_{l=0}^{\infty} \binom{\vec{k}+l}{\vec{k}} \cdot \vec{p}^{\,\vec{k}} \cdot q^l \cdot e^{-\lambda T} \frac{(\lambda T)^{\|\vec{k}\|+l}}{(\|\vec{k}\| + l)!}. \tag{9.16}$$

Consider the generating function

$$\pi(\vec{z}) := \mathbb{E}\left[\prod_{j=1}^{m} z_j^{\nu_j}\right], \quad |z_j| \le 1, \quad j = 1, 2, \ldots, m.$$

Applying Lemma 9.2 with $\hat{p}(z) := \mathbb{E}[z^{X_T}] = e^{\lambda T(z-1)}$, $x_j = p_j$ and $y = q$, we derive

$$\pi(\vec{z}) = \prod_{j=1}^{m} e^{\lambda \tau_j (z_j - 1)}.$$

The latter relation implies independence of the random variables $v_j, (j = 1, 2, \dots, m)$. Therefore, the process $N_t, (0 < t \le T)$ has independent increments. Theorem 9.2 is thus proved.

Remark 9.3 *The conditional uniformity of the (unsorted) arrival moments is a characteristic property of the more general class of mixed Poisson processes (Nawrotzki, 1962; Rolski* et al.*, 1999).*

Now we are in a position to formulate the BS method in the of case of $J = 1$.

Step 1. Generate a random number, n, having the Poisson distribution with parameter λT: $n \sim \text{Pois}(\lambda T)$, and assign $N_T := n$.
Step 2. Generate n uniformly distributed random variables in the interval $[0, T]$, $T_i \sim U([0, T]), i = 1, 2, \dots, n$.
Step 3. Sort the random variables, T_i, in the ascending order.
Step 4. Repeat Steps 1–3 n_{mc} times, where n_{mc} is the required number of scenarios.

Remark 9.4 *The BS method is applicable to the class of mixed Poisson processes with random intensity. It can also be used for simulation of jump processes represented as a time-transformed Poisson process (Feigin, 1979).*

Remark 9.5 *The BS of the Poisson processes can be implemented using quasi–Monte Carlo (QMC) algorithms (see Fox, 1996; Fox and Glynn, 1988). In Fox (1996), in Step 1, the stratified sampling was proposed for generation of the number of arrivals, N_T, and QMC for generation of arrival times. This numerical strategy significantly increases the rate of convergence.*

9.4.2.1 Backward Simulation: $J > 1$

It is easy to extend the BS algorithm to the multivariate Poisson processes. Suppose that marginal distributions of the random vector $\mathbf{N}_* = (N_*^{(1)}, N_*^{(2)}, \dots, N_*^{(J)})$ are Poisson, $N_*^{(j)} \sim \text{Pois}(\lambda_j T)$. Denote the correlation coefficient of $N_*^{(i)}$ and $N_*^{(j)}$ by ρ_{ij}.

Theorem 9.3 *Consider the processes*

$$N_t^{(j)} = \sum_{i=1}^{N_*^{(j)}} \mathbb{1}(T_i^{(j)} \le t), \quad j = 1, 2, \dots, J,$$

where the random variables, $T_i^{(j)}, (i = 1, 2, \dots, X_*^{(j)})$, are mutually independent and uniformly distributed in the interval $[0, T]$. Then, $N_t^{(j)}$ is a multivariate Poisson process in the interval $[0, T]$ and

$$corr(N_t^{(i)}, N_t^{(j)}) = \rho_{ij} t T^{-1}, \quad 0 \le t \le T. \tag{9.17}$$

Proof. We have already proved in Theorem 9.2 that $N_t^{(j)}$ is a Poisson process. Let us now formulate the auxiliary result used in the derivation of (9.17).

Consider a bivariate, integer-valued random vector, $\vec{\zeta} = (\zeta_1, \zeta_2)$, $\vec{\zeta} \in \mathbb{Z}_+^2$, and denote

$$p(k,l) = \mathbb{P}(\zeta_1 = k, \zeta_2 = l), \quad k,l = 0,1,2,\dots,$$

its probability distribution. By $\hat{p}(z,w)$, we denote the generating function

$$\hat{p}(z,w) := \sum_{k=0}^{\infty}\sum_{l=0}^{\infty} p(k,l) z^k w^l, \quad |z| \le 1, \quad |w| \le 1.$$

Consider a random vector, $\xi = (\xi_1, \xi_2)$, such that for all $k = 0,1,2,\dots,k'$, and $l = 0,1,\dots,l'$

$$\mathbb{P}(\xi_1 = k, \xi_2 = l \mid \zeta_1 = k', \zeta_2 = l') = \binom{k'}{k} x^k (1-x)^{k'-k} \cdot \binom{l'}{l} y^l (1-y)^{l'-l}, \tag{9.18}$$

where $0 < x \le 1, 0 < y \le 1$. The components of the random vector ξ are conditionally independent. The joint probability, $q_{k,l} = \mathbb{P}(\xi_1 = k, \xi_2 = l)$, can be written as

$$q_{k,l} = \sum_{m=0}^{\infty}\sum_{n=0}^{\infty} p_{k+m,l+n} \cdot \binom{k+m}{k} x^k (1-x)^m \cdot \binom{l+n}{l} y^l (1-y)^n, \quad k,l = 0,1,2,\dots. \tag{9.19}$$

Lemma 9.3 *Suppose that the variance and the first moment of the random variables, ζ_i, $(i = 1,2)$, are equal:*[6] $\mathbb{E}[\zeta_i] = \sigma^2(\zeta_i)$. *The generating function $\hat{q}(z,w) := \sum_{k=0}^{\infty}\sum_{l=0}^{\infty} q_{k,l} z^k w^l$, satisfies the relation*

$$\hat{q}(z,w) = \hat{p}(1-x+xz, 1-y+yw), \quad |z| \le 1, |w| \le 1. \tag{9.20}$$

The correlation coefficient of the random variables ξ_1 and ξ_2 is

$$\rho(\xi_1, \xi_2) = \sqrt{xy} \cdot \rho(\zeta_1, \zeta_2). \tag{9.21}$$

Lemma 9.3 is proved in the Appendix.

Let us apply Lemma 9.3 with $\vec{\zeta} = (N_*^{(i)}, N_*^{(j)})$ and $\vec{\xi} = (N_t^{(i)}, N_t^{(j)})$. The conditional probabilities, $\mathbb{P}((N_t^{(i)} = k, N_t^{(j)} = l) \mid (N_T^{(i)} = k', N_T^{(j)} = l'))$, satisfy (9.18) with $x = y = t\,T^{-1}$ and $\rho(\zeta_1, \zeta_2) = \mathrm{corr}(N_*^{(i)}, N_*^{(j)})$. Then, Lemma 9.3 implies (9.17).

Theorem 9.4 states the Markovian structure of the process $\vec{N}_t$. If $J = 1$, $\vec{N}_t$ is a Poisson process and, obviously, is Markovian in the natural filtration. If $J > 1$, the Markovian nature of the process is not obvious since it is constructed backward.

Theorem 9.4 *The process $\vec{N}_t = (N_t^{(1)}, N_t^{(2)}, \dots, N_t^{(J)})$ is a Markov process in the interval* $[0,T]$.

Proof. Let $0 = t_0 < t_1 < t_2 < \cdots < t_{m-1} < t_m < T$ be a partition of the interval $[0,T]$. Consider integer vectors, $\vec{n}_j \in \mathbb{Z}_+^m$, $\vec{n}_j = (n_1^{(j)}, n_2^{(j)}, \dots, n_{m-1}^{(j)}, n_m^{(j)})$, $(j = 1,2,\dots,J)$, and denote

$$P(\vec{n}_1, \vec{n}_2, \dots, \vec{n}_m) = \mathbb{P}\left(\bigcap_{k=1}^{m}\bigcap_{j=1}^{J}\{N_{t_k}^{(j)} = n_k^{(j)}\}\right).$$

[6] As in the case of the Poisson distributions.

By construction, the coordinates of $\vec{N}_t$ are conditionally independent, and the arrival moments have conditional uniform distributions. Therefore,

$$\mathbb{P}\left(\bigcap_{k=1}^{m}\bigcap_{j=1}^{J}\left\{N_{t_k}^{(j)}=n_k^{(j)}\right\}\Bigg|\ \vec{N}_T=\vec{k}\right)=\prod_{j=1}^{J}\mathbb{P}\left(\bigcap_{k=1}^{m}\left\{N_{t_k}^{(j)}=n_k^{(j)}\right\}\Bigg|\ N_T^{(j)}=k_j\right) \tag{9.22}$$

$$=\prod_{j=1}^{J}\left(\frac{k_j\ !}{(k_j-n_m^{(j)})!\cdot\prod_{k=1}^{m}\left(\Delta n_k^{(j)}\right)!}\cdot\prod_{k=1}^{m+1}\left(\frac{\tau_k}{T}\right)^{\Delta n_k^{(j)}}\right),$$

where $\Delta n_k^{(j)}=n_k^{(j)}-n_{k-1}^{(j)}$, $\tau_k=t_k-t_{k-1}$, $\tau_{m+1}=T-t_m$ and $n_0^{(j)}=0$.

Consider the conditional probabilities

$$P\left(\vec{n}_m\mid\vec{n}_{m-1},\ldots,\vec{n}_1\right):=\mathbb{P}\left(\bigcap_{j=1}^{J}\left\{N_{t_m}^{(j)}=n_m^{(j)}\right\}\Bigg|\bigcap_{j=1}^{J}\bigcap_{k=1}^{m-1}\left\{N_{t_k}^{(j)}=n_k^{(j)}\right\}\right).$$

We have

$$P(\vec{n}_m\mid\vec{n}_{m-1},\ldots,\vec{n}_1)=\frac{P(\vec{n}_1,\vec{n}_2,\ldots,\vec{n}_m)}{P(\vec{n}_1,\vec{n}_2,\ldots,\vec{n}_{m-1})} \tag{9.23}$$

$$=\frac{P(\vec{n}_{m-2},\ldots,\vec{n}_1\mid\vec{n}_m,\vec{n}_{m-1})P(\vec{n}_m,\vec{n}_{m-1})}{P(\vec{n}_1,\vec{n}_2,\ldots,\vec{n}_{m-1})}.$$

Equation (9.23) implies

$$P(\vec{n}_{m-2},\ldots,\vec{n}_1\mid\vec{n}_m,\vec{n}_{m-1})=P(\vec{n}_{m-2},\ldots,\vec{n}_1\mid\vec{n}_{m-1}). \tag{9.24}$$

Notice that $P(\vec{n}_m,\vec{n}_{m-1})=P(\vec{n}_m\mid\vec{n}_{m-1})P(\vec{n}_{m-1})$. Equations, (9.23) and (9.24) imply

$$P(\vec{n}_m\mid\vec{n}_{m-1},\ldots,\vec{n}_1)=P(\vec{n}_m\mid\vec{n}_{m-1}),$$

as was to be proved.

9.4.2.2 Generation of Poisson Random Vectors

Using the BS method, we reduce simulation of the multivariate Poisson processes to generation of the random integer-valued vector with marginal Poisson distribution. The following approach to generation of the Poisson random vectors, based on the idea of transformation of Gaussian random vectors with properly assigned correlations, is considered in Yahav and Shmueli (2011) and Duch *et al.* (2014).

Let us recall one simple, general property of random variables: if ξ is a continuous random variable with the cumulative distribution function (cdf), $F(x)$, then the random variable, $\zeta=F(\xi)$, has the standard uniform distribution, $U([0,1])$, in the unit interval.

Conversely, given a distribution function, F, and a sample from the uniform distribution, ζ_1, $\zeta_2,\ldots,\zeta_M$, we obtain a new sample, $\xi_m=F^{(-1)}(\zeta_m)$, $(m=1,2,\ldots,M)$, from the distribution F (see Johnson and Kotz, 1969; Johnson *et al.*, 1997).

Consider a random vector, $\vec{\eta} = (\eta_1, \dots \eta_J)$, with a mean-zero multivariate normal distribution, $\mathcal{N}(0, \rho)$, with the correlation matrix, ρ, and unit variances. The random vector, $\vec{\xi} = (\Phi(\eta_1), \dots, \Phi(\eta_m))$, where $\Phi(x)$ is the standard normal cdf, has a multivariate distribution with the standard uniform marginals. Let $[x]$ be the integer part of x, $(x \in \mathbb{R})$. Denote

$$P_\lambda(x) := \begin{cases} \sum_{k=0}^{[x]} e^{-\lambda} \frac{\lambda^k}{k!}, & \text{if } x \geq 0, \\ 0, & \text{otherwise,} \end{cases}$$

the cdf of the Poisson random variable. Applying the inverse Poisson cdf to each coordinate of $\vec{\xi}$, we obtain a random vector $\vec{\zeta} = (\zeta_1, \dots, \zeta_J)$, where $\zeta_j = P_{\lambda_j}^{-1}(\Phi(\eta_j))$ has Poisson marginal distribution with parameter λ_j. The correlation coefficient, $\hat{\rho}_{k,l} := \text{corr}(\zeta_k, \zeta_l)$, is

$$\hat{\rho}_{k,l} = \frac{\mathbb{E}[\zeta_k \cdot \zeta_l] - \lambda_k \lambda_l}{\sqrt{\lambda_k \cdot \lambda_l}}, \quad k, l = 1, 2, \dots, J,$$

and

$$\mathbb{E}[\zeta_k \cdot \zeta_l] = \lambda_k \lambda_l + \hat{\rho}_{k,l} \cdot \sqrt{\lambda_k \cdot \lambda_l}. \tag{9.25}$$

9.4.2.3 Model Calibration

Let us now briefly discuss the calibration problem. In Duch *et al.* (2014), the calibration problem was set as a matching of the intensities, λ_j, and the correlation coefficients, $\hat{\rho}_{ij}$. The intensities, λ_j, can be found from the observations of the number of events in the interval, $[0, T]$. In Duch *et al.* (2014), it is shown that

$$\begin{aligned} \mathbb{E}[\zeta_k \cdot \zeta_l] &= \sum_{m=1}^{\infty} \sum_{n=1}^{\infty} n \cdot m \mathbb{P}(\zeta_k = m, \zeta_l = n) \qquad (9.26) \\ &= \sum_{m=1}^{\infty} \sum_{n=1}^{\infty} n \cdot m \mathbb{P}(u_m^{(k-1)} < \Phi(\eta_k) \leq u_m^{(k)}, u_n^{(l-1)} < \Phi(\eta_l) \leq u_n^{(l)}), \end{aligned}$$

where $u_j^{(i)} = P_{\lambda_i}^{-1}(j)$, $(i = 1, 2, \dots, J, j = 1, 2, \dots)$.

The probabilities, $\mathbb{P}(u_m^{(k-1)} < \Phi(\eta_k) \leq u_m^{(k)}, u_n^{(l-1)} < \Phi(\eta_l) \leq u_n^{(l)})$, can be written as a linear combination of the bivariate normal distribution functions, $\Phi_2(\cdot, \cdot, \rho)$, with the arguments depending on the indices m and n. Then, from (9.25) and (9.26), we obtain an implicit equation for the correlation coefficient, $\hat{\rho}_{k,l}$.

A numerical scheme for the computation of the matrix $\hat{\rho} = \|\hat{\rho}_{k,l}\|$ is considered in detail in Duch *et al.* (2014). There is, however, one delicate problem related to the existence of the correlation matrix $\boldsymbol{\rho}$. The problem is the sufficient conditions, for a given positive semidefinite matrix to be a correlation matrix of a random vector with the Poisson marginal distributions, depend on the parameters, λ_j, of these distributions. The latter property is in contrast with Gaussian random vectors. Thus, the admissibility test is required to verify if the calibration problem has a solution. It is shown in Duch *et al.* (2014) that the calibration problem for multivariate processes is decomposed into a series of bivariate problems for each correlation coefficient. This problem is analyzed in Section 9.5.

9.5 Extreme Joint Distribution

We would like to find the range, $[\rho^*, \rho^{**}]$, of the admissible values of the correlation coefficient, given the intensities, λ and μ, of the random variables, $\mathcal{N}$ and $\mathcal{N}'$, with Poisson marginal distributions. A straightforward observation is that this range is asymmetrical. Indeed, it is not difficult to see that the maximal correlation can be 1 in the case of equal intensities. On the other hand, it is impossible to obtain $\rho = -1$, simply because both $\mathcal{N}$ and $\mathcal{N}'$ are nonnegative, unbounded random variables and a linear relation, $\mathcal{N} = -b\mathcal{N}' + c$ with a constant $b > 0$ and an arbitrary finite c, contradicts the nonnegativity of the random variables.

9.5.1 *Reduction to Optimization Problem*

Denote the marginal probabilities of the vector $(\mathcal{N}, \mathcal{N}')$ by $p(i) := \mathbb{P}(\mathcal{N} = i)$ and $q(j) := \mathbb{P}(\mathcal{N}' = j)$, $(i, j = 0, 1, 2, \dots)$. The results of this subsection are derived in a general case and do not use the Poisson specification of the marginal probabilities.

Let us find the joint distributions of the random vector, $(\mathcal{N}, \mathcal{N}')$, such that the correlation coefficient $\rho = \rho(N, \mathcal{N}')$ takes an extreme value, $\rho \to \text{extr}$, and

$$\begin{aligned} \mathbb{P}(\mathcal{N} = i) &= p(i), i = 0, 1, 2, \dots, \\ \mathbb{P}(\mathcal{N}' = j) &= q(j), j = 0, 1, 2, \dots, \end{aligned} \tag{9.27}$$

where $\sum_{i=0}^{\infty} p(i) = \sum_{j=0}^{\infty} q(j) = 1.$

This problem was considered in Embrechts and Puccetti (2006), Griffiths *et al.* (1979), and Nelsen (1987). The coefficients $\rho^* = \min \rho(N, \mathcal{N}')$ and $\rho^{**} = \max \rho(N, \mathcal{N}')$ were numerically calculated in Griffiths *et al.* (1979) using reduction to the optimization problem discussed below. In that chapter, it was shown that both ρ^* and ρ^{**} are nonmonotone functions of the intensities, λ and μ. It was proved in Griffiths *et al.* (1979), using the nonmonotonicity property, that the distribution of the vector $(\mathcal{N}, \mathcal{N}')$ is not infinitely divisible.

Denote by $\mathfrak{p}$ the matrix of the joint probabilities

$$\mathfrak{p}_{i,j} = \mathbb{P}(\mathcal{N} = i, \mathcal{N}' = j), i = 0, 1, 2, \dots, j = 0, 1, \dots .$$

Let

$$f(\mathfrak{p}) := \sum_{i=0}^{\infty} \sum_{j=0}^{\infty} ij\mathfrak{p}_{i,j}.$$

Then the solution to the problem

$$\begin{aligned} &f(\mathfrak{p}) \to \text{extr} \\ &\sum_{j=0}^{\infty} \mathfrak{p}_{i,j} = p(i), \quad i = 0, 1, \dots \\ &\sum_{i=0}^{\infty} \mathfrak{p}_{i,j} = q(j), \quad j = 0, 1, \dots \end{aligned} \tag{9.28}$$

$$\sum_{i=0}^{\infty}\sum_{j=0}^{\infty} \mathfrak{p}_{i,j} = 1$$

$$\mathfrak{p}_{i,j} \geq 0 \quad i,j = 0,1,\ldots$$

is the joint distribution maximizing (or minimizing) the correlation coefficient given marginal distributions of $\mathcal{N}$ and $\mathcal{N}'$. Note that the problem (9.28) belongs to the class of infinitely dimensional linear programming problems.[7]

9.5.2 Monotone Distributions

In this section, we describe the structure of the extreme joint distributions and derive formulae for the joint probabilities, $\mathfrak{p}^*_{i,j}$ and $\mathfrak{p}^{**}_{i,j}$ (see Theorem 9.5). We introduce two special classes of discrete, bivariate distributions that, according to Frechet (1951), we call co-monotone and anti-monotone distributions, and prove that, given marginal distributions, the solution to the extreme problem (9.28) belongs to these classes. We also demonstrate that in the case of discrete bivariate distributions, Theorem 9.5 is equivalent to the famous Frechet–Hoeffding theorem on extreme distributions, obtained in Hoeffding (1940).

Consider a set of points, (finite or infinite), $\mathcal{S} = \{s_n\}_{n\geq 1}$, where $s_n = (x_n, y_n) \in \mathbb{R}^2$. Consider the subsets $\mathcal{R}_+ = \{(x,y) \in \mathbb{R}^2 : x{\cdot}y \geq 0\}$ and $\mathcal{R}_- = \{(x,y) \in \mathbb{R}^2 : x{\cdot}y \leq 0\}$.

Definition 9.1 *A set $\mathcal{S} = \{s_n\}_{n\geq 1} \in \mathbb{R}^2$ is called co-monotone if $\forall i,j$ the vector $s_i - s_j \in \mathcal{R}_+$. A set $\mathcal{S} \in \mathbb{R}^2$ is called anti-monotone if $\forall i,j$ $s_i - s_j \in \mathcal{R}_-$.*

Figure 9.1 illustrates the notion of the co-monotone and anti-monotone sets introduced in Definition 9.1. Let $z \in \mathbb{R}^N$ and π be a permutation of N elements. The group of permutations of N elements is denoted by $\mathfrak{S}_N$. Finally, we introduce the notation $\pi z := (z_{\pi(1)}, z_{\pi(2)}, \ldots z_{\pi(N)})$.

Suppose that $\mathcal{S} = \{(x_n, y_n)\}_{n=1}^N$ is a finite, co-monotone set in $\mathbb{R}^2$, $N \geq 2$. Consider the vectors, $x = (x_1, x_2, \ldots, x_N)$ and $y = (y_1, y_2, \ldots, y_N)$. Then there exists a permutation, π, such that both vectors πx and πy have monotonically increasing coordinates:

$$x_{\pi(1)} \leq x_{\pi(2)} \leq \cdots \leq x_{\pi(N)} \quad \text{and} \quad y_{\pi(1)} \leq y_{\pi(2)} \leq \cdots \leq y_{\pi(N)}.$$

If $\mathcal{S}$ is an anti-monotone set, then there exists a permutation, τ, such that τx has monotonically increasing coordinates and coordinates of τy are monotonically decreasing.

Let us now introduce the co-monotonicity of two-dimensional random vectors. Consider a random vector, (X, Y), where X and Y are discrete random variables. Denote by $\mathcal{S} := \{s_n\}_{n=1}^{\infty}$, $s_n = (x_n, y_n)$, the support of its distribution. The probabilities, $\mathsf{P}(s_n) = \mathbb{P}(X = x_n, Y = y_n)$, satisfy

$$\sum_{n=1}^{\infty} \mathsf{P}(s_n) = 1 \quad \text{and } \mathsf{P}(s_n) > 0 \text{ for all } n,\ (n = 1, 2, \ldots).$$

Definition 9.2 *We call the distribution, P, co-monotone if its support is a co-monotone set. A discrete, bivariate distribution is called anti-monotone if its support is an anti-monotone set.*

---

[7] These problems are often numerically unstable.

Let a random vector (X, Y) take values on the lattice $\mathbb{Z}_+^{(2)} = \{(i,j) : i \geq 0, j \geq 0\}$. Denote $\mathfrak{p}_{i,j} := \mathbb{P}(X = i, Y = j)$, and introduce marginal cdf's $P_i := \mathbb{P}(X \leq i)$, and $Q_j := \mathbb{P}(Y \leq j)$, $(i,j = 0, 1, 2, \dots)$.

Suppose that $\mathsf{P} = \{p(n)\}_{n=0}^{\infty}$ and $\mathcal{Q} = \{q(n)\}_{n=0}^{\infty}$, are discrete distributions on $\mathbb{Z}_+$, and denote by $\mathfrak{D}(\mathsf{P}, \mathcal{Q})$ the class of discrete bivariate distributions,

$$\mathfrak{D}(\mathsf{P}, \mathcal{Q}) = \{f(i,j) : \sum_{j \in \mathbb{Z}_+} f(i,j) = p(i), \sum_{i \in \mathbb{Z}_+} f(i,j) = q(j), i, j \in \mathcal{S}\},$$

with the marginal distributions, P and Q, having finite first and second moments. If $\mathfrak{p} \in \mathfrak{D}(\mathsf{P}, \mathcal{Q})$ is a distribution of a random vector (X, Y), then $\mathbb{E}[XY] < \infty$. Obviously the Poisson distributions, P and Q, belong to $\mathfrak{D}(\mathsf{P}, \mathcal{Q})$.

The main result of this section is the following.

Theorem 9.5 *The distribution, $\mathfrak{p}^{**}$, maximizing the correlation coefficient of X and Y, given marginal distributions, is co-monotone. If a vector (i,j) belongs to the support of $\mathfrak{p}^{**}$, then the probability, $\mathfrak{p}_{i,j}^{**} := \mathbb{P}(X = i, Y = j)$, satisfies*

$$\mathfrak{p}_{i,j}^{**} = \min(P_i, Q_j) - \max(P_{i-1}, Q_{j-1}), i, j = 0, 1, 2, \dots, \tag{9.29}$$

$$\mathfrak{p}_{0,0}^{**} = \min(P_0, Q_0), \tag{9.30}$$

where the probabilities $P_{-1} = Q_{-1} = 0$.

The distribution, $\mathfrak{p}^{*}$, minimizing the correlation coefficient of X and Y is anti-monotone. If a vector (i,j) belongs to the support of $\mathfrak{p}^{*}$ then

$$\mathfrak{p}_{i,j}^{*} = \min(P_i, \bar{Q}_{j-1}) - \max(P_{i-1}, \bar{Q}_j), i, j = 0, 1, 2, \dots, \tag{9.31}$$

where $\bar{Q}_j = 1 - Q_j$ for $j = 0, 1, 2, \dots$ and $\bar{Q}_{-1} := 1$.

The extreme distributions, $\mathfrak{p}^{*}$ and $\mathfrak{p}^{**}$, are unique.

The rest of this subsection is dedicated to the proof of Theorem 9.5. At first, we recall one important result on the monotone sequences of real numbers, that can be found in Hardy *et al.* (1952). This result motivates introduction of the co-monotone or anti-monotone distributions. After that, we prove that the extreme distributions must be co-monotone and find their support and compute the joint probabilities. The reasoning described below is quite general and does not depend on the specific type of the marginal distributions.

Let $x_1 \leq x_2 \leq \dots \leq x_N$ be a monotone sequence of N real numbers, $x_k \in \mathbb{R}$, $(k = 1, 2, \dots, N)$, and $y_1, y_2, \dots, y_N$ be an arbitrary sequence of real numbers, $y_k \in \mathbb{R}$. Denote the inner product of two vectors, $x \in \mathbb{R}^N$, and $y \in \mathbb{R}^N$, by

$$\langle x, y \rangle := \sum_{k=1}^{N} x_k y_k, \quad x = (x_1, \dots, x_N), y = (y_1, \dots, y_N).$$

Lemma 9.4 (Hardy *et al.*, 1952) *For any monotonically increasing sequence, $x_1 \leq x_2 \leq \dots \leq x_N$, and a vector, $y \in \mathbb{R}^N$, there exist permutations, π_+ and π_-, solving the extreme problems*

$$\langle x, \pi_+ y \rangle = \max_{\pi \in \mathfrak{S}_N} \langle x, \pi y \rangle,$$

and

$$\langle x, \pi_{-}y\rangle = \min_{\pi\in\mathfrak{S}_N} \langle x, \pi y\rangle.$$

The permutation, π_+, sorts the vector y in ascending order; the permutation, π_-, sorts the vector y in descending order.

If x and y are co-monotone, then

$$\langle x, y\rangle = \max_{\pi,\tau\in\mathfrak{S}_N} \langle \pi x, \tau y\rangle. \tag{9.32}$$

Indeed,

$$\sum_{i=1}^{N} x_i y_i = \sum_{i=1}^{N} x_{\pi(i)} y_{\pi(i)}, \forall \pi \in \mathfrak{S}_N.$$

By the co-monotonicity assumption, there exists a permutation, π, such that both πx and πy are monotonically increasing. Then, Lemma 9.4 implies (3.32).

From now on, we will consider only discrete random vectors, (X, Y), defined on $\mathbb{Z}^2_+$, the positive quadrant of the two-dimensional lattice. The joint probabilities will be denoted by $\mathfrak{p}_{i,j} := \mathbb{P}(X = i, Y = j)$, $i, j = 0, 1, \ldots$ in what follows.

Consider two monotone sequences, $\{x_n\}_{n=1}^{\infty}$ and $\{y_n\}_{n=1}^{\infty}$, both containing the set of all non-negative, integer numbers satisfying the inequalities $|x_{n+1} - x_n| \le 1$, $|y_{n+1} - y_n| \le 1$. Obviously, these sequences are co-monotone. Take the set $Z \in \mathbb{Z}^2_+$, $Z = \{(x_n, y_n)\}_{n=1}^{N}$, for some fixed N. Let us connect the nodes (x_n, y_n) and (x_{n+1}, y_{n+1}) by the arrows, $(n = 1, 2,\ldots,N - 1)$. Then we obtain a directed path in $\mathbb{Z}^2_+$ (see Figure 9.1a). This path can be viewed as a graph of a monotonically increasing, multivalued function taking integer values.

In the case of the anti-monotone sequences, the path looks similar to that displayed in Figure 9.1b; it can be viewed as a graph of a monotonically decreasing, multivalued function taking integer values. To emphasize that the introduced path contains support of the distribution, we shall call it the *support path*, or, for the sake of brevity, the $\mathcal{S}$-path.

Lemma 9.5 *Suppose that a distribution $\mathfrak{p}$ solves the problem (9.28), $f(\mathfrak{p}) \to \max$. Then $\mathfrak{p}$ is co-monotone.*

Proof. Let (i, j) belong to the support of $\mathfrak{p}$. We shall prove that $\mathfrak{p}_{i+1,j} \cdot \mathfrak{p}_{i,j+1} = 0$. Indeed, the probabilities, $\mathfrak{p}_{i,j}$, satisfy the constraints

$$\mathfrak{p}_{i,j} + \mathfrak{p}_{i,j+1} = \hat{P}_i, \tag{9.33}$$

$$\mathfrak{p}_{i+1,j} + \mathfrak{p}_{i+1,j+1} = \hat{P}_{i+1}, \tag{9.34}$$

where $\hat{P}_i = \sum_{k\neq j,j+1} \mathfrak{p}_{i,k}$ and $\hat{P}_{i+1} = \sum_{k\neq j,j+1} \mathfrak{p}_{i+1,k}$. On the other hand,

$$\mathfrak{p}_{i,j} + \mathfrak{p}_{i+1,j} = \hat{Q}_j, \tag{9.35}$$

$$\mathfrak{p}_{i,j+1} + \mathfrak{p}_{i+1,j+1} = \hat{Q}_{j+1}, \tag{9.36}$$

where $\hat{Q}_j = \sum_{l\neq i,i+1} \mathfrak{p}_{l,j}$ and $\hat{Q}_{j+1} = \sum_{l\neq i,i+1} \mathfrak{p}_{l,j+1}$. Obviously,

$$\hat{P}_i + \hat{P}_{i+1} = \hat{Q}_j + \hat{Q}_{j+1}. \tag{9.37}$$

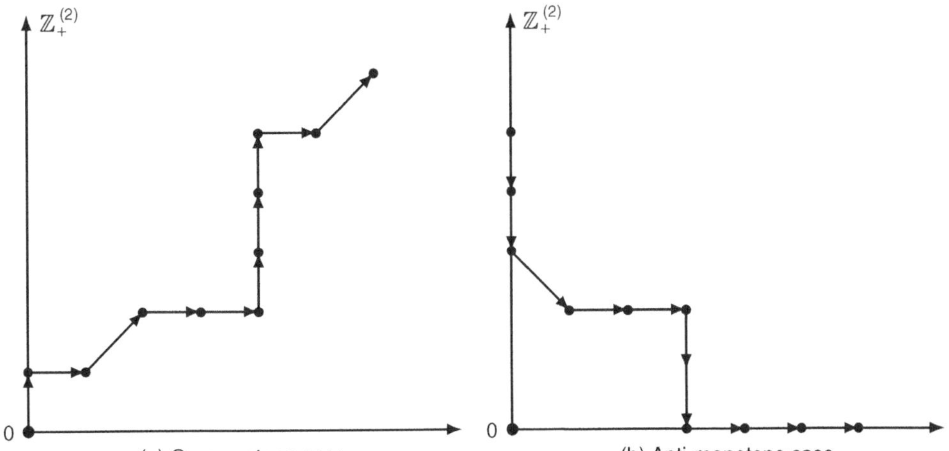

(a) Co-monotone case. (b) Anti-monotone case.

Figure 9.1 Typical monotone paths.

The function $f(\mathfrak{p})$ can be written as

$$f(\mathfrak{p}) = \sum_{\alpha} kl\mathfrak{p}_{k,l} + \sum_{\beta} kl\mathfrak{p}_{k,l},$$

where the set of indices, α, contains only four indices $\alpha = \{(i,j), (i,j+1), (i+1,j), (i+1, j+1)\}$; and β is the complementary set, $\beta = \mathbb{Z}_+^2 \backslash \alpha$.

Denote $t = \mathfrak{p}_{i+1,j+1}$, $t \geq 0$. Equations (9.33–9.37) imply

$$\mathfrak{p}_{i+1,j} = \hat{P}_{i+1} - t,$$
$$\mathfrak{p}_{i,j} = \hat{Q}_j - \hat{P}_{i+1} + t, \text{ and}$$
$$\mathfrak{p}_{i,j+1} = \hat{P}_i - \hat{Q}_j + \hat{P}_{i+1} - t.$$

We have $\max(0, \hat{P}_{i+1} - \hat{Q}_j) \leq t \leq \min(\hat{P}_{i+1}, \hat{Q}_{j+1})$. The objective function can be written as

$$f(\mathfrak{p}) = \sum_{(k,l)\in\beta} kl\mathfrak{p}_{k,l} + ij\cdot(\hat{P}_i + \hat{P}_{i+1}) + j\cdot\hat{P}_{i+1} + i\hat{Q}_{j+1} + t.$$

The solution to the problem, $f(\mathfrak{p}) \to \max$, is attained at the boundary, $t = t_{\max}$. In our case, $t_{\max} = \min(\hat{P}_{i+1}, \hat{Q}_{j+1})$. If $t_{\max} = \hat{P}_{i+1}$, then $\mathfrak{p}_{i+1,j} = 0$. If $t_{\max} = \hat{Q}_{j+1}$, then from (9.37) we obtain $\mathfrak{p}_{i,j+1} = 0$. If $\hat{P}_{i+1} = \hat{Q}_{j+1}$), then $\mathfrak{p}_{i+1,j} = \mathfrak{p}_{i,j+1} = 0$.

A similar statement can be proved for the minimization problem.

Lemma 9.6 *Let $\mathfrak{p}$ be a solution to problem (9.28), $f(\mathfrak{p}) \to \min$. Then $\mathfrak{p}$ is anti-monotone.*

Monotonicity of the distributions is directly related to monotonicity of the samples of random variables from these distributions.

Lemma 9.7 *Consider a finite random sample, $Z = \{Z_n\}_{n=1}^N$, of independent, two-dimensional vectors $Z_n = (X_n, Y_n)'$ from a co-monotone distribution. Then, Z is co-monotone.*

If a random sample of N independent two-dimensional vectors, Z, has co-monotone coordinates for any integer $N \geq 2$, then Z is a sample from a co-monotone distribution.

Proof. Consider an independent, finite sample, Z, from a co-monotone distribution. Let us find a permutation, π, ordering the first coordinate, $\{X_n\}_{n=1}^N$, and apply this permutation to the second coordinate. Then we obtain

$$\hat{X}_1 \leq \hat{X}_2 \leq \cdots \leq \hat{X}_N, \quad \text{where } \hat{X} = \pi X.$$

If the second coordinate, $\hat{Y} = \pi Y$, is also ordered, the proof is finished. Suppose, on the contrary, there is a couple of indices, $i < j$, such that $\hat{Y}_i > \hat{Y}_j$; then the first coordinate must satisfy the relation $\hat{X}_i = \hat{X}_j$, otherwise the co-monotonicity condition of the support is not satisfied. Then the transposition of the elements, $\hat{Y}_i$ and $\hat{Y}_j$, puts these elements in ascending order. After a finite number of transpositions, we find a permutation, τ, ordering the second coordinate and keeping the first coordinate without changes. Then the permutation, $\tau \cdot \pi$, makes Z monotone coordinate-wise.

To prove the converse statement, let us assume that support of the bivariate distribution is not co-monotone. Then there exist two elements of the support, $s_k = (x_k, y_k)$ and $s_l = (x_l, y_l)$, such that $x_k \leq x_l$ but $y_k > y_l$. Since both elements of the support have positive probabilities, there exists sufficiently large N such that the sample of independent vectors contains both vectors, s_k and s_l. Then Z is not co-monotone.

Lemma 9.8 is a direct analog of Lemma 9.7 for the anti-monotone distributions.

Lemma 9.8 *Consider a finite random sample, $Z = (Z_1, \ldots, Z_N)$, of independent, two-dimensional vectors $Z_n = (X_n, Y_n)'$ from an anti-monotone distribution. Then, the set $\{Z_n\}_{n=1}^N$ is anti-monotone. If a random sample of N independent two-dimensional vectors, Z, has anti-monotone coordinates for any integer $N \geq 2$, then Z is a sample from an anti-monotone distribution.*

Let us now find the joint probabilities and compute the support of the extreme distributions. We will use a probabilistic argument based on the Strong Law of Large Numbers (SLLN) (Johnson *et al.*, 1997). Let us start with the distribution $\mathfrak{p}^{**}$, maximizing the correlation coefficient. Consider a sample, $\{(\tilde{X}_n, \tilde{Y}_n)\}_{n=1}^N$, of the size, N, from the distribution $\mathfrak{p}^{**}$. There exists a permutation, π, such that the vectors $X = \pi\tilde{X}$ and $Y = \pi\tilde{X}$ both have monotone coordinates. As $N \to \infty$, we obtain two sequences of increasing length,

$$\begin{array}{ll} X: & \overbrace{0,0,\ldots,0}^{N_X(0)}, \overbrace{1,1,\ldots,1}^{N_X(1)}, \overbrace{2,2,\ldots,2}^{N_X(2)}, \ldots, \overbrace{k,k,\ldots,k}^{N_X(k)}, \ldots \\ Y: & \underbrace{0,\ldots,0}_{N_Y(0)}, \underbrace{1,\ldots,1}_{N_Y(1)}, \underbrace{2,\ldots,2}_{N_Y(2)}, \ldots, \underbrace{k,k,k,\ldots,k,k}_{N_Y(k)}, \ldots, \end{array} \tag{9.38}$$

with the number of elements, $N_X(k)$ and $N_Y(k)$, in the sequences satisfying, by the SLLN, the relations:

$$\lim_{N\to\infty}\frac{N_X(k)}{N}=p_k, k=0,1,2,\ldots, \text{almost surely}, \tag{9.39}$$
$$\lim_{N\to\infty}\frac{N_Y(k)}{N}=q_k, k=0,1,2,\ldots, \text{almost surely}.$$

The probabilities, $P_i=\mathbb{P}(X_k\le i)$ and $Q_j=\mathbb{P}(Y_k\le j)$, satisfy

$$P_i=\sum_{l=0}^{i}p_l,\quad Q_j=\sum_{l=0}^{j}q_l,\quad i,j=0,1,2,\ldots.$$

Let us denote the number of pairs, (i,j), by N_{ij}. In particular, the number of pairs, $(0,0)$, is

$$N_{00}=\min(N_X(0),N_Y(0)). \tag{9.40}$$

Lemma 9.9 *The limits,*

$$\mathfrak{p}_{ij}^{**}=\lim_{N\to\infty}\frac{N_{ij}}{N}$$

exist almost surely for all $i,j=0,1,2,\ldots$, *as* $N\to\infty$. *They satisfy equation (9.29).*

Proof. In the case of the state $(0,0)$, existence of the limit, $\mathfrak{p}_{0,0}^{**}$, follows from the continuity of the function min, equation (9.40), and existence of the limits in (9.39). Let us prove the statement of the lemma for $i>0$ and $j>0$. Consider the sequence of vectors, (X_k,Y_k), $(k=1,2,\ldots)$. It is not difficult to see that N_{ij} satisfies the relation

$$N_{ij}=[\min(M_X(i),M_Y(j))-\max(M_X(i-1),M_Y(j-1))]^+, \tag{9.41}$$

where

$$M_X(i)=\sum_{l=0}^{i}N_X(l)\quad\text{and}\quad M_Y(i)=\sum_{l=0}^{i}N_Y(l),$$

and, as usual, $x^+:=\max(0,x)$. Indeed, the index, k, of the pair, (i,j), in the sequence satisfies the inequalities $k\ge\max(M_X(i-1),M_Y(j-1))$ and $k\le\min(M_X(i),M_Y(j))$. Equation (9.41) follows from these inequalities.

Passing to the limit as $N\to\infty$, we obtain that almost surely

$$\lim_{N\to\infty}\frac{M_X(i)}{N}=P_i,\quad\text{and}\quad\lim_{N\to\infty}\frac{M_Y(i)}{N}=Q_i,\quad i=0,1,\ldots.$$

Existence of the limits, $\mathfrak{p}_{ij}^{**}$, now follows from (9.41) and continuity of the functions $\min(\cdot,\cdot)$ and $\max(\cdot,\cdot)$. From (9.41), we find that the probability, $\mathfrak{p}_{i,j}$, of the state (i,j) satisfies:[8]

$$\mathfrak{p}_{i,j}^{**}=\min(P_i,Q_j)-\max(P_{i-1},Q_{j-1}).$$

Equation (9.29) is thus proved.

[8] If $\min(P_i,Q_j)\le\max(P_{i-1},Q_{j-1})$, then the vector (i,j) does not belong to the support.

In the case of minimal correlation, for any sample, $\{(\tilde{X}_n, \tilde{Y}_n)\}_{n=1}^{N}$, there exists a permutation, π, sorting the first coordinate in ascending order and the second coordinate in descending order:

$$X_1 \leq X_2 \leq \ldots \leq X_N, \quad \text{where } X = \pi\tilde{X},$$

and the second coordinate forms a decreasing sequence:

$$Y_1 \geq Y_2 \geq \ldots \geq Y_N, \quad \text{where } Y = \pi\tilde{X}.$$

As $N \to \infty$, we obtain two sequences,

$$\begin{aligned}
X &: \quad \overbrace{0,\ldots,0}^{N_X(0)},\ldots \overbrace{i-1,\ldots,i-1}^{N_X(i-1)}, \overbrace{i,i,\ldots,i}^{N_X(i)},\ldots, \overbrace{k,k,\ldots,k}^{N_X(k)},\ldots \\
Y &: \quad \ldots, \underbrace{j+1, j+1\ldots, j+1}_{N_Y(j+1)}, \underbrace{j,\ldots,j}_{N_Y(j)}, \underbrace{j-1,\ldots,j-1}_{N_Y(j-1)},\ldots, \underbrace{0,\ldots,0,0}_{N_Y(0)}
\end{aligned} \tag{9.42}$$

with the number of elements, $N_X(k)$ and $N_Y(k)$, satisfying, by the SLLN, equation (9.39) almost surely.

Lemma 9.10 *The limits* $\mathfrak{p}^*_{i,j} = \lim\limits_{N\to\infty} \frac{N_{i,j}}{N}$ *exist almost surely for all* $i, j = 0, 1, 2, \ldots$, *as* $N \to \infty$, *and satisfy equation (9.31).*

Proof. The proof is a minor modification of that of Lemma 9.9. Let for some k, $X_k = i$ and $Y_k = j$. Notice that $k \geq N - M_Y(j)$ (see (9.42)). Then we have

$$N_{i,j} = \min(M_X(i), N - M_Y(j-1)) - \max(M_X(i-1), N - M_Y(j)), \tag{9.43}$$

Passing to the limit in (9.43), as $N \to \infty$, we derive (9.31).

Proof of Lemma 9.10 completes the derivation of Equations (9.29)–(9.31). Uniqueness of the extreme distributions, $\mathfrak{p}^*$ and $\mathfrak{p}^{**}$, follows immediately from (9.29)–(9.31). Theorem 9.5 is thus proved.

9.5.3 *Computation of the Joint Distribution*

Equation (9.29), in principle, allows one to find all the elements of the support of the maximal distribution and to compute the probabilities. In practice, this computation can be done much more efficiently by the algorithm described in this section. The idea of the extreme joint distribution (EJD) algorithm is to walk along the $\mathcal{S}$-path, instead of considering all possible pairs, (i, j), satisfying Equation (9.29).

It is useful to look again at the structure of the sample, represented by (9.38). When N is fixed, we have segments of a random length containing $N_X(k)$ elements equal to k. As $N \to \infty$ the length of the segment, the random variables, $N_X(i)$ and $N_Y(j)$, also tend to infinity, but the ratios (9.39) converge to finite limits.

In the limit, the scaled diagram (9.38) will look like two partitions of the unit interval, [0, 1] with the nodes, $\Pi_X = \{P_0, P_1, \ldots\}$ and $\Pi_Y = \{Q_0, Q_1, \ldots\}$, defined by the cdf's.

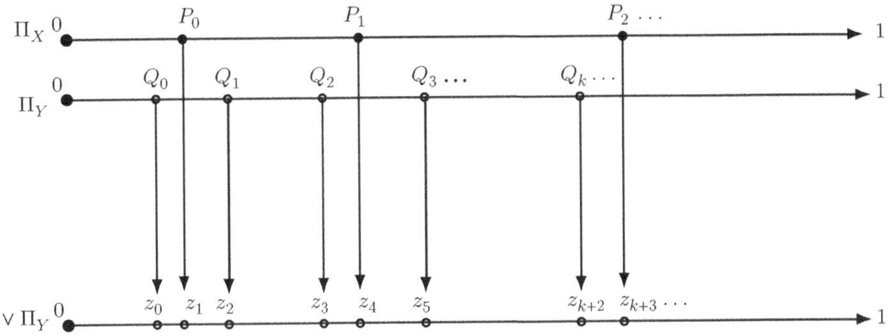

Figure 9.2 Partitions of the unit interval: corr$(X, Y) \to$ max.

The partition, $\Pi_Z = \Pi_X \vee \Pi_Y$, is formed by the union of the nodes of the partitions Π_X and Π_Y; see Figure 9.2. We denote the nodes of the partition, Π_Z, by $z_0 \leq z_1 \leq \ldots$.

It is convenient to assume that each node of Π_Z "remembers" the original partition it came from (Π_X or Π_Y), its index in the original partition, and its cumulative probability. The initial element of the support, $\{(x_k, y_k)\}_{k=0}^{\infty}$, is $(0, 0)$; the probability of this state is $z_0 = \min(P_0, Q_0)$. The algorithm scans the elements of the partition, Π_Z; finds the support of the extreme, bivariate distribution; and computes the joint probabilities, $\mathfrak{p}_{i,j}$.

9.5.3.1 EJD Algorithm

We shall start with the case of maximal correlation.

Step 0. $k := 0$; $x_k := 0, y_k := 0$; $\mathfrak{p}_{0,0} = z_0$.
Step 1. $k := k + 1$.
Step 2. If $z_{k-1} = P_i$ for some i and $z_{k-1} \neq Q_j$ for all j, then $x_k = i + 1$ and $y_k = y_{k-1}$. If $z_{k-1} = Q_j$ for some j and $z_{k-1} \neq P_i$ for all i, then $x_k = x_{k-1}$ and $y_k = j + 1$. If there exist such i and j that $z_{k-1} = P_i = Q_j$, then $x_k = i + 1$ and $y_k = j + 1$.
Step 3. Assign the kth element of the support, (x_k, y_k).
Step 4. $\mathfrak{p}^{**}_{x_k,y_k} := z_k - z_{k-1}$.
Step 5. Go to Step 1.

In the case of minimal correlation, we "flip" the distribution function and use the tail probabilities, $\ldots, 1 - Q_k, 1 - Q_{k-1}, \ldots, 1 - Q_0$, instead of Q_k (see Figure 9.3).

Remark 9.6 *One can think of the EJD algorithm as an observer moving along the $\mathcal{S}$-path in $\mathbb{Z}^2_+$; as such, the time to get to the kth node is z_k. Each time the observer's position is the kth node, (x_k, y_k), of the support, she records the coordinates of the node into the list and marks the node with the probability $\mathfrak{p}^{**}_{x_k,y_k}$.*

9.5.4 *On the Frechet–Hoeffding Theorem*

In the discrete case, the Frechet–Hoeffding theorem can be formulated as follows. Consider a space $\mathfrak{D}(\mathsf{P}, \mathcal{Q})$ of discrete bivariate distributions, with the marginal distributions, P and $\mathcal{Q}$.

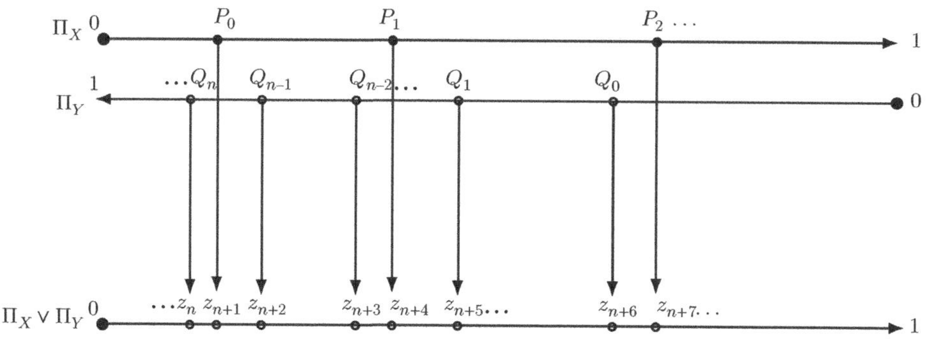

Figure 9.3 Partitions of the unit interval: corr$(X, Y) \to$ min.

Theorem 9.6 (Frechet, 1951; Hoeffding, 1940) *The bivariate cdf,* $H^*(i,j) := \mathbb{P}(X \leq i, Y \leq j)$*, maximizing the correlation of X and Y is*

$$H^*(i,j) = \min(P_i, Q_j),\ i,j \in \mathbb{Z}_+^2. \tag{9.44}$$

The bivariate cdf, $H_*(i,j)$*, minimizing the correlation of X and Y is*

$$H_*(i,j) = \max(0, P_i + Q_j - 1),\ i,j \in \mathbb{Z}_+^2. \tag{9.45}$$

Proposition 9.4 Theorem 9.5 is equivalent to the Frechet–Hoeffding theorem.

Proof. We shall only prove that Theorem 9.5 implies the Frechet–Hoeffding theorem in the case of maximimal correlation. It is enough to prove (9.44) in the case that the node, (i,j), belongs to the $\mathcal{S}$-path. We give the proof by induction on the index of the node.

Let us check the basis of induction. The first node is $(0, 0)$. If $i = j = 0$, then $\mathbb{P}(X \leq 0, Y \leq 0) = \mathbb{P}(X = 0, Y = 0) = \min(P_0, Q_0)$. Thus, equation (9.44) is satisfied.

Suppose that the statement is true for the kth node, (i,j): $\mathbb{P}(X \leq i, Y \leq j) = \min(P_i, Q_j)$. Let us prove the statement for the $(k + 1)$st node.

There are three possibilities: the $(k + 1)$st node is $(i + 1, j)$, the $(k + 1)$st node is $(i + 1, j + 1)$, or the $(k + 1)$st node is $(i, j + 1)$. In the first case, using Equation (9.29), we find

$$\begin{aligned}\mathbb{P}(X \leq i+1, Y \leq j) &= \mathbb{P}(X \leq i, Y \leq j) + \mathfrak{p}_{i+1,j} \\ &= \min(P_i, Q_j) + \min(P_{i+1}, Q_j) - \max(P_i, Q_{j-1}).\end{aligned}$$

Since the $\mathcal{S}$-path contains the arrow from the node (i,j) to $(i + 1, j)$, we have $P_i < Q_j$. On the other hand, since the second coordinate of the node is j, $P_i > Q_{j-1}$. Thus, $\max(P_i, Q_{j-1}) = P_i$. Then we obtain that

$$\min(P_i, Q_j) = \max(P_i, Q_{j-1}) = P_i.$$

Therefore,

$$\mathbb{P}(X \leq i+1, Y \leq j) = \min(P_{i+1}, Q_j),$$

as was to be proved in the first case.

In the second case, the $(k+1)$st node is $(i+1, j+1)$. The diagonal segment in the $\mathcal{S}$-path appears only if $P_i = Q_j$. Then we have

$$\begin{aligned}\mathbb{P}(X \leq i+1, Y \leq j+1) &= \mathbb{P}(X \leq i, Y \leq j) + \mathfrak{p}_{i+1,j+1} \\ &= \min(P_i, Q_j) + \min(P_{i+1}, Q_{j+1}) - \max(P_i, Q_j) \\ &= \min(P_{i+1}, Q_{j+1}).\end{aligned}$$

The last case is analogous to the first one.

9.5.5 Approximation of the Extreme Distributions

For practical computations, the marginal distributions, $\mathsf{P} = \{p_i\}_{i\geq 0}$ and $\mathcal{Q} = \{q_j\}_{j\geq 0}$, with two finite moments can be approximated by the distributions with finite support, $\tilde{\mathsf{P}} = \{\tilde{p}_i\}_{i\geq 0}$ and $\tilde{\mathcal{Q}} = \{\tilde{q}_j\}_{j\geq 0}$, such that

$$\max_{i\leq I_*} |P_i - \tilde{P}_i| \leq \epsilon, \quad 1 - P_{I_*} \leq \epsilon,$$

$$\max_{j\leq J_*} |Q_j - \tilde{Q}_j| \leq \epsilon, \quad 1 - Q_{J_*} \leq \epsilon,$$

where $P_n = \sum_{i=0}^{n} p_i$, $Q_m = \sum_{j=0}^{m} q_j$, $\tilde{P}_n = \sum_{i=0}^{n} \tilde{p}_i$, and $\tilde{Q}_m = \sum_{j=0}^{m} \tilde{q}_j$. Let $\mathfrak{p} \in \mathfrak{D}(\mathsf{P}, \mathcal{Q})$ be an extreme bivariate distribution and $\tilde{\mathfrak{p}} \in \mathfrak{D}(\tilde{\mathsf{P}}, \tilde{\mathcal{Q}})$ be the corresponding extreme bivariate distribution with the marginals, $\tilde{\mathsf{P}}$ and $\tilde{\mathcal{Q}}$. Then it follows from Theorem 9.5 that $\tilde{\mathfrak{p}}$ satisfies the inequality

$$\sup_{\substack{i\geq 0,\\ j\geq 0}} |\mathfrak{p}_{i,j} - \tilde{\mathfrak{p}}_{i,j}| \leq 2\epsilon.$$

Similar inequality can be obtained for the approximation of the correlation coefficient. We shall formulate one sufficient condition that guarantees the approximation.

Proposition 9.5 Suppose that the first and the second moments of the marginal distributions are finite. Then, given $\epsilon > 0$, one can find the integer numbers, I_* and J_*, such that there exists a discrete bivariate distribution $\tilde{\mathfrak{p}}_{i,j}, 0 \leq i \leq I_*, 0 \leq j \leq J_*$ approximating the extreme distribution, $\mathfrak{p}_{i,j}$, in the following sense:

$$\sup_{i\geq 0}\sup_{j\geq 0} |\ \mathfrak{p}_{i,j} - \tilde{\mathfrak{p}}_{i,j}\ | < \epsilon, \tag{9.46}$$

$$\left|\sum_{i=1}^{\infty}\sum_{j=1}^{\infty} ij\cdot(\mathfrak{p}_{i,j} - \tilde{\mathfrak{p}}_{i,j})\right| < 3\epsilon. \tag{9.47}$$

Proof. We consider the case of the distribution, $\mathfrak{p} = \mathfrak{p}^{**}$, maximizing the correlation coefficient. Notice that the first two finite moments of a discrete random variable taking nonnegative, integer values satisfy the relations

$$\mathbb{E}[\xi^2] = 2\sum_{k=1}^{\infty} k\mathbb{P}(\xi > k) + \mathbb{E}[\xi], \tag{9.48}$$

$$\mathbb{E}[\xi] = \sum_{k=0}^{\infty} \mathbb{P}(\xi > k). \tag{9.49}$$

Since $\mathbb{E}[\xi^2] < \infty$, Equations (9.48) and (9.49) imply convergence of the series $\sum_{k=1}^{\infty} k\mathbb{P}(\xi > k)$. Then we can find I_* and J_* such that

$$\sum_{i>I_*} i(1 - P_i) < \epsilon/2, \tag{9.50}$$

$$\sum_{j>J_*} j(1 - Q_j) < \epsilon/2. \tag{9.51}$$

Consider the distribution functions, $\tilde{P}_i$ and $\tilde{Q}_j$:

$$\tilde{P}_i = P_i, i = 0, 1, \ldots, I_* - 1;\ \tilde{P}_i = 1, i \geq I_*, \tag{9.52}$$

$$\tilde{Q}_j = Q_j, j = 0, 1, \ldots, J_* - 1;\ \tilde{Q}_j = 1, j \geq J_*. \tag{9.53}$$

Let $\tilde{\mathfrak{p}}$ be the extreme bivariate distribution, maximizing the correlation, corresponding to the marginals, $\tilde{\mathsf{P}}$ and $\tilde{\mathcal{Q}}$. It is not difficult to see that $\tilde{\mathfrak{p}}_{i,j} = 0$ for $i \geq I_*$ or $j > J_*$.

The first moments of the marginal distributions satisfy the inequalities

$$\left|\sum_{i=0}^{\infty}(1 - P_i) - \sum_{i=0}^{\infty}(1 - \tilde{P}_i)\right| < \frac{\epsilon}{I_*}, \quad \text{and} \quad \left|\sum_{j=0}^{\infty}(1 - Q_j) - \sum_{j=0}^{\infty}(1 - \tilde{Q}_j)\right| < \frac{\epsilon}{J_*}.$$

Indeed,

$$\begin{aligned}\left|\sum_{i=0}^{\infty}(1 - P_i) - \sum_{i=0}^{\infty}(1 - \tilde{P}_i)\right| &= \left|\sum_{i=I_*}^{\infty}(1 - P_i) - \sum_{i=I_*}^{\infty}(1 - \tilde{P}_i)\right| \\ &\leq \frac{1}{I_*}\sum_{i=I_*}^{\infty} i\cdot(1 - P_i).\end{aligned}$$

The second inequality is derived analogously. Consider the expected values

$$\mathcal{E} := \mathbb{E}[XY] = \sum_{i\geq 1}\sum_{j\geq 1} ij\mathfrak{p}^{**}_{i,j} \quad \text{and} \quad \tilde{\mathcal{E}} := \sum_{i\geq 1}\sum_{j\geq 1} ij\tilde{\mathfrak{p}}_{i,j}.$$

We have

$$\mathcal{E} - \tilde{\mathcal{E}} = \sum_{i=I_*}^{\infty}\sum_{j=J_*}^{\infty} ij\mathfrak{p}^{**}_{ij} + \sum_{i=I_*}^{\infty}\sum_{j=1}^{J_*} ij\mathfrak{p}^{**}_{ij} + \sum_{j=J_*}^{\infty}\sum_{i=1}^{I_*} ij\mathfrak{p}^{**}_{ij}$$

The first sum, $S_1 := \sum_{i=I_*}^{\infty} \sum_{j=J_*}^{\infty} ij\mathfrak{p}_{ij}^{**}$, satisfies the inequality

$$S_1^2 \leq \sum_{i=I_*}^{\infty} i^2 \sum_{j=0}^{\infty} \mathfrak{p}_{i,j}^{**} \cdot \sum_{j=J_*}^{\infty} j^2 \sum_{i=0}^{\infty} \mathfrak{p}_{i,j}^{**} < \epsilon^2.$$

The second sum, $S_2 := \sum_{i=I_*}^{\infty} \sum_{j=1}^{J_*} ij\mathfrak{p}_{ij}^{**}$, satisfies the inequality

$$S_2 \leq J_* \sum_{i=I_*}^{\infty} \sum_{j=0}^{\infty} i\mathfrak{p}_{ij}^{**} \leq J_* \frac{\epsilon}{J_*} = \epsilon,$$

and the third sum, $S_3 := \sum_{j=J_*}^{\infty} \sum_{i=1}^{I_*} ij\mathfrak{p}_{ij}^{**}$, also satisfies $S_3 \leq \epsilon$. Putting all the pieces together, we obtain $| \mathcal{E} - \tilde{\mathcal{E}} | \leq 3\epsilon$.

Remark 9.7 *Similar reasoning allows us to prove continuity of the function* $f(\mathfrak{p}) = \sum_{i\geq 1} \sum_{j\geq 1} ij\mathfrak{p}_{i,j}$ *on the space* $\mathfrak{D}(P, Q)$.

9.6 Numerical Results

9.6.1 Examples of the Support

We shall start with the support of the joint distribution maximizing the correlation coefficient when $\lambda = \mu$. In this case, the support is a set of integer points, $\{(k,k)\}_{k=0}^{\infty}$, on the main diagonal; the correlation coefficient is $\rho^{**}(\lambda, \lambda) = 1$.

Figure 9.4 displays the support of the extreme distribution, $\mathfrak{p}^{**}$, with the intensities $\lambda = 3$ and $\mu = 4$. We apply the EJD algorithm with ϵ-approximation[9] of the marginal distributions, where $\epsilon = 10^{-5}$. It allows us to find (numerically) the joint probabilities, $\mathfrak{p}_{ij}^{**}$, and the correlation coefficient,

$$\rho^{**}(\lambda, \mu) = \frac{\sum_{i=1}^{\infty} \sum_{j=1}^{\infty} ij\mathfrak{p}_{ij}^{**} - \lambda\mu}{\sqrt{\lambda\mu}}. \tag{9.54}$$

The error of the approximation of the correlation coefficient $\epsilon(\rho) < 10^{-4}$. The maximal correlation $\rho^{**}(\lambda, \mu) = 0.979$ in this case. One can notice that the support of the joint distribution is located close to the main diagonal, but for large values of the first coordinate, some deviation from the diagonal pattern is observed. The reason for that is inequality of the intensities of the Poisson processes, $\lambda < \mu$, leading to a re-balance of the probability mass of the bivariate distribution maximizing the correlation coefficient.

In the case $\lambda = 3, \mu = 6$, the deviation of the support from the main diagonal increases as $k \to \infty$ (see Figure 9.5). Applying the EJD algorithm, we find the joint probabilities, $\mathfrak{p}_{ij}^{**}$, and from (9.54) the correlation coefficient, $\rho^{**}(3, 6) = 0.977$.

Let us now consider an example of the support for the minimal (negative) correlation of the random variables. We take $\lambda = 3$ and $\mu = 6$. The joint probabilities, $\mathfrak{p}_{ij}^{*}$, can be found by the EJD algorithm. In this case, the minimal value of the correlation coefficient is $\rho_* = -0.944$. The geometric pattern of the support is different in this case because the minimal value of the correlation coefficient is negative.

[9] The same approximation of the marginal distributions is used in all examples considered in this section.

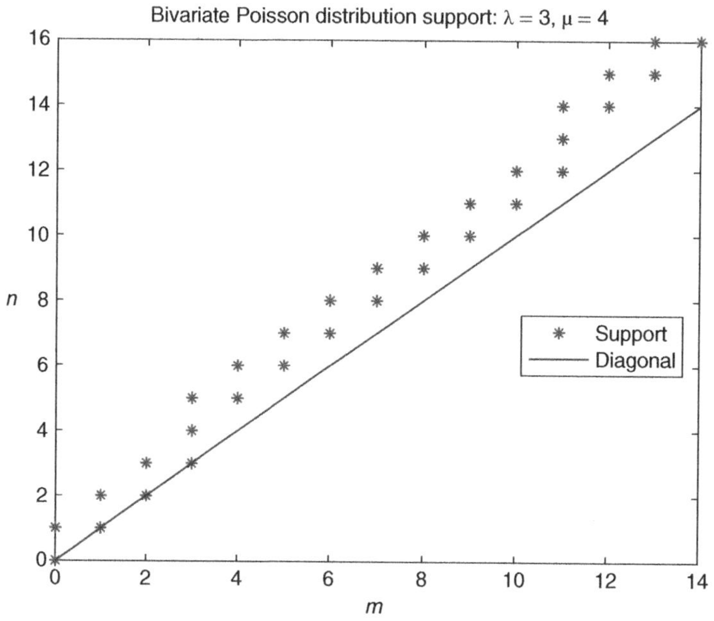

Figure 9.4 Support of the distribution $\mathfrak{p}^{**}$: $\lambda = 3, \mu = 4$.

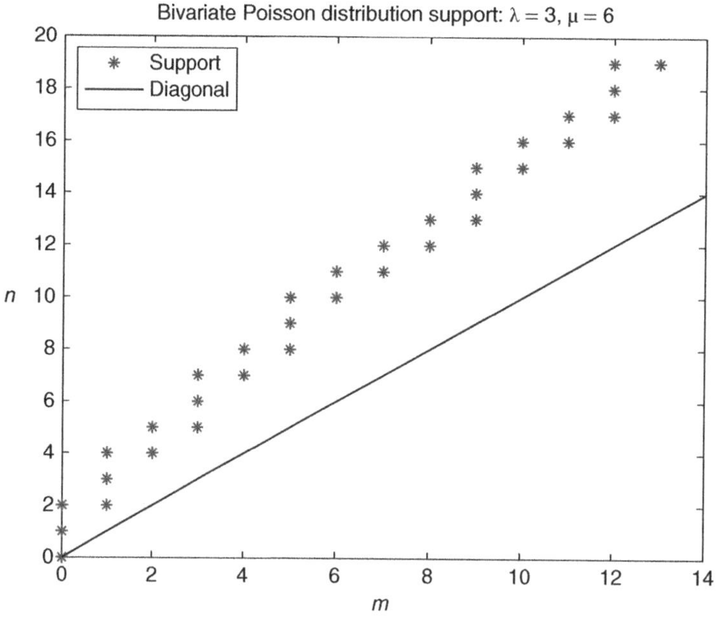

Figure 9.5 Support of the distribution $\mathfrak{p}^{**}$: $\lambda = 3, \mu = 6$.

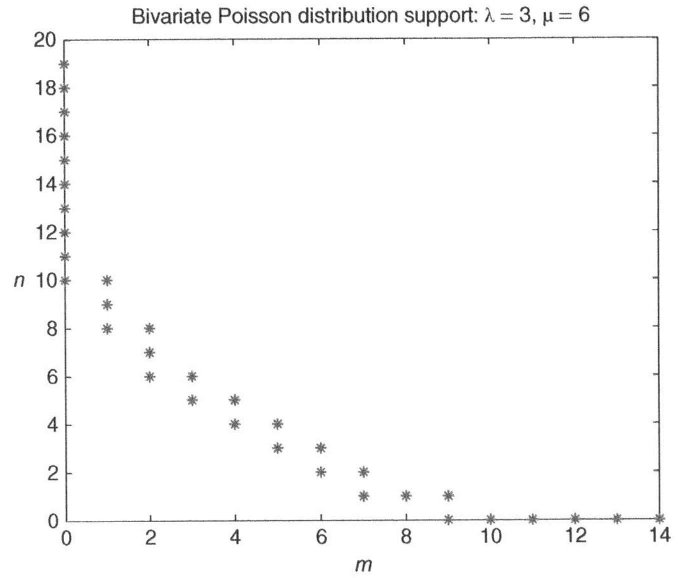

Figure 9.6 Support of the distribution $\mathfrak{p}^*$: $\lambda = 3, \mu = 6$.

9.6.2 *Correlation Boundaries*

The correlation coefficients, $\rho^*(\lambda, \mu)$ and $\rho^{**}(\lambda, \mu)$, are not monotone functions of λ as intensity μ is fixed.[10] The extreme correlation coefficients are shown in Figure 9.7 for $\mu = 3$. We also mention here that the lower and upper boundaries are not symmetrical. It is proved in Griffiths *et al.*, (1979) that the two-dimensional process with variable correlation between the components cannot be infinitely divisible. The latter fact may look surprising because the marginals are the Poisson distributions belonging to the class of infinitely divisible distributions.

Let us now compare the correlation boundaries obtained using the traditional, forward simulation method and the BS method. We shall call the boundaries computed by the forward simulation (FS) the FS boundaries. The boundaries computed by the backward simulation method will be called the BS boundaries.

In Figure 9.8, the correlation boundaries are displayed for the intensities $\lambda = 8$ and $\mu = 4$. The time horizon $T = 3$. Recall that under the FS approach, we have to generate two sequences, ΔT_k and $\Delta T'_k$, such that

$$\mathbb{P}(\Delta T_k \le t) = 1 - e^{-\lambda t}, \quad \mathbb{P}(\Delta T'_k \le t) = 1 - e^{-\mu t}, \quad t \ge 0,$$

and the random variables ΔT_k and $\Delta T'_k$ have minimal possible correlation for each k; at the same time, each of the sequences $\{\Delta T_k\}_{k\ge 0}$ and $\{\Delta T'_k\}_{k\ge 0}$ is formed by independent identically distributed random variables.

[10] This fact is established numerically. The joint probabilities should be computed with the error not exceeding 10^{-4} to provide proof of this fact.

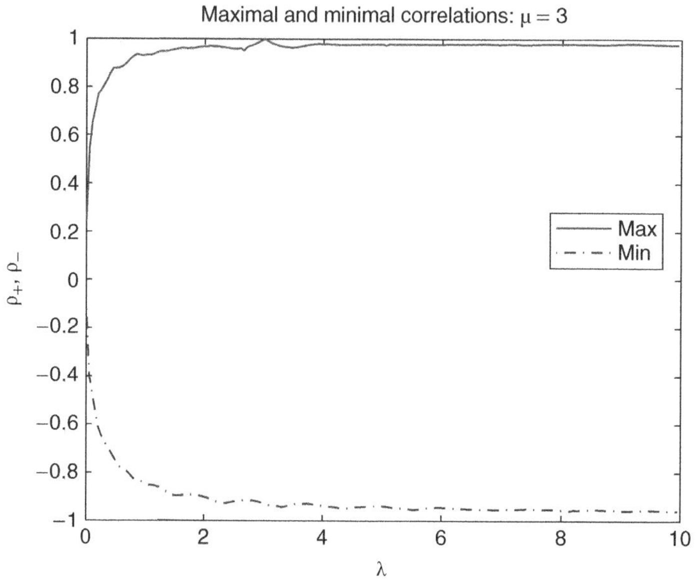

Figure 9.7 Correlation boundaries: $\mu = 3$.

According to the Frechet–Hoeffding theorem, the interarrival times, ΔT_k and $\Delta T'_k$, should satisfy the relation

$$\exp\,(-\lambda\cdot\Delta T_k) + \exp\,(-\mu\cdot\Delta T'_k) = 1, \quad k = 1, 2, \ldots,$$

to provide minimal possible correlation between the interarrival times. The correlation coefficient, $\rho(t) = \mathrm{corr}(N_t, N'_t)$, is computed using a Monte Carlo simulation with 900,000 Monte Carlo samples.

The computation of the FS upper boundary in Figure 9.8 is also based on the Frechet–Hoeffding theorem. According to the latter, the interarrival times, ΔT_k and $\Delta T'_k$, should satisfy the relation

$$\mu\Delta T_k = \lambda\Delta T'_k, k = 1, 2, \ldots,$$

to provide maximal possible correlation between the arrival moments. One can see that the maximal correlation of the processes, N_t and N'_t, under the FS is a constant, while under the BS it is a linear function of time. In the case $\lambda \neq \mu$, the BS allows one to reach stronger correlations of the processes than the FS of the arrival moments.

9.7 Backward Simulation of the Poisson–Wiener Process

The BS approach is applicable to the class of multivariate stochastic process with the components formed by either Poisson or Gaussian processes. In this section, we analyze the basic bivariate process, $\boldsymbol{X}_t = (N_t, W_t)$, where N_t is a Poisson process with intensity μ; and W_t is the Wiener process.

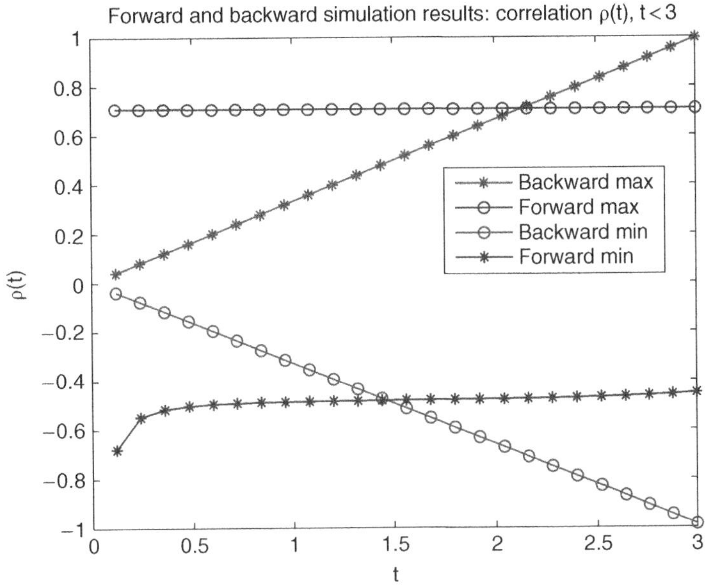

Figure 9.8 Comparison of correlation boundaries: $\lambda = 8, \mu = 4$.

Suppose that the joint distribution of $\boldsymbol{X}_t$ at time $t = T$ is known. Let $\{X(k)\}_{k=1}^{M}, X(k) = (N_T(k), W_T(k))$ be a random sample of the size M from this distribution. For each k, we apply the BS technique to simulate the first coordinate and the Brownian bridge construction to simulate the second coordinate.

More precisely, consider the kth random vector, $X(k) = (N_T(k), W_T(k))$, $(k = 1, 2, \ldots, M)$. Simplifying the notation, denote $n = N_T(k)$ and $x = W_T(k)$. Let us generate the arrival moments of the Poisson process, $0 \le T_1 < T_2 < \ldots < T_n \le T$. As before, the arrival moments are obtained by sorting the sequence of n independent identically distributed random variables, having a uniform distribution in the interval $[0, T]$, in ascending order.

The second coordinate of the process, $\boldsymbol{X}_t$, is generated recursively using the Brownian bridge construction. Suppose that the process W_t has already been generated at times $T_n > T_{n-1} > \ldots > T_k$ and $W_{T_k} = x_k$. Then, at time $t = T_{k-1}$, W_t has a normal distribution, $\mathcal{N}(m(t), \sigma^2(t))$, with the parameters

$$m(T_{k-1}) = \frac{x_k \cdot T_{k-1}}{T_k}, \quad \sigma^2(T_{k-1}) = \sqrt{\frac{T_{k-1} \cdot (T_k - T_{k-1})}{T_k}}.$$

Consider now the correlation function, $\rho(t) = \text{corr}(N_t, W_t)$, of the process, $\boldsymbol{X}_t$, $(0 \le t \le T)$. Our first result is the computation of the extreme correlations of this process at time T. After that we will prove that, as in the case of the BS simulation of the Poisson processes, the correlation coefficient $\rho(t)$ in our case is the linear function of time, $\rho(t) = \frac{t}{T}\rho(T)$.

Let us now find the extreme values of the correlation coefficient, $\rho(T) = \text{corr}(N_T, W_T)$. It is convenient to represent the random variable, $\zeta = N_T$, as a function of a normally distributed random variable, $\eta \sim \mathcal{N}(0, 1)$, $\zeta = P_\lambda^{-1}(\Phi(\eta))$, where $\lambda = \mu T$. The random variable W_T can be represented as $W_T = \sqrt{T}\xi$, where $\xi \sim \mathcal{N}(0,1)$ is a random variable with the standard normal distribution. Then the covariance of the Poisson and Wiener processes at time T is

$$\mathrm{cov}(N_T, W_T) = \sqrt{T}\mathrm{cov}(\xi,\zeta) = \sqrt{T}\cdot\mathbb{E}[\xi\cdot\zeta].$$

Since the variance of W_T is $\sqrt{T}$, we find

$$\rho(T) = \frac{\sqrt{T}\mathbb{E}[\zeta\xi]}{\sqrt{T}\sqrt{\mu T}} = \mathrm{corr}(\xi,\zeta).$$

Denote by ρ_0 the correlation coefficient of ξ and η. Let us represent ξ as $\xi = \rho_0\eta + \sqrt{1-\rho_0^2}\eta'$, where η' is a standard normal random variable independent of η. We have

$$\mathbb{E}[\xi\cdot\zeta] = \rho_0\cdot\Psi(\lambda), \tag{9.55}$$

where $\Psi(\lambda) := \mathbb{E}[\eta P_\lambda^{-1}(\Phi(\eta))]$. Let us compute $\Psi(\lambda)$. The function $P_\lambda(x)$ is piecewise constant: $P_\lambda(x) = P_\lambda([x])$. Define $\beta_k := P_\lambda(k)$, $(k = 0, 1, \dots)$. If $\beta_{k-1} \le \Phi(\eta) < \beta_k$, then $P_\lambda^{-1}(\Phi(\eta)) = k-1$. Denote $\gamma_k(\lambda) := \Phi^{-1}(P_\lambda(k))$. Then we find

$$\Psi(\lambda) = \sum_{k=1}^{\infty}\int_{\gamma_k(\lambda)}^{\gamma_{k+1}(\lambda)} kx\varphi(x)\mathrm{d}x = \sum_{k=1}^{\infty} k\cdot(\varphi(\gamma_k) - \varphi(\gamma_{k+1})),$$

where $\varphi(x)$ is the standard normal density function. The function $\Psi(\lambda)$ is a smooth, monotone function of λ. Then, from (9.55) and the inequality, $|\rho_0| \le 1$, we obtain

Proposition 9.6 The correlation coefficient, $\rho(\xi,\zeta)$, satisfies the inequality

$$-\frac{1}{\sqrt{\lambda}}\Psi(\lambda) \le \rho(\xi,\zeta) \le \frac{1}{\sqrt{\lambda}}\Psi(\lambda), \quad \lambda > 0. \tag{9.56}$$

Figure 9.9 displays the low and upper bounds for the correlation coefficient $\rho(\xi,\zeta)$. Comparing Figures 9.7 and 9.9, we conclude that the correlation boundaries in Proposition 9.6 are smooth, monotone, and symmetric, as opposed to the boundaries in the case of the Poisson processes.

Let us now study the correlation coefficient, $\rho(t) = \mathrm{corr}(N_t, W_t)$.

Proposition 9.7 The correlation coefficient, $\rho(t)$, is a linear function of time:

$$\rho(t) = \frac{t}{T}\cdot\rho(T), 0 \le t \le T. \tag{9.57}$$

Proof. Denote by $p_t(x,k)$ the joint density function of the Brownian motion and the Poisson process at time t. We have

$$\int_{-\infty}^{\infty} p_t(x,k)\mathrm{d}x = e^{-\lambda t}\frac{(\lambda t)^k}{k!}, \quad \sum_{k=0}^{\infty} p_t(x,k) = \frac{e^{-\frac{x^2}{2t}}}{\sqrt{2\pi t}}.$$

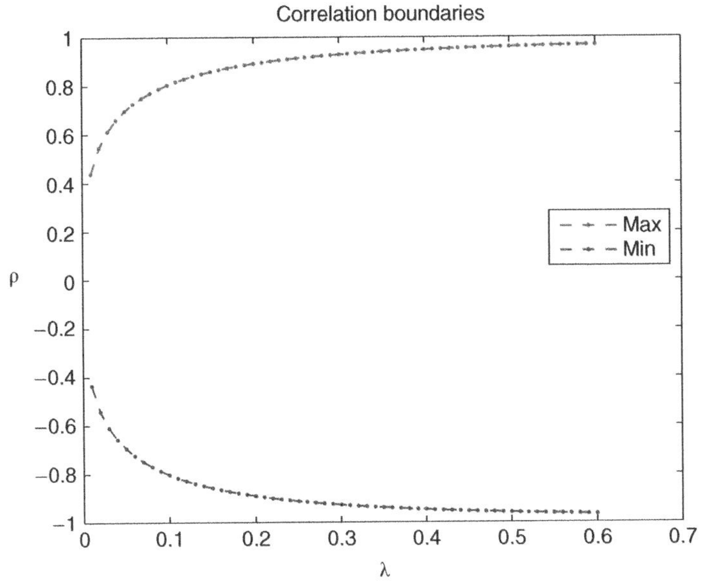

Figure 9.9 Correlation bounds.

Let

$$\hat{p}_t(x,z) := \sum_{n=0}^{\infty} p_t(x,n)z^n, \quad |z| \le 1,$$

be the generating function of the sequence $\{p_t(x,n)\}_{n=0}^{\infty}$. Consider the function

$$\Psi_t(u,z) := \mathbb{E}[\exp(iuW_t)z^{N_t}], \quad u \in \mathbb{R}, |z| \le 1.$$

The function, Ψ_t, is analytical in the area $|z| \le 1, t > 0$; its second derivative satisfies

$$\left.\frac{\partial^2 \Psi_t(u,z)}{\partial u \partial z}\right|_{(u,z)=(0,1)} = i \cdot \mathbb{E}[W_t \cdot N_t].$$

It is convenient to define the function

$$\psi_t(u,z) = -i \cdot \frac{\partial^2 \Psi_t(u,z)}{\partial u \partial z}.$$

Then we have

$$\psi_t(0,1) = \mathbb{E}[W_t \cdot N_t]. \tag{9.58}$$

Taking into account that

$$\mathbb{E}[W_t] = 0, \quad \sigma^2(W_t) = t, \quad \mathbb{E}[N_t] = \lambda t, \quad \sigma^2(N_t) = \lambda t, \quad 0 \le t \le T, \tag{9.59}$$

we derive from (9.58) and (9.59)

$$\psi_t(0,1) = \rho(t) \cdot \sqrt{\lambda} \cdot t. \tag{9.60}$$

Let us apply the BS algorithm to the process (W_t, N_t). The process N_t satisfies the relation

$$\mathbb{P}(N_t = m \mid N_T = n) = \binom{n}{m} q^m (1-q)^{n-m}, \quad m = 0, 1, \dots, n,$$

where $q = t/T$. The relation between $\Psi_t(u, z)$ and $\Psi_T(u, z)$ can be derived as follows. At time t, we have

$$\Psi_t(u, z) = \sum_{n=0}^{\infty} \int_{-\infty}^{\infty} e^{iux} z^n p_t(x, n)\,\mathrm{d}x$$
$$= \int_{-\infty}^{\infty} e^{iux} \hat{p}_t(x, z)\,\mathrm{d}x.$$

Lemma 9.11 *The generating function, $\hat{p}_t$, satisfies the equation*

$$\hat{p}_t(y, z) = \int_{-\infty}^{\infty} \hat{p}_T(x, 1 - q + zq) \frac{1}{\sqrt{2\pi\sigma(t)}} e^{-\frac{(y - xt/T)^2}{2\sigma^2(t)}}\,dx, \quad 0 < t < T, \tag{9.61}$$

where $q = t/T$ and $\sigma(t) = \sqrt{\frac{t\cdot(T-t)}{T}}$.

Proof. The density, $p_t(y, m)$, satisfies

$$p_t(y, m) = \int_{-\infty}^{\infty} \sum_{n=m}^{\infty} \binom{n}{m} q^m (1-q)^{n-m} p_T(x, n) \frac{1}{\sqrt{2\pi\sigma(t)}} e^{-\frac{(y - xt/T)^2}{2\sigma^2(t)}}\,\mathrm{d}x.$$

We have

$$\sum_{m=0}^{\infty} z^m \sum_{n=m}^{\infty} \binom{n}{m} q^m (1-q)^{n-m} p_T(x, n) = \sum_{n=0}^{\infty} p_T(x, n) \sum_{m=0}^{n} \binom{n}{m} (qz)^m (1-q)^{n-m}$$
$$= \sum_{n=0}^{\infty} p_T(x, n)(1 - q + qz)^n$$
$$= \hat{p}_T(x, 1 - q + qz).$$

Therefore, the generating function, $\hat{p}_t(y, z)$, satisfies (9.61), as was to be proved.

Let us finish the proof of Proposition 9.7. The joint generating function, $\Psi_t(u, z)$, is the Fourier transform of $\hat{p}_t$. Therefore,

$$\Psi_t(u, z) = \int_{-\infty}^{\infty} e^{iuy} \hat{p}_t(y, z)\,\mathrm{d}y$$
$$= \int_{-\infty}^{\infty} e^{iuy} \int_{-\infty}^{\infty} \hat{p}_T(x, 1 - q + zq) \frac{1}{\sqrt{2\pi\sigma(t)}} e^{-\frac{(y - xt/T)^2}{2\sigma^2(t)}}\,\mathrm{d}x\,\mathrm{d}y.$$

Changing the order of integration, we obtain

$$\Psi_t(u,z) = \int_{-\infty}^{\infty} e^{iuy} \int_{-\infty}^{\infty} \hat{p}_T(x, 1-q+zq) \frac{1}{\sqrt{2\pi}\sigma(t)} e^{-\frac{(y-xt/T)^2}{2\sigma^2(t)}} \mathrm{d}x\, \mathrm{d}y$$

$$= \int_{-\infty}^{\infty} \hat{p}_T(x, 1-q+zq) \int_{-\infty}^{\infty} e^{iuy} \frac{1}{\sqrt{2\pi}\sigma(t)} e^{-\frac{(y-xt/T)^2}{2\sigma^2(t)}} \mathrm{d}y\, \mathrm{d}x$$

The inner integral

$$\int_{-\infty}^{\infty} e^{iuy} \frac{1}{\sqrt{2\pi}\sigma(t)} e^{-\frac{(y-xt/T)^2}{2\sigma^2(t)}} \mathrm{d}y = e^{ixqu} \cdot e^{-\frac{u^2\sigma^2(t)}{2}}.$$

Therefore,

$$\Psi_t(u,z) = e^{-\frac{u^2\sigma^2(t)}{2}} \cdot \int_{-\infty}^{\infty} e^{ixqu} \cdot \hat{p}_T(x, 1-q+zq)\, \mathrm{d}x, \tag{9.62}$$

where $q = t/T$. Differentiating (9.62) twice and substituting $u = 0$, $z = 1$, we obtain

$$\psi_t(0,1) = \frac{t^2}{T^2} \cdot \int_{-\infty}^{\infty} x \frac{\partial}{\partial z} \hat{p}_T(x, 1) \mathrm{d}x$$

$$= \frac{t^2}{T^2} \cdot \psi_T(0,1).$$

Then, the latter relation and (9.60) immediately imply Equation (9.57).

Remark 9.8 *We have considered the backward simulation of the process with Poisson and Wiener components. It is possible to extend this approach to the class of processes with mean-reverting components instead of Wiener processes. In this case, the process has a nonlinear time structure of correlations.*

9.8 Concluding Remarks

The backward simulation of the multivariate Poisson processes is based on the conditional uniformity of the unsorted arrival moments. The BS allows us to consider in a general framework both Gaussian and Poisson processes describing the dynamics of risk factors. The BS approach to the Poisson processes results in a bigger range of admissible correlations and simpler calibration algorithm.

The computation of the extreme distributions with maximal and minimal correlations of the Poisson processes is solved by the EJD algorithm, proposed in the present chapter, which has a simple probabilistic interpretation. This algorithm has an intimate connection to the class of infinite-dimensional linear optimization problems. It is interesting to understand its role in the analysis of this class of optimization problems.

The boundaries for the correlation coefficients of the multivariate Poisson process delivered by the EJD algorithm give necessary conditions for the existence of the process. It looks natural to try to prove that these conditions are also sufficient. A similar question can be formulated for the multivariate process with the Wiener and Poisson components.

The BS approach is applicable to the class of mixed Poisson processes. These models will be studied in our future publications.

Acknowledgments

I am very grateful to my colleagues Ian Iscoe, Yijun Jiang, Konrad Duch, Helmut Mausser, Oleksandr Romanko, and Asif Lakhany for stimulating discussions: without their support, this chapter would never have appeared.

Andrey Marchenko read the first draft of the paper and made a few important comments and suggestions, especially regarding monotone distributions. I would like to thank Isaac Sonin for his comments pointing at simplification of the proofs and Monique Jeanblanc and Raphael Douady for fruitful discussions during the 8th Bachelier Colloquium. Sebastian Jaimungal and Tianyi Jia made numerous comments and suggestions helping me to write this chapter. Tianyi also verified some of the proofs and derivations. I am very grateful to the referees for their comments. Finally, I would like to thank Dmitry Malioutov for his interest in this project.

Appendix A

A.1 Proof of Lemmas 9.2 and 9.3

A.1.1 Proof of Lemma 9.2

Consider the function, $\pi : \mathbb{Z}_+^m \to \mathbb{R}$,

$$\pi(\vec{k}) = \sum_{l=0}^{\infty} p_{\|\vec{k}\|+l} \binom{\vec{k}+l}{\vec{k}} \cdot \vec{x}^{\,\vec{k}} \cdot y^l, \tag{A.1}$$

and denote by $\hat{\pi}(\vec{z})$ the generating function

$$\hat{\pi}(\vec{z}) := \sum_{\vec{k} \in \mathbb{Z}_+^m} \pi(\vec{k}) \vec{z}^{\,\vec{k}}.$$

We have to show that

$$\hat{\pi}(\vec{z}\,) = \hat{p}\left(1 - \sum_{j=1}^{m} x_j(1 - z_j)\right). \tag{A.2}$$

The generating function $\hat{\pi}(\vec{z}\,)$ can be written as

$$\hat{\pi}(\vec{z}\,) = \sum_{j=1}^{m} \sum_{k_j=0}^{\infty} \sum_{l=0}^{\infty} p_{\|\vec{k}\|+l} \vec{z}^{\,\vec{k}} \binom{\vec{k}+l}{\vec{k}} \cdot \vec{x}^{\,\vec{k}} \cdot y^l. \tag{A.3}$$

Denote $n = \sum_{j=1}^{m} k_j + l$. Let us also introduce the partial sums

$$K_J = \sum_{j=1}^{J} k_j, \quad J = 1, 2, \ldots, m.$$

Then, from Equation (A.3), we find

$$\hat{\pi}(\vec{z}\,) = \sum_{n=0}^{\infty} p_n \cdot \sum_{k_1=0}^{n} \binom{n}{k_1} (x_1 z_1)^{k_1} \sum_{k_2=0}^{n-k_1} \binom{n-k_1}{k_2} (x_2 z_2)^{k_2} \sum_{k_3=0}^{n-K_2} \binom{n-K_2}{k_3} (x_3 z_3)^{k_3} \cdot$$
$$\cdots \sum_{k_m=0}^{n-K_{m-1}} \binom{n-K_{m-1}}{k_m} (x_m z_m)^{k_m} \cdot \left(1 - \sum_{j=1}^{m} x_j\right)^{n-K_m}. \tag{A.4}$$

Financial Signal Processing and Machine Learning, First Edition.
Edited by Ali N. Akansu, Sanjeev R. Kulkarni and Dmitry Malioutov.
© 2016 John Wiley & Sons, Ltd. Published 2016 by John Wiley & Sons, Ltd.

Consider the last sum,

$$S_m = \sum_{k_m=0}^{n-K_{m-1}} \binom{n-K_{m-1}}{k_m} (x_m z_m)^{k_m} \cdot \left(1 - \sum_{j=1}^{m} x_j\right)^{n-K_m}.$$

Since $K_m = K_{m-1} + k_m$, we obtain

$$S_m = \left(1 - \sum_{j=1}^{m-1} x_j - x_m(1 - z_m)\right)^{n-K_{m-1}}.$$

Applying this transformation recursively to the sums over k_j, $(j = m-1, m-2, \ldots, 1)$, in (A.4) we derive

$$\hat{\pi}(\vec{z}\,) = \sum_{n=0}^{\infty} p_n \cdot \left(1 - \sum_{j=1}^{m} x_j(1 - z_j)\right)^n = \hat{p}\left(1 - \sum_{j=1}^{m} x_j(1 - z_j)\right).$$

Lemma 9.2 is proved.

Remark A.1 *Notice that if $\vec{z} = \vec{1}$, $\hat{\pi}(\vec{z}\,) = 1$. The function $\pi(\vec{k}) \geq 0$. Therefore, $\{\pi(\vec{k})\}_{\vec{k} \in \mathbb{Z}_+^m}$ is a probability distribution on the m-dimensional integer lattice.*

A.1.2 Proof of Lemma 9.3

Derivation of Equation (9.20) is completely analogous to that of Equation (9.12) and, for this reason, is omitted. Let us derive (9.21).

From (9.20) we find, by differentiation $\left(\frac{\partial \hat{q}(z,w)}{\partial z} \mid_{z=w=1}, \frac{\partial \hat{q}(z,w)}{\partial w} \mid_{z=w=1}\right)$:

$$\mathbb{E}[\xi_1 \cdot \xi_2] = xy\mathbb{E}[\zeta_1\zeta_2], \quad \mathbb{E}[\xi_1] = x\mathbb{E}[\zeta_1], \quad \mathbb{E}[\xi_2] = y\mathbb{E}[\zeta_2],$$

and

$$\text{cov}(\xi_1, \xi_2) = xy \cdot \text{cov}(\zeta_1, \zeta_2).$$

Then we obtain

$$\sigma^2(\xi_1) = x^2 \cdot \sigma^2(\zeta_1) + \mathbb{E}[\zeta_1] \cdot (x - x^2),$$

and

$$\sigma^2(\xi_2) = y^2 \cdot \sigma^2(\zeta_2) + \mathbb{E}[\zeta_2] \cdot (y - y^2).$$

If the variance and the first moment of the random variables, ζ_i, $(i = 1, 2)$, are equal, then

$$\sigma(\xi_i) = \sqrt{x} \cdot \sigma(\zeta_i), \quad i = 1, 2.$$

Finally, we obtain

$$\rho(\xi_1, \xi_2) = \sqrt{xy} \cdot \rho(\zeta_1, \zeta_2),$$

and Lemma 9.3 is thus proved.

References

Aue, F. and Kalkbrener, M. (2006). LDA at work: Deutsche Banks approach to quantify operational risk. *Journal of Operational Risk*, 1(4), 49–95.

Avellaneda, M. and Zhu, J. (2001). Distance to default. *Risk*, 14(12), 125–129.

Badescu, A., Gang, L., Lin, S. and Tang, D. (2013). A mixture model approach to operational risk management. Working paper. Toronto: University of Toronto, Department of Statistics.

Badescu, A., Gang, L., Lin, S. and Tang, D. (2015). Modeling correlated frequencies with applications in operational risk management. *Journal of Operational Risk*, forthcoming.

Bae, T. (2012). A model for two-way dependent operational losses. Working paper. Regina, SK: University of Regina, Department of Mathematics and Statistics.

Barndorff-Nielsen, O. and Shephard, N. (2001). Non-Gaussian Ornstein–Uhlenbeck-based models and some of their uses in financial economics. *Journal of the Royal Statistical Society*, 63, 167–241.

Barndorff-Nielsen, O. and Yeo, G. (1969). Negative binomial processes. *Journal of Applied Probability*, 6(3), 633–647.

Böcker, K. and Klüppelberg, C. (2010). Multivariate models for operational risk. *Quantitative Finance*, 10(8), 855–869.

Boyarchenko, S. and Levendorski, S. (2002). *Non-Gaussian Merton-Black-Scholes theory*. River Edge, NJ: World Scientific.

Carr, P. and Madan, D. (1998). Option valuation using the fast Fourier transform. *Journal of Computational Finance*, 2, 61–73.

Chavez-Demoulin, V., Embrechts, P. and Nešlehovà, J. (2006). Quantitative models for operational risk: extremes, dependence and aggregation. *Journal of Banking and Finance*, 30(9), 2635–2658.

Cont, R. and Tankov, P. (2003). *Financial modeling with jump processes*. London: Chapman and Hall, CRC Press.

Cont, R. and Tankov, P. (2006). Retrieving Levy processes from option prices: regularization of an ill-posed inverse problem. *SIAM Journal on Control and Optimization*, 45, 1–25.

Duch, K., Kreinin, A. and Jiang, Y. (2014). New approaches to operational risk modeling. *IBM Journal of Research and Development*, 3, 31–45.

Embrechts, P. and Puccetti, G. (2006). Aggregating risk capital with an application to operational risk. *The Geneva Risk and Insurance Review*, 31(2), 71–90.

Feigin, P. (1979). On the characterization of point processes with the order statistic property. *Journal of Applied Probability*, 16(2), 297–304.

Fox, B. (1996). Generating Poisson processes by quasi-Monte Carlo. Working paper. Boulder, CO: SIM-OPT Consulting.

Fox, B. and Glynn, P. (1988). Computing Poisson probabilities. *Communication of the ACM*, 31, 440–445.

Frechet, M. (1951). Sur le tableaux de correlation dont les marges sont donnees. *Annales de l'Université de Lyon*, 14, 53–77.

Griffiths, R., Milne, R. and Wood, R. (1979). Aspects of correlation in bivariate Poisson distributions and processes. *Australian Journal of Statistics*, 21(3), 238–255.

Hardy, G., Littlewood, J. and Polya, G. (1952). *Inequalities*. Cambridge: Cambridge University Press.

Hoeffding, W. (1940). Masstabinvariante korrelations-theorie. *Schriften Math. Inst. Univ. Berlin*, 5, 181–233.

Hull, J. and White, A. (2001). Valuing credit default swaps II: modeling default correlations. *Journal of Derivatives*, 8(3), 29–40.

Iscoe, I. and Kreinin, A. (2006). Recursive valuation of basket default swaps. *Journal of Computational Finance*, 9(3), 95–116.

Iscoe, I. and Kreinin, A. (2007). Valuation of synthetic CDOs. *Journal of Banking and Finance*, 31, 3357–3376.

Iscoe, I., Kreinin, A. and Rosen, D. (1999). Integrated market and credit risk portfolio model. *Algorithmics Research Quarterly*, 2(3), 21–38.

Jaimungal, S., Kreinin, A. and Valov, A. (2013). The generalized Shiryaev problem and Skorokhod embedding. *Theory of Probabilities and Applications*, 58(3), 614–623.

Johnson, N. and Kotz, S. (1969). *Distributions in statistics: discrete distributions*. Boston: Houghton Mifflin.

Johnson, N., Kotz, S. and Balakrishnan, N. (1997). *Discrete multivariate distributions*. New York: John Wiley & Sons.

Kou, S. (2002). A jump diffusion model for option pricing. *Management Sciences*, 48(8), 1086–1101.

Kyprianou, A., Schoutens, W. and Wilmott, P. (2005). *Exotic option pricing and advanced Levy models*. Hoboken, NJ: John Wiley & Sons.

Lindskog, F. and McNeil, A.J. (2001). *Poisson shock models: applications to insurance and credit risk modeling*. Zurich: Federal Institute of Technology ETH Zentrum.

Lipton, A. and Rennie, A. (2008). *Credit correlation: life after copulas*. River Edge, NJ: World Scientific.

Merton, R. (1974). On the pricing of corporate debt: the risk structure of interest rates. *Journal of Finance*, 29, 449–470.

Merton, R. (1976). Option pricing when underlying stock returns are discontinuous. *Journal of Financial Economics*, 3, 125–144.

Musiela, M. and Rutkowsky, M. (2008). *Martingale methods in financial modelling stochastic modelling*. New York: Springer-Verlag.

Nawrotzki, K. (1962). Ein grenwertsatz für homogene züffalige punktfolgen (verallgemeinerung eines satzes von A. Renyi). *Mathematische Nachrichten*, 24, 201–217.

Nelsen, R.B. (1987). Discrete bivariate distributions with givenmarginals and correlation. *Communications in Statistics –Simulation*, 16(1), 199–208.

Nešlehovà, J., Embrechts, P. and Chavez-Demoulin, V. (2006). Infinite mean models and the lda for operational risk. *Journal of Operational Risk*, 1(1), 3–25.

Panjer, H.H. (2006). *Operational risks: modeling analytics*. Hoboken, NJ: John Wiley & Sons.

Peters, G., Shevchenko, P. and Wüthrich, M. (2009). Dynamic operational risk: modeling dependence and combining different data sources of information. *The Journal of Operational Risk*, 4(2), 69–104.

Powojovsky, M., Reynolds, D. and Tuenter, H. (2002). Dependent events and operational risk. *Algorithmics Research Quarterly*, 5(2), 65–74.

Revus, D. and Yor, M. (1991). *Continuous Martingales and Brownian motion: a series of comprehensive studies in mathematics*, 2nd ed. New York: Springer.

Rolski, T., Schmidli, H., Schmidt, V. and Teugles, J. (1999). *Stochastic processes for insurance and finance*. Chichester, UK: John Wiley & Sons.

Shevchenko, P. (2011). *Modeling operational risk using modeling analytics*. New York: Springer.

Shreve, S. (2004). *Stochastic calculus for finance II: continuous-time models –a series of comprehensive studies in mathematics*. New York: Springer-Finance.

Vasicek, O. (1987). *Probability of loss on loan portfolio*. Working paper. San Francisco: KMW.

Vasicek, O. (2002). Loan portfolio value. *Risk*, 15, 160–162.

Whitt, W. (1976). Bivariate distributions with given marginals. *Annals of Statistics*, 4, 1280–1289.

Yahav, I. and Shmueli, G. (2011). On generating multivariate Poisson data in management science applications. *Applied Stochastic Models in Business and Industry*, 28(1), 91–102.

10

CVaR Minimizations in Support Vector Machines

Jun-ya Gotoh[1] and Akiko Takeda[2]
[1]*Chuo University, Japan*
[2]*The University of Tokyo, Japan*

How to measure the riskiness of a random variable has been a major concern of financial risk management. Among the many possible measures, *conditional value at risk* (*CVaR*) is viewed as a promising functional for capturing the characteristics of the distribution of a random variable. CVaR has attractive theoretical properties, and its minimization with respect to involved parameters is often tractable. In portfolio selection especially, the minimization of the empirical CVaR is a linear program. On the other hand, machine learning is based on the so-called regularized empirical risk minimization, where a surrogate of the empirical error defined over the in-sample data is minimized under some regularization of the parameters involved. Considering that both theories deal with empirical risk minimization, it is natural to look at their interaction. In fact, a variant of support vector machine (SVM) known as ν-SVM implicitly carries out a certain CVaR minimization, though the relation to CVaR is not clarified at the time of the invention of ν-SVM.

This chapter overviews the connections between SVMs and CVaR minimization and suggests further interactions beyond their similarity in appearance. Section 10.1 summarizes the definition and properties of CVaR. The authors wish this section to be a quick introduction for those who are not familiar with CVaR. Section 10.2 collects basic formulations of various SVMs for later reference. Those who are familiar with SVM formulations can skip this section and consult it when succeeding sections refer to the formulations therein. Section 10.3 provides CVaR minimization-based representations of the SVM formulations given in Section 10.2. Section 10.4 is devoted to dual formulations of CVaR-based formulations. Section 10.5 describes two robust optimization extensions of these formulations. For further

Financial Signal Processing and Machine Learning, First Edition.
Edited by Ali N. Akansu, Sanjeev R. Kulkarni and Dmitry Malioutov.
© 2016 John Wiley & Sons, Ltd. Published 2016 by John Wiley & Sons, Ltd.

study, Section 10.6 briefly overviews the literature regarding the interaction between risk measure theory and SVM.

10.1 What Is CVaR?

This section introduces CVaR as a risk measure of random variables and describes its relation to VaR and other statistics.

10.1.1 Definition and Interpretations

Let $\tilde{L}$ be a random variable defined on a sample space Ω (i.e., $\tilde{L} : \Omega \to \mathbb{R}$), and let it represent a quantity that we would like to minimize, such as payments, costs, damages, or error. We figuratively refer to such random variables as *losses*.[1] In financial risk management, each element $\omega \in \Omega$ can be interpreted as a future state or scenario. Furthermore, let us suppose that all random variables are associated with a probability measure $\mathbb{P}$ on Ω (and a set of events), satisfying $\mathbb{E}_{\mathbb{P}}[|\tilde{L}|] < +\infty$. In most parts of this chapter, this assumption is satisfied because losses are associated with empirical distributions based on finite observations, and each loss is defined on a finite sample space (i.e., $\Omega = \{\omega_1, \dots, \omega_m\}$).

For a loss $\tilde{L}$, a *risk measure* is a functional that maps $\tilde{L}$ to $\mathbb{R} \cup \{\infty\}$, expressing how risky $\tilde{L}$ is. Among the many risk measures, VaR is popular because it is easy to interpret.

Definition 10.1 (VaR (value-at-risk) or α-quantile of loss) *The VaR of $\tilde{L}$ at a significant level $\alpha \in (0, 1)$ is defined as*

$$VaR_{(\alpha,\mathbb{P})}[\tilde{L}] := \min_c \{c : \mathbb{P}\{\tilde{L} \le c\} \ge \alpha\}.$$

The parameter α is determined by decision makers such as fund managers. To capture the possibility of a large loss that could occur with a small probability, the parameter α is typically set close to 1 in financial risk management (e.g., $\alpha = 0.99$ or 0.999). While VaR can capture the upper tail of $\tilde{L}$, it fails to capture the impact of a loss beyond VaR. In addition, the lack of convexity makes VaR intractable in risk management.[2]

CVaR has gained growing popularity as a convex surrogate to VaR.

Definition 10.2 (CVaR or α-superquantile of loss) *The conditional value-at-risk (CVaR) of $\tilde{L}$ with a significant level $\alpha \in [0, 1)$ is defined as*

$$CVaR_{(\alpha,\mathbb{P})}[\tilde{L}] := \inf_c \left\{ G(c) := c + \frac{1}{1-\alpha}\mathbb{E}_{\mathbb{P}}[\max\{\tilde{L} - c, 0\}] \right\}. \tag{10.1}$$

$\text{CVaR}_{(\alpha,\mathbb{P})}[\tilde{L}]$ is nondecreasing in α since the function G is nondecreasing in α.

[1] In SVM and other statistical learning contexts, the word *loss* takes on a specific meaning. However, in this chapter, we will use this word in a different and more general manner for the sake of consistency with risk measure theory.

[2] A functional $\mathcal{F}$ is *convex* if for all $\tilde{L}, \tilde{L}', \tau \in (0, 1), (1-\tau)\mathcal{F}[\tilde{L}] + \tau\mathcal{F}[\tilde{L}'] \ge \mathcal{F}[(1-\tau)\tilde{L} + \tau\tilde{L}']$.

Note that for $\alpha \in (0,1)$, "inf" in the above formula is replaced with "min." Indeed, for $\alpha \in (0,1)$, Theorem 10 of Rockafellar and Uryasev (2002) shows that

$$\arg\min_{c} G(c) = [\mathrm{VaR}_{(\alpha,\mathbb{P})}[\tilde{L}], \mathrm{VaR}^{+}_{(\alpha,\mathbb{P})}[\tilde{L}]], \tag{10.2}$$

where $\mathrm{VaR}^{+}_{(\alpha,\mathbb{P})}[\tilde{L}] := \inf_{c} \{c : \mathbb{P}\{\tilde{L} \le c\} > \alpha\}$. The relation (10.2) implies that any minimizer in the formula (10.1) can be an approximate VaR and that if the minimizer is unique,[3] it is equal to VaR.

The definition (10.1) of CVaR implies that VaR is always bounded above by CVaR, that is,

$$\mathrm{VaR}_{(\alpha,\mathbb{P})}[\tilde{L}] \le \mathrm{CVaR}_{(\alpha,\mathbb{P})}[\tilde{L}],$$

and $\mathrm{VaR}_{(\alpha,\mathbb{P})}[\tilde{L}] < \mathrm{CVaR}_{(\alpha,\mathbb{P})}[\tilde{L}]$ unless there is no chance of a loss greater than VaR. More specifically, CVaR is represented as a convex combination of VaR and $\mathbb{E}_{\mathbb{P}}[\tilde{L}|\tilde{L} > \mathrm{VaR}_{(\alpha,\mathbb{P})}[\tilde{L}]]$.[4] Indeed, with $t_\alpha := \mathbb{P}\{\tilde{L} > \mathrm{VaR}_{(\alpha,\mathbb{P})}[\tilde{L}]\}/(1-\alpha) \in [0,1]$, it is true that

$$\mathrm{CVaR}_{(\alpha,\mathbb{P})}[\tilde{L}] = t_\alpha \mathbb{E}_{\mathbb{P}}[\tilde{L}|\tilde{L} > \mathrm{VaR}_{(\alpha,\mathbb{P})}[\tilde{L}]] + (1-t_\alpha)\mathrm{VaR}_{(\alpha,\mathbb{P})}[\tilde{L}]. \tag{10.3}$$

When $\tilde{L}$ follows a parametric distribution, CVaR may be explicitly given a closed formula.

Example 10.1 (CVaR under normal distribution) *When $\tilde{L}$ follows a normal distribution $N(\mu, \sigma^2)$, CVaR can be explicitly expressed as*

$$CVaR_{(\alpha,\mathbb{P})}[\tilde{L}] = \mu + \frac{1}{(1-\alpha)\sqrt{2\pi}} \exp\left(-\frac{1}{2}\{\Psi^{-1}(\alpha)\}^2\right)\cdot\sigma, \tag{10.4}$$

where Ψ is the cumulative distribution function of $N(0,1)$ (i.e., $\Psi(z) := \frac{1}{\sqrt{2\pi}}\int_{-\infty}^{z} \exp(-\frac{1}{2}t^2)dt$). On the other hand, $VaR_{(\alpha,\mathbb{P})}[\tilde{L}] = \mu + \Psi^{-1}(\alpha)\cdot\sigma$.

Note that (10.4) is equal to the conditional expectation of a loss exceeding VaR, that is,

$$\mathrm{CVaR}_{(\alpha,\mathbb{P})}[\tilde{L}] = \mathbb{E}_{\mathbb{P}}[\tilde{L}|\tilde{L} > \mathrm{VaR}_{(\alpha,\mathbb{P})}[\tilde{L}]] = \mathbb{E}_{\mathbb{P}}[\tilde{L}|\tilde{L} \ge \mathrm{VaR}_{(\alpha,\mathbb{P})}[\tilde{L}]].$$

This relation also holds for other distributions as long as there is no probability atom at $\mathrm{VaR}_{(\alpha,\mathbb{P})}[\tilde{L}]$.

However, for general loss distributions and arbitrary α, the above equalities only hold in an approximate manner. That is, we have

$$\mathbb{E}_{\mathbb{P}}[\tilde{L}|\tilde{L} \ge \mathrm{VaR}_{(\alpha,\mathbb{P})}[\tilde{L}]] \le \mathrm{CVaR}_{(\alpha,\mathbb{P})}[\tilde{L}] \le \mathbb{E}_{\mathbb{P}}[\tilde{L}|\tilde{L} > \mathrm{VaR}_{(\alpha,\mathbb{P})}[\tilde{L}]].$$

Usually, the loss distributions are not known, so an empirical distribution is used as a surrogate of the true distribution. In such a case, there is a positive probability atom at $\mathrm{VaR}_{(\alpha,\mathbb{P})}[\tilde{L}]$ for all $\alpha \in (0,1)$, and either of the equalities does not hold (see Proposition 5 of Rockafellar and Uryasev (2002) for the details).

[3] The minimizer is unique if and only if $\mathrm{VaR}^{+}_{(\alpha,\mathbb{P})}[\tilde{L}] = \mathrm{VaR}_{(\alpha,\mathbb{P})}[\tilde{L}]$ (i.e., there is no probability atom at $\mathrm{VaR}_{(\alpha,\mathbb{P})}[\tilde{L}]$).

[4] $\mathrm{CVaR}_{(\alpha,\mathbb{P})}[\tilde{L}]$ is also considered to be the mean of the α-tail distribution of $\tilde{L}$. See Proposition 6 of Rockafellar and Uryasev (2002) for the details.

Example 10.2 (CVaR under finite scenarios) *Suppose that $\tilde{L}$ is defined on a finite sample space $\Omega = \{\omega_1, \dots, \omega_m\}$ equipped with $\mathbb{P}\{\omega = \omega_i\} = p_i > 0$, $i = 1, \dots, m$. For $\alpha \in [0, 1)$, CVaR is given by*

$$CVaR_{(\alpha,\mathbb{P})}[\tilde{L}] = CVaR_{(\alpha,p)}(\boldsymbol{L}) := \min_c \left\{ c + \frac{1}{1-\alpha} \sum_{i=1}^{m} p_i \max\{L_i - c, 0\} \right\},$$

where $\boldsymbol{L} := (L_1, \dots, L_m)^\top$ and $\boldsymbol{p} := (p_1, \dots, p_m)^\top$.[5] *In this case, CVaR can be represented with a linear program (LP).*

$$CVaR_{(\alpha,p)}(\boldsymbol{L}) = \left| \begin{array}{ll} \underset{c,z}{\text{minimize}} & c + \frac{1}{1-\alpha} \sum_{i=1}^{m} p_i z_i, \\ \text{subject to} & z_i \geq L_i - c, i = 1, \dots, m, \\ & z_i \geq 0, i = 1, \dots, m. \end{array} \right. \tag{10.5}$$

Note that (10.5) is feasible for any $\boldsymbol{L}$, $\boldsymbol{p}$, and α (e.g., $(c, z_1, \dots, z_m) = (\bar{c}, L_1 - \bar{c}, \dots, L_m - \bar{c})$ with $\bar{c} = \min\{L_1, \dots, L_m\}$ is a feasible solution). The dual problem of (10.5) is derived as

$$\left| \begin{array}{ll} \underset{q}{\text{maximize}} & \sum_{i=1}^{m} q_i L_i, \\ \text{subject to} & \sum_{i=1}^{m} q_i = 1, \\ & 0 \leq q_i \leq \frac{p_i}{1-\alpha}, i = 1, \dots, m. \end{array} \right. \tag{10.6}$$

The solution $\boldsymbol{q} = \boldsymbol{p}$ is feasible for (10.6) for any $\alpha \in [0, 1)$, and therefore, the optimal value of (10.6) is equal to $CVaR_{(\alpha,p)}(\boldsymbol{L})$ because of the strong duality of LP (see, e.g., Vanderbei, 2014).

As will be elaborated on later, (10.6) suggests that CVaR can be interpreted as the worst expected loss over a set of probability measures, $\{\boldsymbol{q} : \sum_{i=1}^{m} q_i = 1, 0 \leq q_i \leq \frac{p_i}{1-\alpha}, i = 1, \dots, m\}$. Note also that (10.6) can be viewed as a variant of the continuous (or fractional) knapsack problem,[6] *and its solution is obtained in a greedy manner as follows:*[7]

1. *Sort the loss scenarios, $L_1, \dots, L_m$, in descending order, and let $L_{(i)}$ denote the i-th largest component (i.e., $L_{(1)} \geq L_{(2)} \geq \dots \geq L_{(m)}$), and $p_{(i)}$, denote the reference probability corresponding to $L_{(i)}$.*
2. *Find the integer k satisfying $\frac{1}{1-\alpha} \sum_{i=1}^{k} p_{(i)} \leq 1 < \frac{1}{1-\alpha} \sum_{i=1}^{k+1} p_{(i)}$. Let $q_{(i)}$ denote the element of the solution vector $\boldsymbol{q}$, corresponding to $L_{(i)}$ in the objective of (10.6), and set $q_{(i)} = \frac{p_{(i)}}{1-\alpha}$ for $i = 1, \dots, k$, $q_{(k+1)} = 1 - \frac{1}{1-\alpha} \sum_{i=1}^{k} p_{(i)}$, and $q_{(i)} = 0$ for $i = k + 2, \dots, m$. Consequently, CVaR is given by the formula*

$$CVaR_{(\alpha,p)}(\boldsymbol{L}) = \frac{1}{1-\alpha} \sum_{i=1}^{k} p_{(i)} L_{(i)} + \left(1 - \frac{1}{1-\alpha} \sum_{i=1}^{k} p_{(i)}\right) L_{(k+1)}. \tag{10.7}$$

---

[5] Since the random variable $\tilde{L}$ and probability measure $\mathbb{P}$ can be expressed as vectors $\boldsymbol{L} := (L_1, \dots, L_m)^\top$ and $\boldsymbol{p} := (p_1, \dots, p_m)^\top$, we denote $\text{CVaR}_{(\alpha,\mathbb{P})}[\tilde{L}]$ by $\text{CVaR}_{(\alpha,p)}(\boldsymbol{L})$.

[6] While the *knapsack problem* is formulated as a binary integer program $\max\{\sum_{i=1}^{m} b_i x_i : \sum_{i=1}^{m} c_i x_i \leq C, \boldsymbol{x} \in \{0, 1\}^m\}$, where $c_i, b_i, C > 0$, the corresponding continuous knapsack problem (CKP) is formulated as an LP: $\max\{\sum_{i=1}^{m} b_i x_i : \sum_{i=1}^{m} c_i x_i = C, \boldsymbol{x} \in [0, 1]^m\}$. With a change of variables $x_i = (1 - \alpha) q_i / p_i$, (10.6) can be rewritten into an LP with $b_i = p_i L_i / (1 - \alpha)$, $c_i = p_i$, and $C = 1 - \alpha$. Strictly speaking, (10.6) is not a CKP since L_i, and thus b_i, can be negative. However, the same greedy algorithm is applicable.

[7] This procedure shows that the complexity of computing the CVaR of a loss is at most on the order of $m \log m$. If $\boldsymbol{p}$ is uniform (i.e., $p_i = 1/m$), we can use the so-called selection algorithm whose complexity is on the order of m.

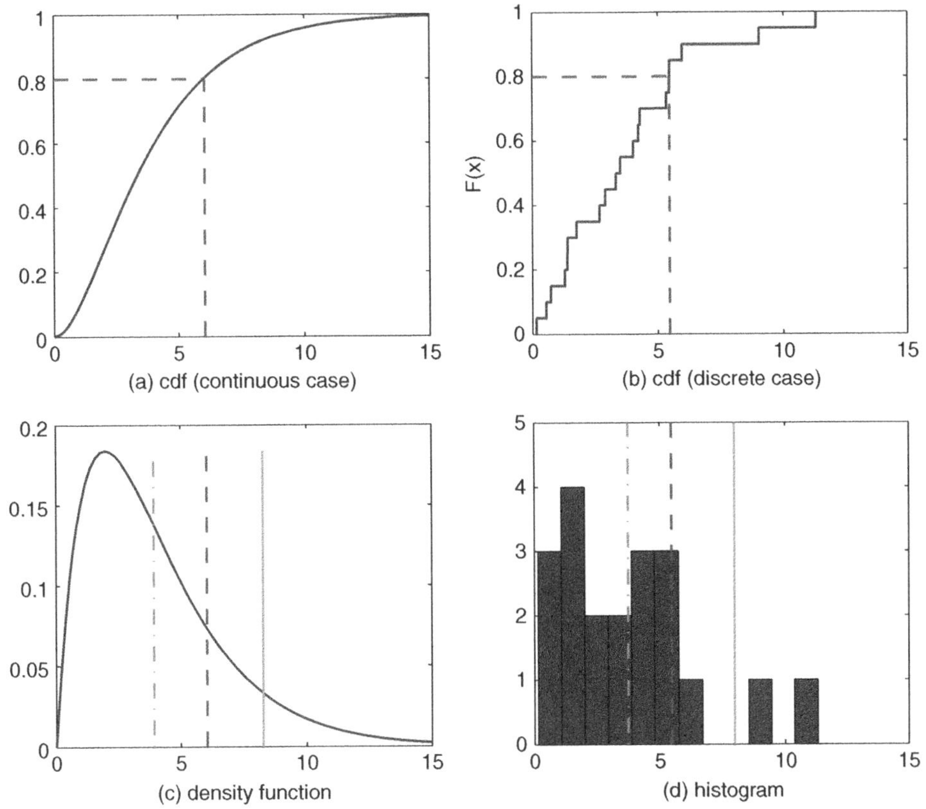

Figure 10.1 CVaR, VaR, mean, and maximum of distribution. (a, c) The cumulative distribution function (cdf) and the density of a continuous loss distribution; (b, d) the cdf and histogram of a discrete loss distribution. In all four figures, the location of VaR with $\alpha = 0.8$ is indicated by a vertical dashed line. In (c) and (d), the locations of CVaR and the mean of the distributions are indicated with vertical solid and dashed-dotted lines. In (b) and (d), the location of the maximum loss is shown for the discrete case.

Formula (10.7) implies that CVaR is approximately equal to the mean of the largest $100(1-\alpha)$ *% losses* $L_{(1)}, \ldots, L_{(k)}$ *if* $\sum_{i=1}^{k} p_{(i)} \approx 1-\alpha$ *(see Figures 10.1b and 10.1d).*

Now let us consider the complementarity condition. Namely, for any optimal solutions $(c^*, \boldsymbol{z}^*)$ *and* $\boldsymbol{q}^*$ *to (10.5) and (10.6), respectively, the condition*

$$\begin{cases} (z_i^* - L_i + c^*)q_i^* = 0, & i = 1, \ldots, m, \\ z_i^*\{p_i - (1-\alpha)q_i^*\} = 0, & i = 1, \ldots, m, \end{cases}$$

implies $z_{(i)}^* = L_{(i)} - c^*$, $i = 1, \ldots, k$, *and* $z_{(i)}^* = 0$, $i = k+1, \ldots, m$, *where* $z_{(i)}^*$ *is the element of the optimal solution corresponding to* $L_{(i)}$ *and* $p_{(i)}$. *Using this for the objective of (10.5) and* *comparing the result with (10.7), we find that* $c^* = L_{(k+1)}$. *Note that*

$$L_{(k+1)} = \begin{cases} VaR_{(\alpha,\boldsymbol{p})}(\boldsymbol{L}) & \text{if} \quad \sum_{i=1}^{k} p_{(i)} < 1-\alpha < \sum_{i=1}^{k+1} p_{(i)}, \\ VaR^{+}_{(\alpha,\boldsymbol{p})}(\boldsymbol{L}) & \text{if} \quad \sum_{i=1}^{k} p_{(i)} = 1-\alpha, \end{cases}$$

where $VaR^{+}_{(\alpha,\boldsymbol{p})}(\boldsymbol{L})$ *is* $VaR^{+}_{(\alpha,\mathbb{P})}$ *of* $\boldsymbol{L}$ *with* $\mathbb{P} = \boldsymbol{p}$, *that is,* $\min_c\{c : \sum_{i=1}^{m} p_i 1_{\{L_i \leq c\}} > \alpha\}$, *where* $1_{\{cond\}}$ *is the 0–1 indicator function (i.e.,* $1_{\{\mathrm{cond}\}} = 1$ *if cond is true, 0 otherwise). Accordingly, formula (10.7) is an expression (10.3) for the discrete distribution case.*

CVaR can also be considered to be a generalization of the average and the maximum of the underlying random variable.

- With $\alpha = 0$, CVaR is equal to the mean of the loss, that is, $\mathrm{CVaR}_{(0,\mathbb{P})}[\tilde{L}] = \mathbb{E}_{\mathbb{P}}[\tilde{L}]$ (or $\mathrm{CVaR}_{(0,\boldsymbol{p})}(\boldsymbol{L}) = \mathbb{E}_{\boldsymbol{p}}(\boldsymbol{L}) := \boldsymbol{p}^{\top}\boldsymbol{L}$ for the discrete distribution case).
- With α close to 1, it approximates the largest loss. Indeed, in the case of a discrete distribution, it is true that $\mathrm{CVaR}_{(\alpha',\boldsymbol{p})}(\boldsymbol{L}) = \max\{L_1, \dots, L_m\} = L_{(1)}$ for $\alpha' > 1 - p_{(1)}$.

As a result of the nondecreasing property with respect to α, CVaR is typically between the maximum and mean for an intermediate $\alpha \in (0, 1 - p_{(1)})$ (see Figure 10.1d).

10.1.2 Basic Properties of CVaR

The convexity of a risk functional often makes the associated risk minimization tractable and enables us to exploit duality theory. Those advantages of the convexity of CVaR are worth emphasizing again.

Let us consider a probabilistic representation. Letting B be the target level of loss and $\alpha \in (0, 1)$ be the significant level, define

$$\mathbb{P}\{\tilde{L} \geq B\} \leq 1 - \alpha. \tag{10.8}$$

In general, the set of losses $\tilde{L}$ satisfying (10.8) is nonconvex. To avoid this nonconvexity, the left side of (10.8) is often approximated with a convex upper bound of the form $\mathbb{E}_{\mathbb{P}}[f(\tilde{L})]$, where f is a convex function on $\mathbb{R}$ such that $f(L) \geq 1_{\{L \geq B\}}$ for all $L \in \mathbb{R}$. Note that $\mathbb{E}_{\mathbb{P}}[f(\tilde{L})] \leq 1 - \alpha$ implies $\mathbb{P}\{\tilde{L} \geq B\} \leq 1 - \alpha$ since $\mathbb{P}\{\tilde{L} \geq B\} = \mathbb{E}_{\mathbb{P}}[1_{\{\tilde{L} \geq B\}}] \leq \mathbb{E}_{\mathbb{P}}[f(\tilde{L})]$. The expression $\mathbb{E}_{\mathbb{P}}[f(\tilde{L})] \leq 1 - \alpha$ is thus called a *conservative approximation* of (10.8). To tighten the bound, it is enough to consider a piecewise linear function $f(L) = \max\{(L - C)/(B - C), 0\}$ with some C such that $B > C$ (see Figure 10.2). Namely, (10.8) can be replaced with $\mathbb{E}_{\mathbb{P}}[\max\{(\tilde{L} - C)/(B - C), 0\}] \leq 1 - \alpha$, which becomes

$$C + \frac{1}{1-\alpha}\mathbb{E}_{\mathbb{P}}[\max\{\tilde{L} - C, 0\}] \leq B \quad \text{for some } C.$$

Noting that this is equivalent to $\mathrm{CVaR}_{(\alpha,\mathbb{P})}[\tilde{L}] \leq B$, we can see that CVaR provides a tight convex conservative approximation of the probabilistic condition (10.8).

CVaR has three properties that are useful in financial risk management; CVaR is

1. *monotonic*: $\mathrm{CVaR}_{(\alpha,\mathbb{P})}[\tilde{L}] \geq \mathrm{CVaR}_{(\alpha,\mathbb{P})}[\tilde{L}']$ when $\tilde{L} \geq \tilde{L}'$;

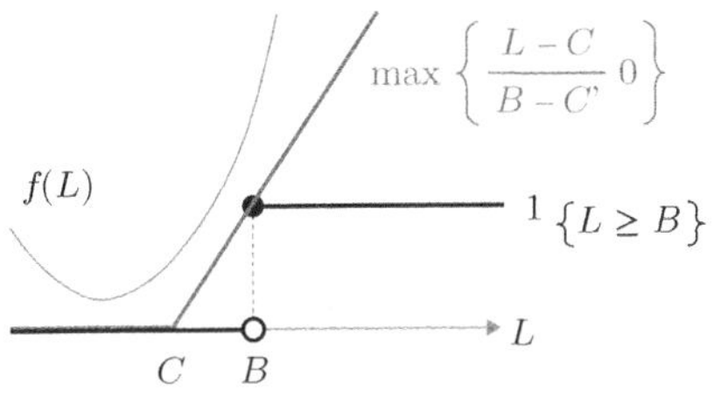

Figure 10.2 Convex functions dominating $1_{\{L \geq B\}}$.

2. *translation invariant*: $\text{CVaR}_{(\alpha,\mathbb{P})}[\tilde{L} + \tau] = \text{CVaR}_{(\alpha,\mathbb{P})}[\tilde{L}] + \tau$ for all $\tau \in \mathbb{R}$; and
3. *positively homogeneous*: $\text{CVaR}_{(\alpha,\mathbb{P})}[\tau\tilde{L}] = \tau\text{CVaR}_{(\alpha,\mathbb{P})}[\tilde{L}]$ for all $\tau > 0$.

If a convex risk functional satisfies all three of these properties, it is said to be *coherent* (Artzner *et al.*, 1999).[8] Monotonicity is a useful property in machine learning contexts, whereas translation invariance and positive homogeneity exist for technical reasons rather than for intuitive reasons. However, each of these properties plays roles in tractability and in compatibility with regularization terms (Gotoh and Uryasev, 2013).

In general, monotonicity, translation invariance, and positive homogeneity can be characterized in a dual manner. To avoid unnecessary technicalities, we will assume a finite sample space $\Omega = \{\omega_1, \dots, \omega_m\}$ (i.e., which implies that risk functionals $\mathcal{F}$ are functions on $\mathbb{R}^m$).[9] Let us define $\mathcal{F}^*$ to be the conjugate of $\mathcal{F}$, that is, $\mathcal{F}^*(\boldsymbol{\lambda}) := \sup_{\boldsymbol{L}}\{\boldsymbol{L}^\top \boldsymbol{L} - \mathcal{F}(\boldsymbol{L})\}$. Furthermore, let us define dom $\mathcal{F}$ as the effective domain of $\mathcal{F}$, that is, dom $\mathcal{F} := \{\boldsymbol{L} \in \mathbb{R}^m : \mathcal{F}(\boldsymbol{L}) < +\infty\}$.

Theorem 10.1 (dual characterization of risk functional properties (Ruszczyński and Shapiro, 2006)) *Suppose that $\mathcal{F} : \mathbb{R}^m \to (-\infty, \infty)$ is an l.s.c.,*[10] *proper,*[11] *and convex function. Accordingly:*

1. *$\mathcal{F}$ is monotonic if and only if dom $\mathcal{F}^*$ is in the nonnegative orthant.*
2. *$\mathcal{F}$ is translation invariant if and only if $\forall \boldsymbol{\lambda} \in$ dom $\mathcal{F}^*$, $\mathbf{1}_m^\top \boldsymbol{\lambda} = 1$.*
3. *$\mathcal{F}$ is positively homogeneous if and only if $\mathcal{F}$ can be represented in the form*

$$\mathcal{F}(\boldsymbol{L}) = \sup_{\boldsymbol{\lambda}} \{\boldsymbol{L}^\top \boldsymbol{\lambda} : \boldsymbol{\lambda} \in \textit{dom } \mathcal{F}^*\}. \tag{10.9}$$

See Ruszczyński and Shapiro (2006) for the proof.

[8] CVaR is also *law invariant* (i.e., if the distribution functions of $\tilde{L}$ and $\tilde{L}'$ are identical, $\text{CVaR}_{(\alpha,\mathbb{P})}[\tilde{L}] = \text{CVaR}_{(\alpha,\mathbb{P})}[\tilde{L}']$) and *co-monotonically additive* (i.e., $\text{CVaR}_{(\alpha,\mathbb{P})}[\tilde{L} + \tilde{L}'] = \text{CVaR}_{(\alpha,\mathbb{P})}[\tilde{L}] + \text{CVaR}_{(\alpha,\mathbb{P})}[\tilde{L}']$ for any $\tilde{L}, \tilde{L}'$ satisfying $(\tilde{L}(\omega) - \tilde{L}(\omega'))(\tilde{L}'(\omega) - \tilde{L}'(\omega')) \geq 0$ for any $\omega, \omega' \in \Omega$). A coherent risk measure that has these two properties is called a *spectral (or distortion) risk measure*. See, for example, Acerbi (2002) for details.

[9] The results hold true in a more general setting. See, for example, Rockafellar and Uryasev (2013) and Ruszczyński and Shapiro (2006) for more general statements.

[10] $\text{CVaR}_{(\alpha,\mathbb{P})}$ is *lower semicontinuous (l.s.c.)*, that is, $\text{CVaR}_{(\alpha,\mathbb{P})}[\tilde{L}] \leq \liminf_{k\to\infty} \text{CVaR}_{(\alpha,\mathbb{P})}[\tilde{L}^k]$ for any $\tilde{L}$ and any sequence $\tilde{L}^1, \tilde{L}^2, \dots$, converging to $\tilde{L}$.

[11] $\text{CVaR}_{(\alpha,\mathbb{P})}$ is *proper*, that is, $\text{CVaR}_{(\alpha,\mathbb{P})}[\tilde{L}] > -\infty$ for all $\tilde{L}$, and $\text{CVaR}_{(\alpha,\mathbb{P})}[\tilde{L}'] < \infty$ for some $\tilde{L}'$.

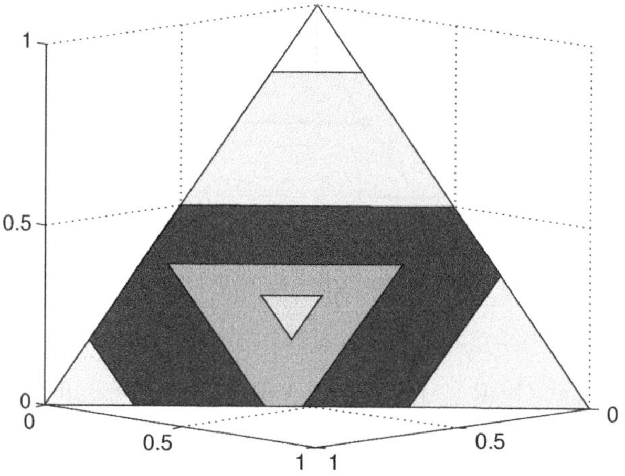

Figure 10.3 Illustration of $\mathcal{Q}_{\mathrm{CVaR}(\alpha,p)}$ in a discrete distribution on IR^3 with $(p_1, p_2, p_3) = (5/12, 4/12, 3/12)$. This figure shows how $\mathcal{Q}_{\mathrm{CVaR}(\alpha,p)}$ varies depending on α ($\alpha = 0.1, 0.3, 0.5, 0.7$). As α approaches 1, $\mathcal{Q}_{\mathrm{CVaR}(\alpha,p)}$ approaches the unit simplex Π^3. The risk envelope shrinks to the point $(p_1, p_2, p_3) = (5/12, 4/12, 3/12)$ as α decreases to 0.

In particular, from (10.9), we can see that any l.s.c. proper positively homogeneous convex risk functional can be characterized by the effective domain of its conjugate, which is referred to as the *risk envelope*. Let us denote the risk envelope of CVaR by $\mathcal{Q}_{\mathrm{CVaR}(\alpha,p)}$ (i.e., $\mathcal{Q}_{\mathrm{CVaR}(\alpha,p)} := \mathrm{dom}\ \mathrm{CVaR}^*_{(\alpha,p)}$).

Noting that the dual LP (10.6) is written in the form of (10.9), the risk envelope of CVaR is

$$\mathcal{Q}_{\mathrm{CVaR}(\alpha,p)} := \{\boldsymbol{q} \in \mathrm{IR}^m : \mathbf{1}_m^\top \boldsymbol{q} = 1, \mathbf{0} \le \boldsymbol{q} \le \boldsymbol{p}/(1-\alpha)\}. \tag{10.10}$$

Figure 10.3 illustrates an example of the risk envelope of CVaR with $m = 3$. The conjugate of CVaR is given by

$$\mathrm{CVaR}^*_{(\alpha,p)}(\boldsymbol{\lambda}) = \delta_{\mathcal{Q}_{\mathrm{CVaR}(\alpha,p)}}(\boldsymbol{\lambda}),$$

where δ_C is the indicator function of a set C (i.e., $\delta_C(\xi) := 0$ if $\xi \in C$, and $+\infty$ otherwise). Since $\mathcal{Q}_{\mathrm{CVaR}(\alpha,p)} \subset \Pi^m := \{\boldsymbol{q} \in \mathrm{IR}^m : \mathbf{1}_m^\top \boldsymbol{q} = 1, \boldsymbol{q} \ge \mathbf{0}\}$, the dual LP (10.6) is symbolically represented as the worst-case expected loss over a set of probabilities, that is,

$$\mathrm{CVaR}_{(\alpha,p)}(\boldsymbol{L}) = \max_{\boldsymbol{q}} \{\mathbb{E}_{\boldsymbol{q}}(\boldsymbol{L}) : \boldsymbol{q} \in \mathcal{Q}_{\mathrm{CVaR}(\alpha,p)}\}. \tag{10.11}$$

Indeed, any coherent function on IR^m can be characterized by using a non-empty closed convex set in Π^m in place of $\mathcal{Q}_{\mathrm{CVaR}(\alpha,p)}$ (see, e.g., Artzner *et al.*, 1999).

10.1.3 Minimization of CVaR

Now, let us consider the case where the loss is defined with one or more parameters, and find the parameter values that minimize CVaR.

Let the loss $\tilde{L}$ be parametrized with $\boldsymbol{\theta} \in \mathbb{R}^n$ (i.e., $\tilde{L}(\boldsymbol{\theta})$), and suppose that the probability is independent of $\boldsymbol{\theta}$. Rockafellar and Uryasev (2002) prove the following theorem.

Theorem 10.2 (CVaR minimization) *Let* $G(\boldsymbol{\theta}, c) := c + \frac{1}{1-\alpha}\mathbb{E}_{\mathbb{P}}[\max\{\tilde{L}(\boldsymbol{\theta}) - c, 0\}]$, *and let* $\Theta \subset \mathbb{R}^n$ *denote the set of admissible* $\boldsymbol{\theta}$, *and* $(\boldsymbol{\theta}^\star, c^\star) \in \arg\min_{\boldsymbol{\theta}\in\Theta, c} G(\boldsymbol{\theta}, c)$. *Then,*

1. $\min_{\boldsymbol{\theta}\in\Theta} CVaR_{(\alpha,\mathbb{P})}[\tilde{L}(\boldsymbol{\theta})] = G(\boldsymbol{\theta}^\star, c^\star) = \min_{\boldsymbol{\theta}\in\Theta, c} G(\boldsymbol{\theta}, c)$.
2. $\boldsymbol{\theta}^\star \in \arg\min_{\boldsymbol{\theta}\in\Theta} CVaR_{(\alpha,\mathbb{P})}[\tilde{L}(\boldsymbol{\theta})]$ *and* $c^\star \in [VaR_{(\alpha,\mathbb{P})}[\tilde{L}(\boldsymbol{\theta}^\star)], VaR^+_{(\alpha,\mathbb{P})}[\tilde{L}(\boldsymbol{\theta}^\star)]]$.
3. *Furthermore, if* $\tilde{L}(\boldsymbol{\theta})$ *is convex with respect to* $\boldsymbol{\theta}$, *then so are both* $CVaR_{(\alpha,\mathbb{P})}[\tilde{L}(\boldsymbol{\theta})]$ *and* $G(\boldsymbol{\theta}, c)$.
4. *If* $\tilde{L}(\boldsymbol{\theta})$ *is homogeneous with respect to* $\boldsymbol{\theta}$ *(i.e., for any* $a \in \mathbb{R}$, $\tilde{L}(a\boldsymbol{\theta}) = a\tilde{L}(\boldsymbol{\theta})$*), then both* $CVaR_{(\alpha,\mathbb{P})}[\tilde{L}(\boldsymbol{\theta})]$ *and* $G(\boldsymbol{\theta}, c)$ *are positively homogeneous with respect to* $(\boldsymbol{\theta}, c)$.

The first property states that the minimization of CVaR reduces to simultaneous minimization of the function $G(\boldsymbol{\theta}, c)$ in $\boldsymbol{\theta}$ and c. The second statement guarantees that the interpretation of the variable c as an approximate VaR (i.e., α-quantile of L) remains valid even in the case of CVaR minimization. The third property states that the associated CVaR minimization is a convex minimization if $\tilde{L}(\boldsymbol{\theta})$ is convex in $\boldsymbol{\theta}$. The fourth property, which is not exactly stated in Rockafellar and Uryasev (2002), is that the (positive) homogeneity of the loss propagates to that of CVaR in terms of the involved parameters. As will be discussed in Section 10.3, this property plays a role in analyzing the form of a regularized empirical CVaR minimization.

Example 10.3 (CVaR-minimizing portfolio selection) *Let* $\tilde{R}_j$ *denote a random rate of return of an investable asset* j, *and suppose that* $\tilde{\boldsymbol{R}} := (\tilde{R}_1, \dots, \tilde{R}_n)$ *follows a discrete distribution satisfying*

$$(\tilde{R}_1, \dots, \tilde{R}_n)(\omega_i) = (R_{i,1}, \dots, R_{i,n}) \quad \text{and} \quad p_i := \mathbb{P}\{\omega = \omega_i\} > 0, i = 1, \dots, m.$$

Let θ_j *denote the investment ratio of asset* j. *To make the investment self-financing, we impose a constraint* $\sum_{j=1}^n \theta_j = 1$. *In addition, to meet the investor's requirements, several constraints are imposed on* $\boldsymbol{\theta}$. *We will impose, for example, a restriction of the form* $\boldsymbol{0} \le \boldsymbol{\theta} \le \boldsymbol{u}$ *with upper bounds* $\boldsymbol{u} \in \mathbb{R}^n$ *and* $\boldsymbol{r}^\top\boldsymbol{\theta} \ge \tau$ *using the expected return* $\boldsymbol{r} = \mathbb{E}_p[\tilde{\boldsymbol{R}}]$ *and the minimum target return* $\tau > 0$. *The problem of determining a portfolio* $(\theta_1, \dots, \theta_n)$ *that minimizes CVaR defined with* $\tilde{L}(\boldsymbol{\theta}) = -\tilde{\boldsymbol{R}}^\top\boldsymbol{\theta}$ *(or, equivalently,* $\boldsymbol{L}(\boldsymbol{\theta}) = -(\boldsymbol{R}_1, \dots, \boldsymbol{R}_m)^\top\boldsymbol{\theta}$ *with* $\boldsymbol{R}_i = (R_{i,1}, \dots, R_{i,n})^\top$*) is an LP:*

$$\begin{array}{|ll}
\underset{\theta,c,z}{\text{minimize}} & c + \frac{1}{1-\alpha}\sum_{i=1}^m p_i z_i \\
\text{subject to} & z_i \ge -\sum_{j=1}^n R_{ij}\theta_j - c, i = 1, \dots, m, \\
& z_i \ge 0, i = 1, \dots, m, \\
& \sum_{j=1}^n \theta_j = 1, \quad \sum_{j=1}^n r_j\theta_j \ge \tau, \quad 0 \le \theta_j \le u_j, j = 1, \dots, n.
\end{array}$$

Typically, a discrete distribution is obtained from historical observations, for example, periodic asset returns (e.g., daily, weekly, or monthly) in real markets, and $p_i = 1/m$ *is used unless there is particular information about* $\boldsymbol{p}$.[12]

Example 10.4 (CVaR-based passive portfolio selection) *Strategies seeking to mimic market indexes such as the S&P 500 (i.e., a certain average of asset prices) are popular in portfolio management.*[13] *Let* $\tilde{I}$ *denote the return of an index, and assume that* $(\tilde{R}_1, \dots, \tilde{R}_n, \tilde{I})$ *follows a discrete distribution satisfying* $(\tilde{R}_1, \dots, \tilde{R}_n, \tilde{I})(\omega_i) = (R_{i,1}, \dots, R_{i,n}, I_i)$ *and* $p_i := \mathbb{P}\{\omega = \omega_i\} > 0, i = 1, \dots, m$. *Measure the deviation of the portfolio return* $\tilde{\boldsymbol{R}}^\top \boldsymbol{\theta}$ *from the benchmark return* $\tilde{I}$ *by using the CVaR associated with the loss* $\tilde{L}(\boldsymbol{\theta}) = |\tilde{I} - \tilde{\boldsymbol{R}}^\top \boldsymbol{\theta}|$ *(or, equivalently,* $\boldsymbol{L}(\boldsymbol{\theta}) = (|I_1 - \boldsymbol{R}_1^\top \boldsymbol{\theta}|, \dots, |I_m - \boldsymbol{R}_m^\top \boldsymbol{\theta}|)^\top$ *with* $\boldsymbol{R}_i = (R_{i,1}, \dots, R_{i,n})^\top$*). The problem of finding a portfolio mimicking the index can then be formulated as*

$$
\left|
\begin{array}{ll}
\underset{\boldsymbol{\theta}, c, \boldsymbol{z}}{\text{minimize}} & c + \frac{1}{1-\alpha} \sum_{i=1}^{m} p_i z_i \\
\text{subject to} & z_i \geq I_i - \sum_{j=1}^{n} R_{ij}\theta_j - c, i = 1, \dots, m, \\
& z_i \geq -I_i + \sum_{j=1}^{n} R_{ij}\theta_j - c, i = 1, \dots, m, \\
& z_i \geq 0, i = 1, \dots, m, \\
& \sum_{j=1}^{n} \theta_j = 1, \quad \sum_{j=1}^{n} r_j\theta_j \geq \tau, \quad 0 \leq \theta_j \leq u_j, j = 1, \dots, n.
\end{array}
\right.
$$

10.2 Support Vector Machines

SVMs are one of the most successful supervised learning methods that can be applied to classification or regression. This section introduces several SVM formulations, whose relation to CVaR minimization will be discussed in the succeeding sections.

10.2.1 Classification

Suppose that we have m samples $(\boldsymbol{x}_i, y_i), i = 1, \dots, m$, where $\boldsymbol{x}_i := (x_{i1}, \dots, x_{in})^\top \in \mathbb{R}^n$ denotes the vector of the attributes of sample i and $y_i \in \{-1, +1\}$ denotes its binary label, $i = 1, \dots, m$. SVM classification (or SVC, for short) finds a hyperplane, $\boldsymbol{w}^\top \boldsymbol{x} = b$, that separates the training samples as much as possible. The labels of the new (unknown) samples can be predicted on the basis of which side of the hyperplane they fall on.

By using the so-called kernel trick, SVC constructs a nonlinear classifier, a hyperplane in a high (possibly, infinite) dimensional space. Namely, it implicitly uses a mapping $\boldsymbol{\phi} : \mathbb{R}^n \to \mathbb{R}^N$ (N can be infinite) and obtains a hyperplane $\boldsymbol{w}^\top \boldsymbol{\phi}(\boldsymbol{x}) = b$ and a decision function $d(\boldsymbol{x}) = \text{sign}(\boldsymbol{w}^\top \boldsymbol{\phi}(\boldsymbol{x}) - b)$, where sign($z$) is 1 if $z \geq 0$ and -1 otherwise.

---

[12] If $(\tilde{R}_1, \dots, \tilde{R}_n)$ follows a multivariate normal distribution $N(\boldsymbol{r}, \boldsymbol{\Sigma})$, the portfolio return (i.e., $\tilde{\boldsymbol{R}}^\top \boldsymbol{\theta}$), follows a normal distribution $N(\boldsymbol{r}^\top \boldsymbol{\theta}, \boldsymbol{\theta}^\top \boldsymbol{\Sigma} \boldsymbol{\theta})$. With the loss $\tilde{L}(\boldsymbol{\theta}) = -\tilde{\boldsymbol{R}}^\top \boldsymbol{\theta}$, the CVaR minimization reduces to a second-order cone program, that is, $\min_{\boldsymbol{\theta} \in \Theta} -\boldsymbol{r}^\top \boldsymbol{\theta} + C_\alpha \sqrt{\boldsymbol{\theta}^\top \boldsymbol{\Sigma} \boldsymbol{\theta}}$ with $C_\alpha := \exp(-\{\Psi^{-1}(\alpha)\}^2/2)/\{(1-\alpha)\sqrt{2\pi}\}$ (see formula (10.4)). This is equivalent to the so-called mean–variance criterion (Markowitz, 1952) for a specific trade-off parameter.

[13] This type of investment strategy is called *passive*, while those seeking to beat the market average are called *active*.

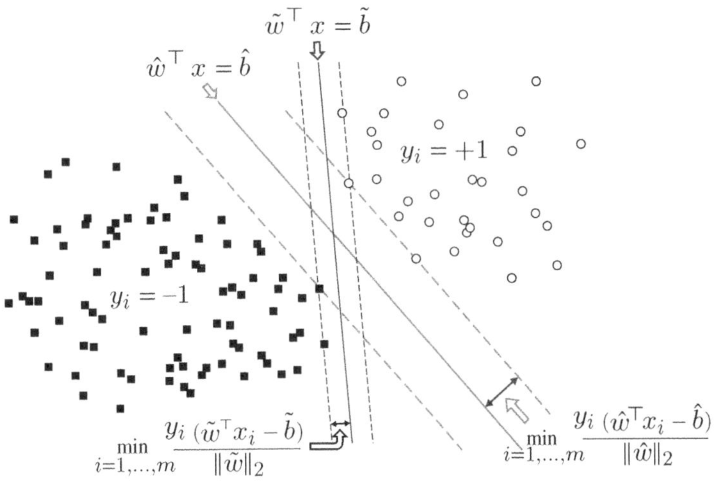

Figure 10.4 Two separating hyperplanes and their geometric margins. The dataset is said to be *linearly separable* if there exist $\boldsymbol{w} \neq \boldsymbol{0}$ and b such that $y_i(\boldsymbol{w}^\top \boldsymbol{x}_i - b) > 0$ for all $i = 1, \ldots, m$. If the dataset is linearly separable, there are infinitely many hyperplanes separating the dataset. According to generalization theory (Vapnik, 1995), the hyperplane $\hat{\boldsymbol{w}}^\top \boldsymbol{x} = \hat{b}$ is preferable to $\tilde{\boldsymbol{w}}^\top \boldsymbol{x} = \tilde{b}$. The optimization problem (10.12) (or, equivalently, (10.13)) finds a hyperplane that separates the datasets with the largest margin.

The Vapnik–Chervonenkis theory shows that a large geometric margin classifier has a small generalization error (Vapnik, 1995). Namely, the search for a hyperplane that has the largest distance to the nearest data points decreases the upper bound of the out-of-sample error. Motivated by this theoretical result, Boser *et al.* (1992) developed an algorithm for finding a hyperplane $(\boldsymbol{w}, b)$ with the maximum geometric margin, which is formulated as

$$\underset{\boldsymbol{w},b}{\text{maximize}} \min_{i=1,\ldots,m} \frac{y_i(\boldsymbol{w}^\top \boldsymbol{\phi}(\boldsymbol{x}_i) - b)}{\|\boldsymbol{w}\|_2} = -\underset{\boldsymbol{w},b}{\text{minimize}} \max_{i=1,\ldots,m} \frac{-y_i(\boldsymbol{w}^\top \boldsymbol{\phi}(\boldsymbol{x}_i) - b)}{\|\boldsymbol{w}\|_2}, \tag{10.12}$$

where $\|\cdot\|_2$ denotes the ℓ_2-norm (or the Euclidean norm), that is, $\|\boldsymbol{w}\|_2 := \sqrt{\boldsymbol{w}^\top \boldsymbol{w}}$.

If the data samples are *linearly separable*, that is, there exists a hyperplane that separates the samples $\boldsymbol{x}_i$ such that $y_i = +1$ from those such that $y_i = -1$, as in Figure 10.4, the fractional optimization (10.12) can be rewritten as the following quadratic program:

$$\left|\begin{array}{ll} \underset{\boldsymbol{w},b}{\text{minimize}} & \frac{1}{2}\|\boldsymbol{w}\|_2^2 \\ \text{subject to} & y_i(\boldsymbol{w}^\top \boldsymbol{\phi}(\boldsymbol{x}_i) - b) \geq 1, \ i = 1, \cdots, m. \end{array}\right. \tag{10.13}$$

This is called *hard-margin SVC*. Note that (10.13) is valid only when the training samples are linearly separable.

10.2.1.1 *C*-Support Vector Classification

Cortes and Vapnik (1995) extend the SVC algorithm to linearly nonseparable cases and trade off the margin size with the data separation error. More precisely, by introducing slack variables

$z_1, \ldots, z_m$ and adding their sum to the objective function, the hard-margin SVC formulation can be modified into

$$f_{\text{CSVC}} := \left| \begin{array}{ll} \underset{w,b,z}{\text{minimize}} & \frac{1}{2}\|\boldsymbol{w}\|_2^2 + C\sum_{i=1}^m z_i, \\ \text{subject to} & z_i \geq -y_i(\boldsymbol{w}^\top \boldsymbol{\phi}(\boldsymbol{x}_i) - b) + 1, i = 1, \ldots, m, \\ & z_i \geq 0, i = 1, \ldots, m, \end{array} \right. \tag{10.14}$$

where $C > 0$ is a user-defined parameter. Formulation (10.14) is often viewed as a correction that adds the so-called *hinge loss* $\sum_{i=1}^m \max\{L_i + 1, 0\}$ with $L_i = -y_i(\boldsymbol{w}^\top \boldsymbol{\phi}(\boldsymbol{x}_i) - b)$, as a surrogate of the *0–1 loss* $\sum_{i=1}^m 1_{\{L_i \geq 0\}}$, which would otherwise involve nonconvexity.[14] Formulation (10.14) is usually referred to as *C-SVC*,[15] and it has been shown to work very well in various real-world applications (see, e.g., Schölkopf and Smola, 2002).

The (Lagrangian) dual formulation of (10.14) is derived as

$$f_{\text{CSVC}} = \left| \begin{array}{ll} \underset{\lambda}{\text{maximize}} & \sum_{i=1}^m \lambda_i - \frac{1}{2}\sum_{h=1}^m \sum_{i=1}^m \lambda_i \lambda_h y_i y_h k(\boldsymbol{x}_i, \boldsymbol{x}_h) \\ \text{subject to} & \sum_{i=1}^m y_i \lambda_i = 0, \quad 0 \leq \lambda_i \leq C, i = 1, \ldots, m. \end{array} \right. \tag{10.15}$$

Here, $k(\boldsymbol{x}_i, \boldsymbol{x}_h) = \boldsymbol{\phi}(\boldsymbol{x}_i)^\top \boldsymbol{\phi}(\boldsymbol{x}_h)$ is a *kernel function* defined directly on the inputs of $\boldsymbol{x}_i$ and $\boldsymbol{x}_h$. The use of a kernel function is preferable to that of the explicit mapping $\boldsymbol{\phi}(\cdot)$, because we can treat a highly nonlinear mapping without bothering about how large the dimension N of the mapped space should be.

Moreover, by using the optimality condition, we can recover a dual solution from a primal solution.[16] With an optimal solution $(\boldsymbol{w}^\star, b^\star, \boldsymbol{z}^\star, \boldsymbol{\lambda}^\star)$, the decision function is given by

$$d(\boldsymbol{x}) = \text{sign}(\sum_{i=1}^m \lambda_i^\star y_i k(\boldsymbol{x}_i, \boldsymbol{x}) - b^\star).$$

10.2.1.2 ν-Support Vector Classification

ν-SVC is another formulation of soft-margin SVC (Schölkopf *et al.*, 2000),

$$f_{\nu\text{-SVC}} := \left| \begin{array}{ll} \underset{w,b,\rho,z}{\text{minimize}} & \frac{1}{2}\|\boldsymbol{w}\|_2^2 - \nu\rho + \frac{1}{m}\sum_{i=1}^m z_i, \\ \text{subject to} & z_i \geq -y_i(\boldsymbol{w}^\top \boldsymbol{\phi}(\boldsymbol{x}_i) - b) + \rho, i = 1, \ldots, m, \\ & z_i \geq 0, i = 1, \ldots, m, \end{array} \right. \tag{10.16}$$

where $\nu \in (0, 1]$ is a user-defined parameter. The dual of (10.16) is described as

$$f_{\nu\text{-SVC}} = \left| \begin{array}{ll} \underset{\lambda}{\text{maximize}} & -\frac{1}{2}\sum_{h=1}^m \sum_{i=1}^m \lambda_i \lambda_h y_i y_h k(\boldsymbol{x}_i, \boldsymbol{x}_h) \\ \text{subject to} & \sum_{i=1}^m \lambda_i = 1, \sum_{i=1}^m y_i \lambda_i = 0, \\ & 0 \leq \lambda_i \leq \frac{1}{\nu m}, i = 1, \ldots, m. \end{array} \right. \tag{10.17}$$

---

[14] More precisely, hinge loss is viewed as a special case of the convex upper bound, $\max\{(L - C)/(B - C), 0\}$ with $B = 0$ and $C = -1$, of the 0–1 loss, as shown in Figure 10.2.

[15] To make a contrast with (10.13), formulations of this type are sometimes referred to as *soft-margin SVCs*.

[16] Strong duality also holds (i.e., the optimal values of (10.14) and (10.15) approach the same f_{CSVC}).

Formulation (10.17) indicates that the optimal value of (10.16) as well as (10.17) is non-increasing with respect to ν. Moreover, the optimal value is nonpositive (or unbounded) because $(\boldsymbol{w}, b, \rho, z) = \boldsymbol{0}$ is feasible for (10.16).

Note that (10.17) is not necessarily well defined for any ν between 0 and 1 (Chang and Lin, 2001; Crisp and Burges, 2000). Let m_+ (resp. m_-) denote the number of samples with positive (resp. negative) labels. When ν is larger than $\nu_{\max} := 2\min\{m_+, m_-\}/m$, we can show that the primal ν-SVC (10.16) is unbounded and the dual ν-SVC (10.17) becomes infeasible. On the other hand, when ν is smaller than some threshold $\nu_{\min}$, ν-SVC produces a trivial solution satisfying $(\boldsymbol{w}, b) = \boldsymbol{0}$ (Chang and Lin, 2001). The lower threshold $\nu_{\min}$ is defined as the smallest upper bound of ν with which the optimal value of ν-SVC becomes zero.[17]

Schölkopf *et al.*, (2000) show that the relation between ν-SVC and C-SVC is as follows.

Theorem 10.3 (Schölkopf *et al.* 2000) *Suppose that (10.16) has an optimal solution $(\boldsymbol{w}, b, \rho, z)$ with $\rho > 0$. Then (10.14) with $C = 1/(\rho m)$ provides the same decision function as (10.16) does.*

Crisp and Burges (2000) show that an optimal solution of ν-SVC (10.16) satisfies $\rho \geq 0$. In the above sense, ν-SVC and C-SVC are equivalent except for the case of $\rho = 0$.

10.2.1.3 Extended ν-Support Vector Classification

Recall that for $\nu \in (0, \nu_{\min})$, ν-SVC produces a trivial solution satisfying $(\boldsymbol{w}, b) = \boldsymbol{0}$. To prevent this, Perez-Cruz *et al.* (2003) require the norm of $\boldsymbol{w}$ to be unity:

$$\left|\begin{array}{ll}\underset{\boldsymbol{w},b,\rho,z}{\text{minimize}} & -\rho + \frac{1}{\nu m}\sum_{i=1}^{m} z_i,\\ \text{subject to} & z_i \geq -y_i(\boldsymbol{w}^{\top}\boldsymbol{\phi}(\boldsymbol{x}_i) - b) + \rho, i = 1, \ldots, m,\\ & z_i \geq 0, i = 1, \ldots, m,\\ & \|\boldsymbol{w}\|_2 = 1.\end{array}\right. \tag{10.18}$$

As a result of this modification, a nontrivial solution can be obtained even for $\nu \in (0, \nu_{\min})$. This modified formulation is called *extended ν-SVC (Eν-SVC)*.

Problem (10.18) is nonconvex because of the equality-norm constraint $\|\boldsymbol{w}\|_2 = 1$.[18] For $\nu \in [\nu_{\min}, \nu_{\max}]$, E$\nu$-SVC has the same optimal solutions as ν-SVC does and can be reduced to ν-SVC. Perez-Cruz *et al.* (2003) experimentally show that the out-of-sample performance of Eν-SVC with $\nu \in (0, \nu_{\min}]$ is often better than that with $\nu \in (\nu_{\min}, \nu_{\max}]$.

10.2.1.4 One-class ν-Support Vector Classifications

Next, we will consider a problem that has been referred to as *outlier/novelty detection*, *high-density region estimation*, or *domain description*.[19]

---

[17] ν-SVC with $\nu = \nu_{\min}$ may result in a nontrivial solution, whereas ν-SVC with $\nu \in (0, \nu_{\min})$ always results in the trivial solution. The computation of $\nu_{\min}$ will be discussed in Section 10.4.

[18] Perez-Cruz *et al.* (2003) propose an iterative algorithm for computing a solution. It goes as follows. First, for some $\bar{\boldsymbol{w}}$ satisfying $\|\bar{\boldsymbol{w}}\|_2^2 = 1$, define an LP by replacing $\|\boldsymbol{w}\|_2^2 = 1$ by $\bar{\boldsymbol{w}}^{\top}\boldsymbol{w} = 1$, and solve it. Then, use the obtained solution $\hat{\boldsymbol{w}}$ to update $\bar{\boldsymbol{w}}$, and repeat this procedure until convergence.

[19] This class of problems is sometimes referred to as one-class classification or, more broadly, unsupervised learning.

Let $X := \{\boldsymbol{x}_1, \ldots, \boldsymbol{x}_m\} \subset \mathbb{R}^n$ be a given dataset. The one-class problem is to define (possible) outliers in X. An outlier detection model known as a *one-class ν-support vector machine* (Schölkopf and Smola, 2002) is formulated as

$$\left|\begin{array}{ll} \underset{\boldsymbol{w},\rho,\boldsymbol{z}}{\text{minimize}} & \frac{1}{2}\|\boldsymbol{w}\|_2^2 - \nu\rho + \frac{1}{m}\sum_{i=1}^m z_i, \\ \text{subject to} & z_i \geq -\boldsymbol{w}^\top \boldsymbol{\phi}(\boldsymbol{x}_i) + \rho, i = 1, \ldots, m, \\ & z_i \geq 0, i = 1, \ldots, m, \end{array}\right. \tag{10.19}$$

where $\nu \in (0, 1]$ is a user-defined parameter. With an optimal solution $(\boldsymbol{w}^\star, \rho^\star, \boldsymbol{z}^\star)$, a sample $\boldsymbol{x}$ satisfying ${\boldsymbol{w}^\star}^\top \boldsymbol{\phi}(\boldsymbol{x}) < \rho^\star$ is regarded as an "outlier." Or, equivalently, we can define a high-density region to be the set $\{\boldsymbol{x} \in \mathbb{R}^n : {\boldsymbol{w}^\star}^\top \boldsymbol{\phi}(\boldsymbol{x}) \geq \rho^\star\}$. We will see in Section 10.3.1 that the (10.19) formulation can be interpreted on the basis of CVaR.

Support vector domain description (SVDD) (Tax and Duin, 1999) is a variant of the one-class problem. It detects (possible) outliers on the basis of the quadratically constrained optimization problem,

$$\left|\begin{array}{ll} \underset{\boldsymbol{\gamma},R,\boldsymbol{z}}{\text{minimize}} & R^2 + C\sum_{i=1}^m z_i, \\ \text{subject to} & z_i \geq \|\boldsymbol{\phi}(\boldsymbol{x}_i) - \boldsymbol{\gamma}\|_2^2 - R^2, i = 1, \ldots, m, \\ & z_i \geq 0, i = 1, \ldots, m, \end{array}\right. \tag{10.20}$$

where $C > 0$ is a user-defined parameter. SVDD defines outliers as points $\boldsymbol{x}_i$ satisfying $\|\boldsymbol{\phi}(\boldsymbol{x}_i) - \boldsymbol{\gamma}^\star\|_2 > R^\star$ by using an optimal solution $(\boldsymbol{\gamma}^\star, R^\star, \boldsymbol{z}^\star)$ of (10.20). Since the high-density region of $\boldsymbol{x}$ is defined as $\{\boldsymbol{x} \in \mathbb{R}^n : \|\boldsymbol{\phi}(\boldsymbol{x}) - \boldsymbol{\gamma}^\star\|_2 \leq R^\star\}$, this type of problem is called *high-density region estimation* or *domain description*. The high-density region is compact even when $\boldsymbol{\phi}$ is a linear mapping, whereas one-class ν-SVM (10.19) is not compact then.

10.2.2 Regression

Following its success in classification, SVC was extended so that it could handle real-valued outputs (Drucker *et al.*, 1997), (Schölkopf *et al.*, 2000). The *support vector regression (SVR)* method performs well in regression analysis and is a popular data analysis tool in machine learning and signal processing.

Let us consider the regression problem of obtaining a model $y = \boldsymbol{w}^\top \boldsymbol{\phi}(\boldsymbol{x}) + b$ using m training samples, $(\boldsymbol{x}_i, y_i), i = 1, \cdots, m$, where $\boldsymbol{x}_i \in \mathbb{R}^n$ is an input and $y_i \in \mathbb{R}$ is the corresponding output value.

10.2.2.1 ε-Support Vector Regression

In the ε-SVR framework (Drucker *et al.*, 1997), the model, or equivalently, $(\boldsymbol{w}, b)$, is determined so that the following regularized empirical risk functional is minimized:

$$\underset{\boldsymbol{w},b}{\text{minimize}} \quad \frac{1}{2}\|\boldsymbol{w}\|_2^2 + C\sum_{i=1}^m \max\{|y_i - \boldsymbol{w}^\top \boldsymbol{\phi}(\boldsymbol{x}_i) - b| - \varepsilon, 0\},$$

where C and ε are positive constants. Among the two parameters, $C > 0$ is a regularization constant that controls the trade-off between the goodness-of-fit and the complexity of the model. The parameter ε controls the sensitivity to the residuals (i.e., $|y_i - \boldsymbol{w}^\top \boldsymbol{\phi}(\boldsymbol{x}_i) - b|$). A potential weakness of the ε-SVR formulation is that the choice of ε is not intuitive.[20]

10.2.2.2 ν-Support Vector Regression

Another formulation of SVR, called *ν-SVR*, was proposed by Schölkopf *et al.* (2000); it uses a parameter $\nu \in (0, 1)$, instead of ε in ε-SVR. The optimization formulation of ν-SVR is

$$\left|\begin{array}{ll} \underset{\boldsymbol{w},b,c,\boldsymbol{z}}{\text{minimize}} & \frac{H}{2}\|\boldsymbol{w}\|_2^2 + c + \frac{1}{\nu m}\sum_{i=1}^m z_i, \\ \text{subject to} & z_i \geq y_i - \boldsymbol{w}^\top \boldsymbol{\phi}(\boldsymbol{x}_i) - b - c, i = 1, \dots, m, \\ & z_i \geq -y_i + \boldsymbol{w}^\top \boldsymbol{\phi}(\boldsymbol{x}_i) + b - c, i = 1, \dots, m, \\ & z_i \geq 0, i = 1, \dots, m, \end{array}\right. \tag{10.21}$$

where $H > 0$ is a user-defined constant. By setting $\nu = 1$ and restricting $c = \epsilon$, the (10.21) formulation reduces to the ϵ-SVR formulation.

10.3 ν-SVMs as CVaR Minimizations

In this section, we reformulate several SVMs in terms of CVaR minimization. We classify the CVaR minimizations into two cases: Case 1, where the loss $\boldsymbol{L}(\boldsymbol{\theta})$ is homogeneous with respect to the involving parameters $\boldsymbol{\theta}$ (i.e., for any $a \in \mathbb{R}$, $\boldsymbol{L}(a\boldsymbol{\theta}) = a\boldsymbol{L}(\boldsymbol{\theta})$); and Case 2, where the loss $\boldsymbol{L}(\boldsymbol{\theta})$ is not homogeneous.

10.3.1 ν-SVMs as CVaR Minimizations with Homogeneous Loss

We can formulate various machine-learning methods by using different types of loss. Let us begin with a binary classification problem defined with the positively homogeneous loss.

10.3.1.1 ν-SVC as a CVaR Minimization

Using the notation of CVaR with a linear loss $L_i(\boldsymbol{w}, b) = -y_i(\boldsymbol{w}^\top \boldsymbol{\phi}(\boldsymbol{x}_i) - b)$ and $p_i = 1/m$, $i = 1, \dots, m$, the quadratic program (10.16) can be symbolically rewritten as

$$\underset{\boldsymbol{w},b}{\text{minimize}} \quad \frac{1}{2}\|\boldsymbol{w}\|_2^2 + C\cdot\text{CVaR}_{(1-\nu,\mathbf{1}_m/m)}(-\boldsymbol{Y}(\boldsymbol{X}\boldsymbol{w} - \mathbf{1}_m b)), \tag{10.22}$$

with $C = \nu$, where

$$\boldsymbol{Y} := \text{diag}(\boldsymbol{y}) := \begin{pmatrix} y_1 & & \\ & \ddots & \\ & & y_m \end{pmatrix} \in \mathbb{R}^{m\times m}, \quad \boldsymbol{X} := \begin{pmatrix} \boldsymbol{\phi}(\boldsymbol{x}_1)^\top \\ \vdots \\ \boldsymbol{\phi}(\boldsymbol{x}_m)^\top \end{pmatrix} \in \mathbb{R}^{m\times N}.$$

[20] The function $\max\{|y_i - \boldsymbol{w}^\top \boldsymbol{\phi}(\boldsymbol{x}_i) - b| - \varepsilon, 0\}$ is called *Vapnik's ε-insensitive loss function.*

Note that (10.22) is a regularized empirical risk minimization in which CVaR is used as the empirical risk (see Figure 10.5). Note also that the empirical risk part in (10.22) is homogeneous in $(\boldsymbol{w}, b)$. To deal with the trade-off between the regularization term and the empirical CVaR, we may be able to perform another form of optimization,

$$\underset{\boldsymbol{w},b}{\text{minimize}} \quad \frac{1}{2}\|\boldsymbol{w}\|_2^2 \quad \text{subject to} \quad \text{CVaR}_{(1-\nu,\mathbf{1}_m/m)}(-\boldsymbol{Y}(\boldsymbol{X}\boldsymbol{w} - \mathbf{1}_m b)) \leq -D, \tag{10.23}$$

and[21]

$$\underset{\boldsymbol{w},b}{\text{minimize}} \quad \text{CVaR}_{(1-\nu,\mathbf{1}_m/m)}(-\boldsymbol{Y}(\boldsymbol{X}\boldsymbol{w} - \mathbf{1}_m b)) \quad \text{subject to} \quad \|\boldsymbol{w}\|_2 \leq E, \tag{10.24}$$

where C, D, and E are positive parameters for reconciling the trade-off. Under a mild assumption, the above three regularized empirical risk minimizations are equivalent for any positive parameters C, D, and E (see Tsyurmasto *et al.*, 2013).[22] Accordingly, C, D, and E can be restricted to 1. For example, (10.24) can be restricted to

$$\left|\begin{array}{ll} \underset{\boldsymbol{w},b}{\text{minimize}} & \text{CVaR}_{(1-\nu,\mathbf{1}_m/m)}(-\boldsymbol{Y}(\boldsymbol{X}\boldsymbol{w} - \mathbf{1}_m b)) \\ \text{subject to} & \|\boldsymbol{w}\|_2 \leq 1. \end{array}\right. \tag{10.25}$$

It is worth emphasizing that the equivalence of (10.22), (10.23), and (10.24) (or (10.25)) relies on the homogeneity of the empirical CVaR with respect to $(\boldsymbol{w}, b)$. Conversely, such an equivalence also holds true for any positively homogeneous risk functionals $\mathcal{F}(\cdot)$ in combination with a homogeneous loss $\boldsymbol{L}(\boldsymbol{\theta})$, that is, $\mathcal{F}(\boldsymbol{L}(\boldsymbol{\theta}))$. On the other hand, with the hinge loss employed in C-SVC (10.14), a risk-constrained variant like (10.23) is infeasible for any $D > 0$, and a parallel equivalence is no longer valid.

10.3.1.2 Eν-SVC as the Geometric Margin-based CVaR Minimization

Eν-SVC (10.18) can be symbolically rewritten as

$$\left|\begin{array}{ll} \underset{\boldsymbol{w},b}{\text{minimize}} & \text{CVaR}_{(1-\nu,\mathbf{1}_m/m)}(-\boldsymbol{Y}(\boldsymbol{X}\boldsymbol{w} - \mathbf{1}_m b)) \\ \text{subject to} & \|\boldsymbol{w}\|_2 = 1. \end{array}\right. \tag{10.26}$$

Comparing (10.25) and (10.26), we can see that ν-SVC (10.25) is a convex relaxation of Eν-SVC (10.26).

Note that when ν is in the range $(\nu_{\min}, \nu_{\max})$, the optimal value of (10.25) (i.e., the minimum CVaR) is negative, and $\|\boldsymbol{w}\|_2 = 1$ is attained at optimality because of the homogeneity of the objective function. In other words, when $\nu > \nu_{\min}$, the equality constraint $\|\boldsymbol{w}\|_2 = 1$ in

---

[21] Based on the discussion at the end of Section 10.1, the CVaR-constraint in (10.23) can be regarded as a convex conservative approximation of the chance constraint $\mathbb{P}\{\tilde{L}(\boldsymbol{w}, b) \geq -D\} \leq \nu$.

[22] This equivalence holds in the sense that these models provide the same optimal decision functions $d(\boldsymbol{x}) = \text{sign}(\boldsymbol{w}^\top \boldsymbol{\phi}(\boldsymbol{x}) - b)$ for any C, D, and E. Schölkopf *et al.* (2000) use the optimality condition to show the independence of the resulting classifiers of the parameter C. On the other hand, Tsyurmasto *et al.* (2013) show equivalence only on the basis of functional properties of CVaR such as positive homogeneity and continuity.

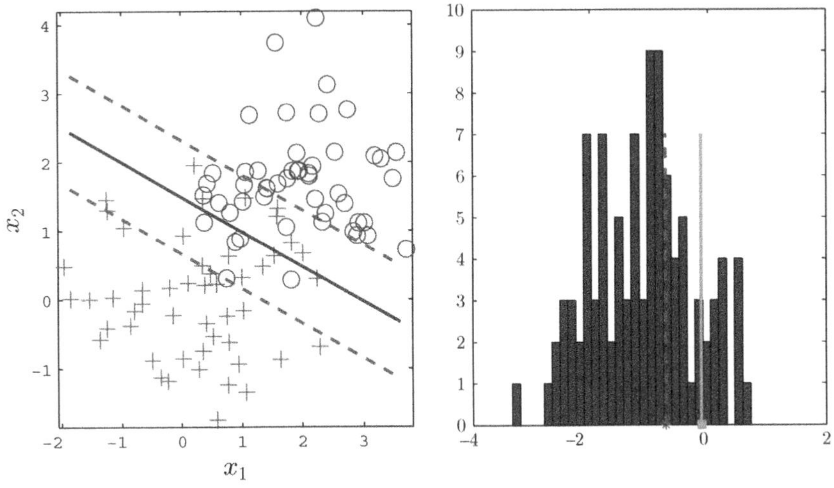

Figure 10.5 ν-SVC as a CVaR minimization. The figure on the left shows an optimal separating hyperplane $w_1^\star x_1 + w_2^\star x_2 = b^\star$ given by ν-SVC ($\nu = 0.3$). The one on the right is a histogram of the optimal distribution of the negative margin, $-y_i(w_1^\star x_{1i} + w_2^\star x_{2i} - b^\star), i = 1, \ldots, 100$. The locations of the minimized CVaR (solid line) and the corresponding VaR (broken line) are indicated in the histogram.

(10.26) can be relaxed to $\|\boldsymbol{w}\|_2 \leq 1$ without changing the optimal solution. On the other hand, when $\nu < \nu_{\min}$, ν-SVC (10.25) results in a trivial solution satisfying $(\boldsymbol{w}, b) = \mathbf{0}$.[23] Therefore, to obtain a solution to Eν-SVM (10.26) for $\nu < \nu_{\min}$, a nonconvex optimization method needs to be applied (Gotoh and Takeda, 2005; Perez-Cruz *et al.*, 2003; Takeda and Sugiyama, 2008). Figure 10.6 illustrates the relation between the sign of the optimal value of Eν-SVC (10.26) and ν.

Note that (10.18) can be equivalently rewritten as

$$\underset{\boldsymbol{w},b,c}{\text{minimize}} \quad c + \frac{1}{\nu m}\sum_{i=1}^{m} \max\left\{\frac{-y_i(\boldsymbol{w}^\top \boldsymbol{\phi}(\boldsymbol{x}_i) - b)}{\|\boldsymbol{w}\|_2} - c, 0\right\}, \tag{10.27}$$

(Takeda and Sugiyama, 2008). Namely, Eν-SVC (10.18) can be described as another CVaR minimization problem,

$$\underset{\boldsymbol{w},b}{\text{minimize}} \quad \text{CVaR}_{(1-\nu,\mathbf{1}_m/m)}\left(-\frac{\boldsymbol{Y}(\boldsymbol{X}\boldsymbol{w} - \mathbf{1}_m b)}{\|\boldsymbol{w}\|_2}\right),$$

by adopting the negative geometric margin as the loss, that is, $L_i(\boldsymbol{w}, b) = -y_i(\boldsymbol{w}^\top \boldsymbol{\phi}(\boldsymbol{x}_i) - b)/\|\boldsymbol{w}\|_2, i = 1, \ldots, m$. Since CVaR includes the maximum loss as a special limiting case (see Section 10.1.1), formulation (10.27) is a generalization of the maximum margin formulation (10.12). Figure 10.7 summarizes the relations among the four CVaR minimizations.

[23] In this case, $\|\boldsymbol{w}\|_2 = 1$ of Eν-SVC (10.26) can be relaxed to $\|\boldsymbol{w}\|^2 \geq 1$, but the resulting optimization problem is still nonconvex.

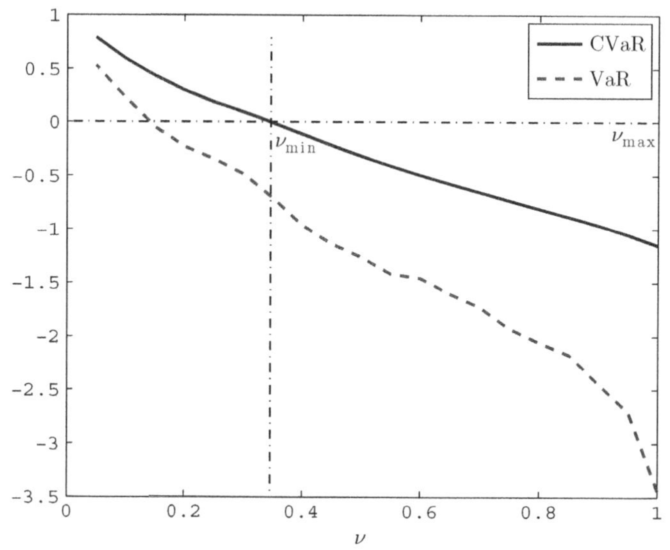

Figure 10.6 Minimized CVaR and corresponding VaR with respect to ν. CVaR indicates the optimal value of Eν-SVC (10.26) for binary classification. $\nu_{\min}$ is the value of ν at which the optimal value becomes zero. For $\nu > \nu_{\min}$, Eν-SVC (10.26) reduces to ν-SVC (10.25). For $\nu < \nu_{\min}$, ν-SVC (10.25) results in a trivial solution, while Eν-SVC (10.26) still attains a nontrivial solution with the positive optimal value.

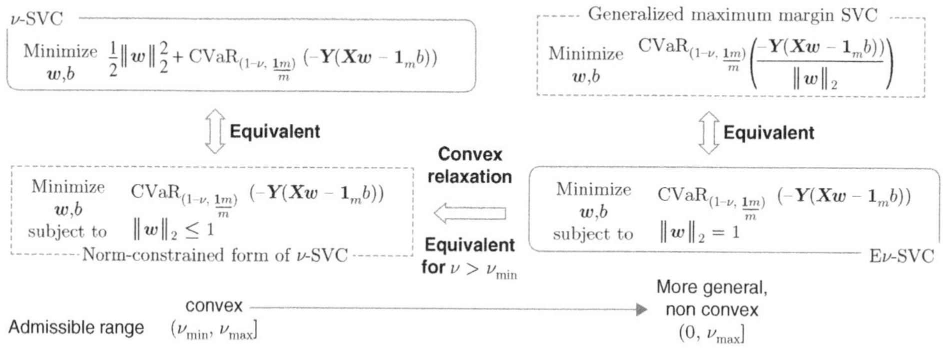

Figure 10.7 Relations among four classification formulations. The two formulations on the left are equivalent to the standard ν-SVC (10.16), while those on the right are equivalent to Eν-SVC (10.18). By resolving the nonconvexity issues that arise from the equality constraint, Eν-SVC provides a classifier that cannot be attained by ν-SVC.

10.3.1.3 One-class ν-SVC as a CVaR Minimization

With $L_i(\boldsymbol{w}) = -\boldsymbol{w}^\top \boldsymbol{\phi}(\boldsymbol{x}_i)$ and regularization term $\frac{1}{2}\|\boldsymbol{w}\|_2^2$, we obtain a CVaR minimization,

$$\underset{\boldsymbol{w}}{\text{minimize}} \quad \text{CVaR}_{(1-\nu,\mathbf{1}_m/m)}(-\boldsymbol{X}\boldsymbol{w}) + \frac{1}{2}\|\boldsymbol{w}\|_2^2, \tag{10.28}$$

which is equivalent to one-class ν-SVC (10.19).[24] Namely, outliers found by (10.19) are viewed as points $\boldsymbol{x}_i$ having values $-\boldsymbol{w}^\top \boldsymbol{\phi}(\boldsymbol{x}_i)$ greater than the 100ν-percentile under the CVaR minimizer $\boldsymbol{w}$.

10.3.2 ν-SVMs as CVaR Minimizations with Nonhomogeneous Loss

Next, we consider cases where the loss is not homogeneous with respect to the involved parameters.

10.3.2.1 CVaR-based Regression and ν-SVR

Assuming a regression model $y = b + \boldsymbol{w}^\top \boldsymbol{\phi}(\boldsymbol{x})$, we can employ a loss of the form $L_i(b, \boldsymbol{w}) = \epsilon(y_i - \{b + \boldsymbol{w}^\top \boldsymbol{\phi}(\boldsymbol{x}_i)\})$, where $\epsilon : \mathbb{R} \to [0, +\infty]$. Note that this is no longer homogeneous in $(b, \boldsymbol{w})$. By using this loss and $\boldsymbol{p}$, we can readily attain a regression version of CVaR minimization:

$$\underset{b,\boldsymbol{w}}{\text{minimize}} \quad \text{CVaR}_{(1-\nu,\boldsymbol{p})}(\epsilon(y_1 - \{b + \boldsymbol{w}^\top \boldsymbol{\phi}(\boldsymbol{x}_1)\}), \ldots, \epsilon(y_m - \{b + \boldsymbol{w}^\top \boldsymbol{\phi}(\boldsymbol{x}_m)\})). \tag{10.29}$$

In particular, when $\nu = 1$, (10.29) includes a number of popular regression formulations. For example:

- With $\boldsymbol{p} = \mathbf{1}_m/m$ and $\epsilon(z) = z^2$, (10.29) is equivalent to *ordinary least squares* (*OLS*).
- With arbitrary $\boldsymbol{p} \in \Pi^m$ and $\epsilon(z) = z^2$, it is equivalent to a weighted least square with a weight vector $\boldsymbol{p}$.
- With $\boldsymbol{p} = \mathbf{1}_m/m$ and $\epsilon(z) = az$ for $z \geq 0$ and $\epsilon(z) = -(1-a)z$ for $z < 0$ with some $a \in (0, 1)$, (10.29) is equivalent to *quantile regression* (Koenker and Bassett, 1978).

By adding a certain regularizer, we can attain more generalized formulations:

$$\underset{b,\boldsymbol{w}}{\text{minimize}} \quad \text{"Objective of (10.29)"} + Cg(\|\boldsymbol{w}\|), \tag{10.30}$$

where $C \geq 0$ is a constant, $g : [0, \infty) \to (-\infty, +\infty]$ a nondecreasing function, and $\|\cdot\|$ a norm. For example:

- With $\nu = 1, \boldsymbol{p} = \mathbf{1}_m/m$, $\epsilon(z) = z^2$, and $g(\|\boldsymbol{w}\|) = \frac{1}{2}\|\boldsymbol{w}\|_2^2$, it is equivalent to *ridge regression* (Hoerl and Kennard, 1970).

---

[24] Since the loss $L_i(\boldsymbol{w}) = -\boldsymbol{w}^\top \boldsymbol{\phi}(\boldsymbol{x}_i)$ is positively homogeneous with respect to $\boldsymbol{w}$, we can show that (10.19) is equivalent to $\min\{\frac{1}{2}\|\boldsymbol{w}\|_2^2 : \text{CVaR}_{(1-\nu,\mathbf{1}_m/m)}(\boldsymbol{L}(\boldsymbol{w})) \leq -1\}$ and $\min\{\text{CVaR}_{(1-\nu,\mathbf{1}_m/m)}(\boldsymbol{L}(\boldsymbol{w})) : \|\boldsymbol{w}\|_2 \leq 1\}$ in the sense that all of them provide the same decision function, as in two-class ν-SVC.

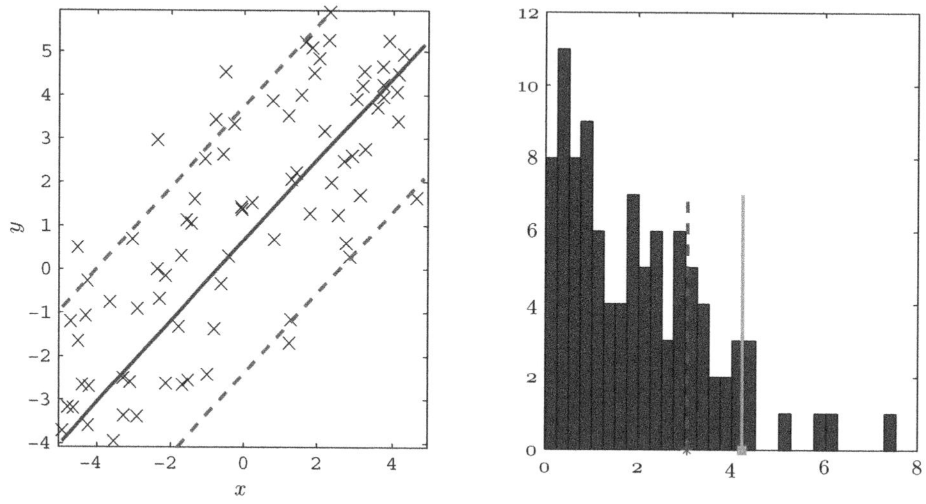

Figure 10.8 ν-SVR as a CVaR minimization. The left figure shows the regression model $y = w^{\star}x + b^{\star}$ given by ν-SVR ($\nu = 0.2$). The right one shows the histogram of the optimal distribution of the residual $|y_i - w^{\star}x_i - b^{\star}|, i = 1, \ldots, 100$. The locations of the minimized CVaR (green solid line) and the corresponding VaR (red dashed line) are indicated in the histogram.

- With $\nu = 1, \boldsymbol{p} = \mathbf{1}_m/m$, $\epsilon(z) = z^2$, and $g(\|\boldsymbol{w}\|) = \|\boldsymbol{w}\|_1$, it is equivalent to *lasso regression* (Tibshirani, 1996).

See Chapter 11 written by Uryasev for the definitions and a general look at ridge regression and lasso.

If $\nu < 1$, (10.30) is related to ν-SVR. Indeed, with $\boldsymbol{p} = \mathbf{1}_m/m$, $\epsilon(z) = |z|$, and $g(\|\boldsymbol{w}\|) = \frac{1}{2}\|\boldsymbol{w}\|_2^2$, it is equivalent to ν-SVR (10.21). Figure 10.8 shows the results of ν-SVR and the distribution of the residual.

Different from formulations (10.16) and (10.19) for the classification problem, formulation (10.21) depends on a trade-off parameter H in addition to ν. Indeed, the value of H changes the decision function $d(x)$ of (10.21). (10.21) is also different from the norm-constrained formulation:

$$\left|\begin{array}{ll} \underset{\boldsymbol{w},b,c,z}{\text{minimize}} & c + \frac{1}{\nu m}\sum_{i=1}^{m} z_i, \\ \text{subject to} & z_i \geq y_i - \boldsymbol{w}^{\top}\boldsymbol{\phi}(\boldsymbol{x}_i) - b - c, i = 1, \ldots, m, \\ & z_i \geq -y_i + \boldsymbol{w}^{\top}\boldsymbol{\phi}(\boldsymbol{x}_i) + b - c, i = 1, \ldots, m, \\ & z_i \geq 0, i = 1, \ldots, m, \\ & \|\boldsymbol{w}\|_2 \leq E, \end{array}\right. \tag{10.31}$$

unless E and H are appropriately set. This dependence on the parameters is due to the lack of positive homogeneity of the loss $\boldsymbol{L}(b, \boldsymbol{w})$. On the other hand, the CVaR minimization (10.29) has an optimal solution for any $\nu \in (0, 1)$, unlike ν-SVC.

10.3.2.2 Domain Description Problems as CVaR Minimizations

If we use a loss of the form $L_i(\gamma) = g(\|\phi(x_i) - \gamma\|)$ with a nondecreasing convex function g defined over $(0, \infty)$ and a norm $\|\cdot\|$, we arrive at another CVaR minimization,

$$\underset{\gamma}{\text{minimize}} \quad \text{CVaR}_{(1-\nu,p)}(g(\|\phi(x_1) - \gamma\|), \dots, g(\|\phi(x_m) - \gamma\|)),$$

which can be explicitly rewritten as a convex optimization problem,

$$\left|\begin{array}{ll} \underset{\gamma,c,z}{\text{minimize}} & c + \frac{1}{\nu}\sum_{i=1}^{m} p_i z_i, \\ \text{subject to} & z_i \geq g(\|\phi(x_i) - \gamma\|) - c, i = 1, \dots, m, \\ & z_i \geq 0, i = 1, \dots, m. \end{array}\right. \tag{10.32}$$

In particular, when we employ $g(\|\cdot\|) = \|\cdot\|_2^2$, $C = 1/\nu$, and $p = \mathbf{1}_m/m$, it is equivalent to SVDD (10.20). Namely, SVDD minimizes the CVaR of the distribution of the squared Euclidean distance $\|\phi(x_i) - \gamma\|^2$ from a center γ.

Formulation (10.32) can be considered to be a generalized version of the so-called *minimum enclosing ball*. Indeed, let us suppose that g is an increasing function. When $\nu < \min_i p_i$, formulation (10.32) becomes the optimization for obtaining a minimum ball enclosing the set of m points $\phi(x_1), \dots, \phi(x_m)$:

$$\left|\begin{array}{ll} \underset{\gamma,r}{\text{minimize}} & r, \\ \text{subject to} & r \geq \|\phi(x_i) - \gamma\|, i = 1, \dots, m. \end{array}\right.$$

On the other hand, when $\nu = 1$, (10.32) becomes

$$\underset{\gamma}{\text{minimize}} \sum_{i=1}^{m} p_i g(\|\phi(x_i) - \gamma\|),$$

which is known to characterize various centers of points $\phi(x_1), \dots, \phi(x_m)$ depending on the norm $\|\cdot\|$ employed.[25]

10.3.3 Refining the ν-Property

So far, we have shown that ν-SVMs can be viewed as CVaR minimizations, each being associated with a certain loss function. This fact enables us to look at SVMs on the basis of the distribution of the loss, $(L_1, \dots, L_m)$, as described in Section 10.1.

10.3.3.1 ν-property

An advantage of ν-SVM over C-SVM is that ν can be interpreted on the basis of the so-called *ν-property*.

---

[25] For example, with $g(\|\cdot\|) = \|\cdot\|_2^2$, the optimal γ is equal to the weighted average of $\phi(x_1), \dots, \phi(x_m)$; with $g(\|\cdot\|) = \|\cdot\|_1$, it is equal to the median center of $\phi(x_1), \dots, \phi(x_m)$; with $g(\|\cdot\|) = \|\cdot\|_2$, it is equal to the geometric median (or one-median) of $\phi(x_1), \dots, \phi(x_m)$. When $\phi(x_i) \in \mathbb{R}^2$ or $\mathbb{R}^3$, the optimal γ is sometimes called the Fermat–Weber point.

The ν-property is usually defined using the Karush–Kuhn–Tucker (KKT) condition (see, e.g., Vanderbei (2014) for the KKT condition). More precisely, let us consider ν-SVC (10.16). Given an optimal solution $(\boldsymbol{w}^\star, b^\star, c^\star, \boldsymbol{z}^\star)$ to (10.16) and $\boldsymbol{\lambda}^\star$ to (10.17), let us denote the set of samples that contribute to the margin error and the set of *support vectors (SVs)* by

$$\text{Err} := \{i \in I : z_i^\star > 0\}, \quad \text{SV} := \{i \in I : \lambda_i^\star > 0\},$$

where $I := \{1, \ldots, m\}$. Accordingly, the *margin error* and number of SVs can be expressed as $|\text{Err}|$ and $|\text{SV}|$. Note that $\text{Err} \subset \text{SV}$ and

$$|\text{SV}| - |\text{Err}| = |\{i \in I : -y_i(\boldsymbol{x}_i^\top \boldsymbol{w}^\star - b^\star) = c^\star\}|. \tag{10.33}$$

Proposition 10.1 ***(ν-property (Schölkopf et al., 2000))*** Any KKT solution to ν-SVC (10.16) or (10.17) satisfies

$$\frac{|\text{Err}|}{m} \leq \nu \leq \frac{|\text{SV}|}{m}.$$

This proposition says that ν is an upper bound of the fraction of margin errors and a lower bound of the fraction of SVs.[26]

Because of (10.33), we can see that the number of SVs is bounded above by a number depending on ν, as well. Indeed, we have

$$m\nu \leq |\text{SV}| \leq m\nu + |\{i \in I : -y_i(\boldsymbol{x}_i^\top \boldsymbol{w}^\star - b^\star) = c^\star\}|.$$

10.3.3.2 Quantile-based ν-property

The ν-property described above depends on the KKT condition of the optimization problems (10.16) and (10.17). However, the interpretation of ν is independently obtained from the definition of the quantile (i.e., VaR).

Let us refine the margin errors and support vectors with the notion of VaR. Denoting

$$\text{Err}_{(\nu,\boldsymbol{p})}(\boldsymbol{\theta}) := \{i \in I : L_i(\boldsymbol{\theta}) > \text{VaR}_{(1-\nu,\boldsymbol{p})}(\boldsymbol{L}(\boldsymbol{\theta}))\},$$

we can define the *quantile-based margin error* as $|\text{Err}_{(\nu,\boldsymbol{p})}(\boldsymbol{\theta})|$. Furthermore, we can denote the set of *quantile-based support vectors* by

$$\text{SV}_{(\nu,\boldsymbol{p})}(\boldsymbol{\theta}) := \{i \in I : L_i(\boldsymbol{\theta}) \geq \text{VaR}_{(1-\nu,\boldsymbol{p})}(\boldsymbol{L}(\boldsymbol{\theta}))\}.$$

Note that the difference between $\text{Err}_{(\nu,\boldsymbol{p})}(\boldsymbol{\theta})$ and $\text{SV}_{(\nu,\boldsymbol{p})}(\boldsymbol{\theta})$ is only in the equality in the above definitions. Indeed, we have $|\text{SV}_{(\nu,\boldsymbol{p})}(\boldsymbol{\theta})| - |\text{Err}_{(\nu,\boldsymbol{p})}(\boldsymbol{\theta})| = |\{i \in I : L_i(\boldsymbol{\theta}) = \text{VaR}_{(1-\nu,\boldsymbol{p})}(\boldsymbol{L}(\boldsymbol{\theta}))\}|$, and the following proposition is straightforward.

Proposition 10.2 The following holds for any $\boldsymbol{\theta}$:

$$\frac{|\text{Err}_{(\nu,\boldsymbol{p})}(\boldsymbol{\theta})|}{m} \leq \nu \leq \frac{|\text{SV}_{(\nu,\boldsymbol{p})}(\boldsymbol{\theta})|}{m}.$$

[26] Note that Eν-SVC also has this property because this proposition only relies on the KKT condition.

Similarly to the standard notion of SVs, the following holds:

$$m\nu \leq |\mathrm{SV}_{(\nu,\boldsymbol{p})}(\boldsymbol{\theta})| \leq m\nu + |\{i \in I : L_i(\boldsymbol{\theta}) = \mathrm{VaR}_{(1-\nu,\boldsymbol{p})}(\boldsymbol{L}(\boldsymbol{\theta}))\}|. \tag{10.34}$$

Note that the inequalities in Proposition 10.2 are valid for any $\boldsymbol{\theta}$, whereas the ordinary ν-property of Proposition 10.1 is shown for a KKT solution (i.e., an optimal solution). By separately defining the risk functional and optimization as in Sections 10.1.1 and 10.1.3, we can introduce the ν-property independently of the optimality condition. Accordingly, the above relation (10.34) suggests that the number of SVs can be reduced by making ν small.

10.3.3.3 Generalization Bounds for ν-SVC and Eν-SVC

The goal of learning methods is to obtain a classifier or a regressor that has a small generalization error. As mentioned in Section 10.2.1, the maximum margin hyperplane of hard-margin SVC minimizes an upper bound of the generalization error. This is considered as the reason why a high generalization performance can be obtained by hard-margin SVC for linearly separable datasets. Here, we give generalization error bounds based on the CVaR risk measure for ν-SVC and Eν-SVC and show that minimizing CVaR leads to a lower generalization bound, which will explain why a high generalization performance can be obtained by ν-SVC and Eν-SVC for linearly nonseparable datasets.

A classifier $d(\boldsymbol{x}) = \mathrm{sign}(\boldsymbol{w}^\top \boldsymbol{\phi}(\boldsymbol{x}) - b)$ is learned on a training set. Here, we assume that such training samples are drawn from an unknown independent and identically distributed (i.i.d.) probability distribution $P(\boldsymbol{x}, y)$ on $\mathrm{IR}^n \times \{\pm 1\}$. The goal of the classification task is to obtain a classifier d (precisely, $(\boldsymbol{w}, b)$ of d) that minimizes the generalization error defined as

$$\mathrm{GE}[d] := \int 1_{\{d(\boldsymbol{x}) \neq y\}} \,\mathrm{d}P(\boldsymbol{x}, y),$$

which corresponds to the misclassification rate for unseen test samples, but unfortunately, GE[d] cannot be computed since P is unknown. A bound on the generalization error is derived, as discussed further here, and used for theoretical analysis of the learning model.

We begin with the case of $\nu \in (\nu_{\min}, \nu_{\max})$, where E$\nu$-SVC is equivalent to ν-SVC.

Theorem 10.4 (Takeda and Sugiyama, 2008) *Let $\nu \in (\nu_{\min}, \nu_{\max}]$ and $\boldsymbol{L}(\boldsymbol{w}, b) = -\boldsymbol{Y}(\boldsymbol{X}\boldsymbol{w} - \mathbf{1}_m b)$. Suppose that $P(\boldsymbol{x}, y)$ has support in a ball of radius R around the origin. Then, for all $(\boldsymbol{w}, b)$ such that $\|\boldsymbol{w}\|_2 \leq 1$ and $CVaR_{(1-\nu,\mathbf{1}_m/m)}(\boldsymbol{L}(\boldsymbol{w}, b)) < 0$, there exists a positive constant c such that the following bound holds with a probability of at least $1 - \delta$:*

$$\begin{aligned} GE[d] &\leq \frac{|Err_{(\nu,\mathbf{1}_m/m)}(\boldsymbol{w}, b)|}{m} + \Gamma_c(VaR_{(1-\nu,\mathbf{1}_m/m)}(\boldsymbol{L}(\boldsymbol{w}, b))) \\ &\leq \nu + \Gamma_c(VaR_{(1-\nu,\mathbf{1}_m/m)}(\boldsymbol{L}(\boldsymbol{w}, b))), \end{aligned} \tag{10.35}$$

where

$$\Gamma_c(\gamma) := \sqrt{\frac{2}{m}\left(\frac{4c^2(R^2+1)^2}{\gamma^2}\log_2(2m) - 1 + \log\frac{2}{\delta}\right)}.$$

The generalization error bound in (10.35) is furthermore upper-bounded as

$$GE[d] \leq \nu + \Gamma_c(CVaR_{(1-\nu,\mathbf{1}_m/m)}(\boldsymbol{L}(\boldsymbol{w}, b))).$$

The function $\Gamma(\gamma)$ decreases as $|\gamma|$ increases. Note also that for $\nu \in (\nu_{\min}, \nu_{\max}]$, ν-SVC and, equivalently, Eν-SVC attain a negative minimum CVaR, that is, $\text{CVaR}_{(1-\nu,\mathbf{1}_m/m)}(\boldsymbol{L}(\boldsymbol{w}^\star, b^\star)) < 0$ (see Section 10.3.1). Accordingly, Theorem 10.4 implies that the minimum generalization bound regarding $(\boldsymbol{w}, b)$ is attained at an optimal solution of ν-SVC (10.25), which minimizes $\text{CVaR}_{(1-\nu,\mathbf{1}_m/m)}(\boldsymbol{L}(\boldsymbol{w}, b))$ subject to $\|\boldsymbol{w}\|_2 \leq 1$. That is, it is expected that the classifier of ν-SVC has a small generalization (i.e., out-of-sample) error.

Next, we consider the case of $\nu \in (0, \nu_{\min}]$, for which ν-SVC results in a trivial solution satisfying $\boldsymbol{w} = \mathbf{0}$, but E$\nu$-SVC leads to a reasonable solution. The discussion below depends on the sign of VaR (see Figure 10.6, where the range $(0, \nu_{\min})$ is divided into two subranges corresponding to a negative VaR or a nonnegative VaR).

Theorem 10.5 (Takeda and Sugiyama, 2008) *Let* $\nu \in (0, \nu_{\min})$*, and let* $(\boldsymbol{w}, b)$ *satisfy* $\|\boldsymbol{w}\|_2 = 1$.

- *Additionally, if* $(\boldsymbol{w}, b)$ *satisfies* $VaR_{(1-\nu,\mathbf{1}_m/m)}(\boldsymbol{L}(\boldsymbol{w}, b)) < 0$*, there exists a positive constant c such that the following bound holds with a probability of at least* $1 - \delta$*:*

$$GE[d] \leq \nu + \Gamma_c(VaR_{(1-\nu,\mathbf{1}_m/m)}(\boldsymbol{L}(\boldsymbol{w}, b))).$$

- *On the other hand, if* $(\boldsymbol{w}, b)$ *satisfies* $VaR_{(1-\nu,\mathbf{1}_m/m)}(\boldsymbol{L}(\boldsymbol{w}, b)) > 0$*, there exists a positive constant c such that the following bound holds with a probability of at least* $1 - \delta$*:*

$$GE[d] \geq \nu - \Gamma_c(VaR_{(1-\nu,\mathbf{1}_m/m)}(\boldsymbol{L}(\boldsymbol{w}, b))).$$

This theorem implies that the upper bound or lower bound of GE[d] can be lowered if $\text{VaR}_{(1-\nu,\mathbf{1}_m/m)}(\boldsymbol{L}(\boldsymbol{w}, b))$ is reduced; indeed, minimizing $\text{VaR}_{(1-\nu,\mathbf{1}_m/m)}(\boldsymbol{L}(\boldsymbol{w}, b))$ with respect to $(\boldsymbol{w}, b)$ subject to $\text{VaR}_{(1-\nu,\mathbf{1}_m/m)}(\boldsymbol{L}(\boldsymbol{w}, b)) < 0$ minimizes the upper bound, while minimizing $\text{VaR}_{(1-\nu,\mathbf{1}_m/m)}(\boldsymbol{L}(\boldsymbol{w}, b))$ subject to $\text{VaR}_{(1-\nu,\mathbf{1}_m/m)}(\boldsymbol{L}(\boldsymbol{w}, b)) > 0$ minimizes the lower bound. Recalling that VaR is upper-bounded by CVaR and that Eν-SVC (10.26) is a CVaR minimization subject to $\|\boldsymbol{w}\|_2 = 1$, Eν-SVC is expected to reduce the generalization error through minimization of the upper bound or lower bound.

10.4 Duality

In this section, we present the dual problems of the CVaR-minimizing formulations in the Section 10.3. As mentioned in Section 10.2, dual representations expand the range of algorithms and enrich the theory of SVM.

10.4.1 Binary Classification

As explained in Section 10.1.3, optimization problems (10.22) with $C = 1$ and (10.25) (i.e., (10.24) with $E = 1$), are equivalent formulations of ν-SVC because of the positive homogeneity of CVaR and the loss $L(w,b) = -Y(Xw - 1_m b)$. Correspondingly, the dual formulations of the CVaR-based representations (10.22) with $C = 1$ and (10.25) can be derived as

$$\left|\begin{array}{ll}\underset{\lambda}{\text{maximize}} & -\frac{1}{2}\|X^\top Y\lambda\|_2^2 \\ \text{subject to} & y^\top\lambda = 0, \lambda \in \mathcal{Q}_{\text{CVaR}(1-\nu,p)},\end{array}\right. \tag{10.36}$$

and

$$\left|\begin{array}{ll}\underset{\lambda}{\text{maximize}} & -\|X^\top Y\lambda\|_2 \\ \text{subject to} & y^\top\lambda = 0, \lambda \in \mathcal{Q}_{\text{CVaR}(1-\nu,p)},\end{array}\right. \tag{10.37}$$

with $p = 1_m/m$.[27]

By using a kernel function $k(x,\xi) = \phi(x)^\top\phi(\xi)$, we can readily obtain a kernelized nonlinear classification. Indeed, letting $K := XX^\top$ and replacing the objective functions of (10.36) and (10.37) with $-\frac{1}{2}\lambda^\top YKY\lambda$ and $-\sqrt{\lambda^\top YKY\lambda}$, respectively, each of them becomes a kernelized formulation.

10.4.2 Geometric Interpretation of ν-SVM

Observe that $\sum_{i=1}^m y_i\lambda_i = 0$ can be rewritten as $\sum_{i\in I_+}\lambda_i = \sum_{i\in I_-}\lambda_i$ and that

$$\lambda^\top YKY\lambda = \|X^\top Y\lambda\|_2^2 = (\sum_{i=1}^m y_i\phi(x_i)\lambda_i)^2 = (\sum_{i\in I_+}\phi(x_i)\lambda_i - \sum_{i\in I_-}\phi(x_i)\lambda_i)^2,$$

where $I_+ := \{i \in \{1,\dots,m\} : y_i = +1\}$, and $I_- := \{1,\dots,m\}\backslash I_+$. With a change of variables $\mu := \lambda/2$, the duals (10.36) and (10.37) can be rewritten as

$$-\left|\begin{array}{ll}\underset{\xi^+,\xi^-}{\text{minimize}} & \frac{1}{8}\|\xi^+ - \xi^-\|_2^2 \\ \text{subject to} & \xi^+ \in D_+^\nu, \xi^- \in D_-^\nu,\end{array}\right. \quad \text{and} \quad -\left|\begin{array}{ll}\underset{\xi^+,\xi^-}{\text{minimize}} & \frac{1}{2}\|\xi^+ - \xi^-\|_2 \\ \text{subject to} & \xi^+ \in D_+^\nu, \xi^- \in D_-^\nu,\end{array}\right.$$

where

$$D_\bullet^\nu := \{\xi \in \mathbb{R}^N : \xi = \sum_{i\in I_\bullet}\phi(x_i)\mu_i, \mu \in \mathcal{Q}_{\text{CVaR}(1-\nu,2p^\bullet)}\}, \quad \bullet \in \{+,-\},$$

where $p^+ \in \mathbb{R}^{m_+}$ and $p^- \in \mathbb{R}^{m_-}$ denote vectors whose elements come from p corresponding to $i \in I_+$ and $i \in I_-$, respectively.[28] Accordingly, the dual problems can be interpreted as ones of finding the nearest two points each belonging to D_+^ν and D_-^ν (see Figure 10.9). These sets, D_+^ν and D_-^ν, are referred to as *reduced convex hulls* or *soft convex hulls* in Bennett and Bredensteiner (2000) and Crisp and Burges (2000). Indeed, for $\nu < \nu_{\min}$, they are equivalent to the convex hulls of $\{\phi(x_i) : i \in I_+\}$ and $\{\phi(x_i) : i \in I_-\}$, respectively; the size of set $D_\bullet^\nu$ monotonically decreases in ν, that is, $D_\bullet^{\nu_1} \subset D_\bullet^{\nu_2}$ for $\nu_1 > \nu_2$; for $\nu \geq \nu_{\max}$, they shrink to their centers $\sum_{i\in I_\bullet} p_i^\bullet\phi(x_i)$.

[27] Obviously, the difference between (10.36) and (10.37) is only in the objective, and in practice, (10.36) is easier to solve since a quadratic program is more stably solvable than a second-order cone program.

[28] $m_+ = |I_+|$ and $m_- = |I_-|$.

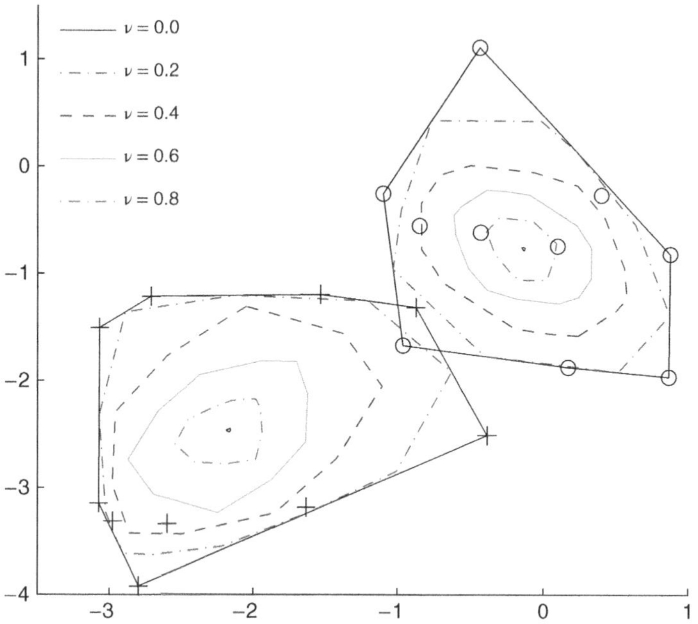

Figure 10.9 Two-dimensional examples of reduced convex hulls. Here, '+' and '∘' represent the data samples. As ν increases, the size of each reduced convex hull shrinks. The reduced convex hull is a single point for $\nu = 1$, whereas it is equal to the convex hull for ν sufficiently close to 0. For linearly inseparable datasets, the corresponding convex hulls (or the reduced convex hulls for a small ν) intersect, and the primal formulation (10.25) results in a trivial solution satisfying $(\boldsymbol{w}, b) = \mathbf{0}$.

10.4.3 Geometric Interpretation of the Range of ν for ν-SVC

Crisp and Burges (2000) pointed out that $\nu_{\min}$ is the largest ν such that two reduced convex hulls D_+^ν and D_-^ν intersect. Namely, $\nu_{\min}$ is the value such that $D_+^{\nu_{\min}}$ and $D_-^{\nu_{\min}}$ touch externally. Indeed, $\nu_{\min}$ can be computed by solving the following optimization problem:

$$\frac{1}{\nu_{\min}} := \left| \begin{array}{ll} \underset{\eta, \boldsymbol{\mu}^+, \boldsymbol{\mu}^-}{\text{minimize}} & \eta \\ \text{subject to} & \sum\limits_{i \in I_+} \boldsymbol{\phi}(\boldsymbol{x}_i)\mu_i^+ = \sum\limits_{j \in I_-} \boldsymbol{\phi}(\boldsymbol{x}_j)\mu_j^-, \\ & \boldsymbol{\mu}^+ \in \mathcal{Q}_{\mathrm{CVaR}(1-\frac{1}{\eta}, \frac{2}{m}\mathbf{1}_{m_+})}, \quad \boldsymbol{\mu}^- \in \mathcal{Q}_{\mathrm{CVaR}(1-\frac{1}{\eta}, \frac{2}{m}\mathbf{1}_{m_-})}. \end{array} \right.$$

Note that this problem reduces to an LP.

On the other hand, if ν is smaller than $\nu_{\min}$, D_+^ν and D_-^ν intersect, and thus, ν-SVC attains zero optimal value. This is the geometric interpretation of the trivial solution mentioned in Section 10.2.1.

However, to make D_+^ν and D_-^ν non-empty, we need to choose ν satisfying $\nu \leq \frac{2m_+}{m}$ and $\nu \leq \frac{2m_-}{m}$. Consequently, $\nu_{\max}$ is defined as $\min\{\frac{2m_+}{m}, \frac{2m_-}{m}\}$.

10.4.4 Regression

The dual problems for regression (10.21) can be derived in a similar way as the SVC formulations. Using the notation $|\lambda| := (|\lambda_1|, \ldots, |\lambda_m|)^\top$, the (kernelized) dual formulation of (10.21) can be symbolically represented by

$$\left| \begin{array}{ll} \underset{\lambda}{\text{maximize}} & -\frac{1}{2H}\lambda^\top \boldsymbol{K}\lambda + \boldsymbol{y}^\top\lambda \\ \text{subject to} & \boldsymbol{1}_m^\top\lambda = 0, \quad |\lambda| \in \mathcal{Q}_{\text{CVaR}(1-\nu, \boldsymbol{1}_m/m)}, \end{array} \right. \tag{10.38}$$

where $\boldsymbol{K} = (k(\boldsymbol{x}_i, \boldsymbol{x}_h))_{i,h}$ with a kernel function $k(\boldsymbol{x}, \xi) = \boldsymbol{\phi}(\boldsymbol{x})^\top\boldsymbol{\phi}(\xi)$. On the other hand, the dual of the norm-constrained version (10.31) can be symbolically represented by

$$\left| \begin{array}{ll} \underset{\lambda}{\text{maximize}} & \boldsymbol{y}^\top\lambda - E\sqrt{\lambda^\top \boldsymbol{K}\lambda} \\ \text{subject to} & \boldsymbol{1}_m^\top\lambda = 0, \quad |\lambda| \in \mathcal{Q}_{\text{CVaR}(1-\nu, \boldsymbol{1}_m/m)}. \end{array} \right. \tag{10.39}$$

10.4.5 One-class Classification and SVDD

The (kernelized) dual formulations of the CVaR-based one-class classification (10.28) are given by

$$\left| \begin{array}{ll} \underset{\lambda}{\text{maximize}} & -\lambda^\top \boldsymbol{K}\lambda \\ \text{subject to} & \lambda \in \mathcal{Q}_{\text{CVaR}(1-\nu, \boldsymbol{1}_m/m)}. \end{array} \right. \tag{10.40}$$

On the other hand, the dual of the CVaR-based SVDD (10.32) is derived as

$$\left| \begin{array}{ll} \underset{\lambda}{\text{maximize}} & \sum_{i=1}^m \sqrt{k(\boldsymbol{x}_i, \boldsymbol{x}_i)}\lambda_i - \sum_{i=1}^m \sum_{h=1}^m k(\boldsymbol{x}_i, \boldsymbol{x}_h)\lambda_i\lambda_h \\ \text{subject to} & \lambda \in \mathcal{Q}_{\text{CVaR}(1-\nu, \boldsymbol{p})}. \end{array} \right.$$

10.5 Extensions to Robust Optimization Modelings

The assumptions of machine-learning theory do not always fit real situations.[29] Some modifications can be made to bridge the gap between theory and practice. Among them is the *robust optimization* modeling (Ben-Tal *et al.*, 2009) option. In this section, we show that two kinds of robust modeling of the CVaR minimization for binary classification are tractable.

10.5.1 Distributionally Robust Formulation

The uniform distribution $\boldsymbol{p} = \boldsymbol{1}_m/m$ is reasonable as long as the dataset is i.i.d. sampled. However, if this does not hold true, the choice $\boldsymbol{p} = \boldsymbol{1}_m/m$ may not be the best. Let us tackle such

[29] For example, the i.i.d. assumption is often violated. In such a situation, it may be better to choose a non-uniform reference probability.

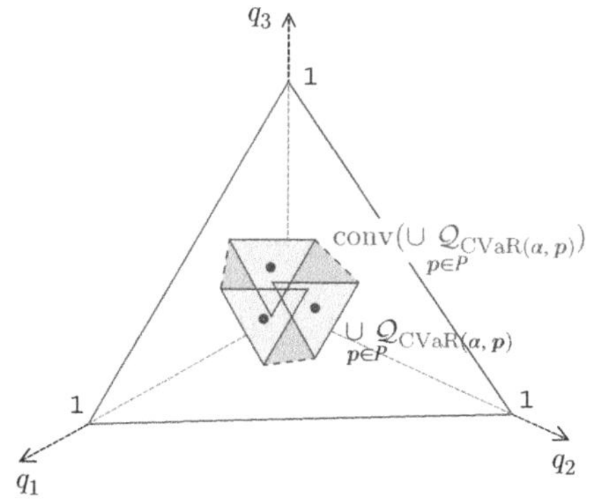

Figure 10.10 Convex hull of the union of risk envelopes ($m = 3$).

uncertainties in the choice of the reference probability $\boldsymbol{p}$ with a min-max strategy. Let $P \subset \mathrm{IR}^m$ be the set of possible $\boldsymbol{p}$, and let us call it the *uncertainty set*. For simplicity, we will assume that P is closed and bounded. For some P, we replace $\mathrm{CVaR}_{(1-\nu,\boldsymbol{p})}(\boldsymbol{L})$ with the distributionally worst-case CVaR, defined by $\max_{\boldsymbol{p}\in P} \mathrm{CVaR}_{(1-\nu,\boldsymbol{p})}(\boldsymbol{L})$.

Recalling (10.11), the worst-case CVaR is represented by

$$\begin{aligned}\max_{\boldsymbol{p}\in P,\boldsymbol{q}\in\mathcal{Q}_{\mathrm{CVaR}(1-\nu,\boldsymbol{p})}} \boldsymbol{q}^\top \boldsymbol{L} &= \max_{\boldsymbol{q}} \{\boldsymbol{q}^\top \boldsymbol{L} : \boldsymbol{q} \in \cup_{\boldsymbol{p}\in P} \mathcal{Q}_{\mathrm{CVaR}(1-\nu,\boldsymbol{p})}\} \\ &= \max_{\boldsymbol{q}} \{\boldsymbol{q}^\top \boldsymbol{L} : \boldsymbol{q} \in \mathrm{conv}(\cup_{\boldsymbol{p}\in P} \mathcal{Q}_{\mathrm{CVaR}(1-\nu,\boldsymbol{p})}\}.\end{aligned}$$

The second equality holds because of the linearity of the function $\boldsymbol{q}^\top \boldsymbol{L}$ (see Figure 10.10 for an illustration of a convex hull of the union of risk envelopes). Note that the above expression shows that the distributionally worst-case CVaR is coherent as long as $\cup_{\boldsymbol{p}\in P}\mathcal{Q}_{\mathrm{CVaR}(1-\nu,\boldsymbol{p})} \subset \Pi^m$. (Recall the dual representation of the coherent risk measure described in Section 10.1.2.) Consequently, the distributionally robust ν-SVC can be written as

$$f_P^\star := \left| \begin{array}{ll} \underset{\boldsymbol{w},b}{\text{minimize}} & \max_{\boldsymbol{q}} \{-\boldsymbol{q}^\top \boldsymbol{Y}(\boldsymbol{X}\boldsymbol{w} - \boldsymbol{1}_m b) : \boldsymbol{q} \in \mathrm{conv}(\cup_{\boldsymbol{p}\in P} \mathcal{Q}_{\mathrm{CVaR}(1-\nu,\boldsymbol{p})})\} \\ \text{subject to} & \|\boldsymbol{w}\| \le 1. \end{array} \right. \tag{10.41}$$

Similar to the case of (10.37), the (kernelized) dual form of (10.41) is derived as[30]

$$(f_P^\star)^2 = \left| \begin{array}{ll} \underset{\lambda}{\text{maximize}} & -\lambda^\top \boldsymbol{YKY}\lambda \\ \text{subject to} & \boldsymbol{y}^\top \lambda = 0, \lambda \in \mathrm{conv}(\cup_{\boldsymbol{p}\in P} \mathcal{Q}_{\mathrm{CVaR}(1-\nu,\boldsymbol{p})}). \end{array} \right. \tag{10.42}$$

For some P, problem (10.42) becomes a tractable convex optimization. Here are special cases of the examples given in Gotoh and Uryasev (2013).

[30] The objective function of (10.42) is squared: $-\lambda^\top \boldsymbol{YKY}\lambda = -\lambda^\top \boldsymbol{YXX}^\top \boldsymbol{Y}\lambda = -\|\boldsymbol{X}^\top \boldsymbol{Y}\lambda\|_2^2$.

Example 10.5 (Finite-scenario uncertainty) *If we employ an uncertainty set defined by* $P = \{\boldsymbol{p}^1, \dots, \boldsymbol{p}^J\}$*, with J candidates* $\boldsymbol{p}^1, \dots, \boldsymbol{p}^J \in \Pi^m$*, (10.42) can be rewritten as*

$$\left|\begin{array}{ll} \underset{\boldsymbol{\lambda},\boldsymbol{\pi},\boldsymbol{\tau}}{\text{maximize}} & -\boldsymbol{\lambda}^\top \boldsymbol{YKY\lambda} \\ \text{subject to} & \boldsymbol{y}^\top \boldsymbol{\lambda} = 0, \boldsymbol{1}_m^\top \boldsymbol{\lambda} = 1, \boldsymbol{0} \le \boldsymbol{\lambda} \le \boldsymbol{\pi}/\nu, \\ & \boldsymbol{\pi} = \sum_{k=1}^{J} \tau_k \boldsymbol{p}_k, \boldsymbol{1}_J^\top \boldsymbol{\tau} = 1, \boldsymbol{\tau} \ge \boldsymbol{0}. \end{array}\right.$$

This formulation was first presented in Wang (2012), which extends the robust formulation to a multiclass classification setting.

Example 10.6 (Distance-based uncertainty) *If we use an uncertainty set defined by* $P = \{\boldsymbol{q} \in \Pi^m : \boldsymbol{q} = \boldsymbol{p} + \boldsymbol{A\zeta}, \|\boldsymbol{\zeta}\| \le 1\}$*, with* $\boldsymbol{p} \in \Pi^m$*, the symmetric positive definite matrix* $\boldsymbol{A} \in I\!R^{m\times m}$*, and* $\|\cdot\|$ *a norm in* $I\!R^m$*, (10.42) can be rewritten as*

$$\left|\begin{array}{ll} \underset{\boldsymbol{\lambda},\boldsymbol{\pi},\boldsymbol{\zeta}}{\text{maximize}} & -\boldsymbol{\lambda}^\top \boldsymbol{YKY\lambda} \\ \text{subject to} & \boldsymbol{y}^\top \boldsymbol{\lambda} = 0, \boldsymbol{1}_m^\top \boldsymbol{\lambda} = 1, \boldsymbol{0} \le \boldsymbol{\lambda} \le \boldsymbol{\pi}/\nu, \\ & \boldsymbol{\pi} = \boldsymbol{p} + \boldsymbol{A\zeta}, \boldsymbol{1}_m^\top \boldsymbol{A\zeta} = 0, \|\boldsymbol{\zeta}\| \le 1. \end{array}\right.$$

Example 10.7 (Entropy-based uncertainty) *If we use an uncertainty set defined by* $P = \left\{\boldsymbol{q} \in \Pi^m : \sum_{i=1}^m q_i \ln(q_i/p_i) \le t\right\}$*, with* $t > 0$ *and* $\boldsymbol{p} \in \Pi^m$ *such that* $p_i > 0$*, (10.42) can be rewritten as*

$$\left|\begin{array}{ll} \underset{\boldsymbol{\lambda},\boldsymbol{\pi}}{\text{maximize}} & -\boldsymbol{\lambda}^\top \boldsymbol{YKY\lambda} \\ \text{subject to} & \boldsymbol{y}^\top \boldsymbol{\lambda} = 0, \boldsymbol{1}_m^\top \boldsymbol{\lambda} = 1, \boldsymbol{0} \le \boldsymbol{\lambda} \le \boldsymbol{\pi}/\nu, \\ & \sum_{i=1}^m \pi_i \ln(\pi_i/p_i) \le t, \boldsymbol{1}_m^\top \boldsymbol{\pi} = 1. \end{array}\right.$$

These convex optimization problems can be solved using off-the-shelf nonlinear programming solver software packages.

10.5.2 Measurement-wise Robust Formulation

Aside from the distributionally robust formulation in Section 10.5.1, another form of uncertainty can be introduced to CVaR defined with a linear loss such as $\boldsymbol{L}(\boldsymbol{w}, b) = -\boldsymbol{Y}(\boldsymbol{Xw} - \boldsymbol{1}_m b)$ and $\boldsymbol{L}(\boldsymbol{w}) = -\boldsymbol{YXw}$.

Recall that classifiers obtained by SVCs depend on the support vectors, which typically make up a small subset of samples. Accordingly, they are likely to be susceptible to the measurement error of a sample, that is, $\boldsymbol{\phi}(\boldsymbol{x}_i) - \Delta\boldsymbol{\phi}(\boldsymbol{x}_i)$. To mitigate the effect of such a perturbation, we may be able to use the min-max strategy, similarly to the case of distributionally robust optimization.

Let us define the set of perturbations as

$$\mathcal{S} := \{(\boldsymbol{\delta}_1, \dots, \boldsymbol{\delta}_m) : \|\boldsymbol{\delta}_i\| \le C, i = 1, \dots, m\},$$

where $\|\cdot\|$ is some norm in IR^n. Given $\mathcal{S}$, the worst-case CVaR is defined as

$$\mathrm{WCVaR}^{\mathcal{S}}_{(\alpha,p)}(-Y(Xw-\mathbf{1}_m b)) := \max_{\Delta\in\mathcal{S}} \mathrm{CVaR}_{(\alpha,p)}(-Y\{(X-\Delta)w-\mathbf{1}_m b\}).$$

This worst-case CVaR enables us to consider variants of ν-SVC. For example, CVaR minimization (10.25) can be modified into

$$\left|\begin{array}{ll}\underset{w,b}{\text{minimize}} & \mathrm{WCVaR}^{\mathcal{S}}_{(\alpha,p)}(-Y(Xw-\mathbf{1}_m b))\\ \text{subject to} & \|w\|_2 \le 1,\end{array}\right. \tag{10.43}$$

whereas the usual ν-SVC (10.22) can be modified into

$$\underset{w,b}{\text{minimize}} \quad \frac{1}{2}\|w\|_2^2 + \mathrm{WCVaR}^{\mathcal{S}}_{(\alpha,p)}(-Y(Xw-\mathbf{1}_m b)). \tag{10.44}$$

Note that (10.43) and (10.44) are formulated as min-max optimizations. However, we can represent a robust optimization in a tractable manner by using the following formula.[31]

Proposition 10.3 (Gotoh and Uryasev, 2013) For any (w,b), we have

$$\mathrm{WCVaR}^{\mathcal{S}}_{(\alpha,p)}(-Y(Xw-\mathbf{1}_m b)) = C\|w\|^\circ + \mathrm{CVaR}_{(\alpha,p)}(-Y(Xw-\mathbf{1}_m b)), \tag{10.45}$$

where $\|\cdot\|^\circ$ is the dual norm of $\|\cdot\|$, that is, $\|x\|^\circ := \max\{x^\top z : \|z\| \le 1\}$.

On the basis of formula (10.45), the robust ν-SVC formulations (10.43) and (10.44) can be rewritten as

$$\left|\begin{array}{ll}\underset{w,b}{\text{minimize}} & C\|w\|^\circ + \mathrm{CVaR}_{(\alpha,p)}(-Y(Xw-\mathbf{1}_m b))\\ \text{subject to} & \|w\|_2 \le 1,\end{array}\right.$$

and

$$\left|\;\underset{w,b}{\text{minimize}} \quad \tfrac{1}{2}\|w\|_2^2 + C\|w\|^\circ + \mathrm{CVaR}_{(\alpha,p)}(-Y(Xw-\mathbf{1}_m b)).\right. \tag{10.46}$$

Note that in defining $\mathcal{S}$, we do not have to limit the norm $\|\cdot\|$ to being the ℓ_2-norm. Indeed, if $\|\cdot\| = \|\cdot\|_\infty$ is used, formulation (10.46) virtually has a regularization term of the form $\frac{1}{2}\|w\|_2^2 + C\|w\|_1$, since $\|\cdot\|_\infty^\circ = \|\cdot\|_1$. In this sense, the min-max strategy yields a justification of the modified regularization term used in the *elastic net* (Zou and Hastie, 2005).[32]

10.6 Literature Review

In this final section, we briefly summarize the related literature and further extensions to the results described in this chapter.

[31] A similar result for C-SVC is derived in Xu *et al.* (2009), in which a smaller perturbation set is used in place of $\mathcal{S}$.

[32] Proposition 10.3 relies on the monotonicity and the translation invariance of CVaR, but we cannot say that the elastic net is underpinned by the same reasoning. Note that the risk functional of the elastic net satisfies neither of these properties. However, we can say that DrSVM of Wang *et al.* (2006), which is a C-SVC with an elastic net-type regularizer, is underpinned by the reasoning in Xu *et al.* (2009).

10.6.1 CVaR as a Risk Measure

The term CVaR was originally introduced by Rockafellar and Uryasev (2000), and *tail VaR* and *expected shortfall* are sometimes used for signifying the same notion. Rockafellar and Uryasev (2000) show that its minimization can be represented by an LP and the optimal $c^\star$ in the minimization formula yields an approximate VaR. Rockafellar and Uryasev (2002) extensively analyze the basic properties of CVaR and its minimization for general distributions.

On the other hand, the empirical CVaR minimization requires a large number of scenarios to make an accurate estimation. Indeed, Lim *et al.* (2011) demonstrate by simulation that CVaR minimization leads to poor out-of-sample performance.

10.6.2 From CVaR Minimization to SVM

Gotoh and Takeda (2005) were the first to describe the connection between CVaR minimization and ν-SVM (actually, Eν-SVC), while Takeda and Sugiyama (2008) were the first to point out that the model of Gotoh and Takeda (2005) is equivalent to Eν-SVC.

Regarding the robustification of CVaR minimization, Zhu and Fukushima (2009) proposed a distributionally robust portfolio selection. They consider finite scenarios and norm-based uncertainty and formulate convex optimization problems without the dual representation of CVaR. Wang (2012) uses a similar robust optimization modeling to formulate a distributionally robust multiclass ν-SVC. Gotoh *et al.* (2014) and Gotoh and Uryasev (2013) extend this robust modeling to the cases of coherent and convex functionals. Indeed, they extend the examples in Section 10.5.2 to non-CVaR functionals.

As for the measurement-wise robust ν-SVC, Gotoh and Uryasev (2013) show that any monotonic and translation-invariant functional results in a regularized empirical functional, as in Proposition 10.3.

10.6.3 From SVM to CVaR Minimization

On the other hand, the notions used in SVMs can be used in portfolio selection. Indeed, the two examples of portfolio optimization at the end of Section 10.1.3 are constrained versions of one-class ν-SVC and ν-SVR. Gotoh and Takeda (2011) present a regularized CVaR minimization by placing a norm constraint on the portfolio vector. They use regularization in the same way as in DeMiguel *et al.* (2009), where the variance-minimizing portfolio is coupled with a norm constraint. They develop generalization bounds for a portfolio selection based on the norm-constrained VaR or CVaR minimization. Gotoh and Takeda (2012) also develop generalization error bounds for portfolio selection and devise a fractional optimization problem on the basis of the empirical VaR and CVaR. El Karoui *et al.* (2012) use the asymptotic variance of the empirical CVaR as a regularizer.

10.6.4 Beyond CVaR

CVaR is the most popular *coherent measure of risk* (Artzner *et al.* 1999). The coherence of CVaR was proven by Pflug (2000) as well as by Rockafellar and Uryasev (2002). It is easy to see how such a generalized class of risk measures can be incorporated in SVMs. Gotoh *et al.* (2014) apply a coherent measure of risk to SVMs in place of the CVaR of the geometric margin. By making a straightforward extension of the maximum margin SVC, they can use a negative geometric margin as the loss and deal with nonconvex optimization. On the other hand, Gotoh and Uryasev (2013) study SVC formulations on the basis of convex functionals and discuss what properties of the risk measure affect the SVC formulation. Tsyurmasto *et al.* (2013) focus on the positive homogeneity of risk measures.

A generalized risk functional of the form $\mathcal{F}[\tilde{L}] = \inf_{c} \{c + \mathbb{E}_{\mathbb{P}}[v(\tilde{L} - c)]\}$, where v is a nondecreasing convex function on $\mathbb{R}$, was first studied by Ben-Tal and Teboulle (1986). This functional is named *optimized certainty equivalent (OCE)*. Note that OCE can be viewed as a generalization of CVaR, since CVaR is equivalent to OCE when $v(z) = \max\{z, 0\}/(1 - \alpha)$. Kanamori *et al.* (2013) (unintentionally) apply OCE to SVC as an extension of ν-SVC. Gotoh and Uryasev (2013) point out that the use of OCE-based SVC is related to the Csiszár f-divergence (Csiszár, 1967).

Rockafellar and Uryasev (2013) provide a systematic view of risk functionals, named the *risk quadrangle*, within which a wide class of convex functionals used in risk management, statistics, economic theory, and so on are shown to be related to each other. Indeed, CVaR is associated with quantile regression (Koenker and Bassett, 1978) within the quantile-based quadrangle.

References

Acerbi, C. (2002). Spectral measures of risk: a coherent representation of subjective risk aversion. *Journal of Banking and Finance*, 26 1505–1518.

Artzner, P., Delbaen, F., Eber, J.M. and Heath, D. (1999). Coherent measures of risk. *Mathematical Finance*, 9 (3) 203–228.

Bennett, K.P. and Bredensteiner, E.J. (2000). Duality and geometry in SVM classifiers. *Proceedings of International Conference on Machine Learning*, 57–64.

Ben-Tal, A. and El-Ghaoui, L. and Nemirovski, A. (2009). *Robust optimization*. Princeton: Princeton University Press.

Ben-Tal, A. and Teboulle, M. (1986). Expected utility, penalty functions, and duality in stochastic nonlinear programming. *Management Science*, 32, 1445–1466.

Boser B.E., Guyon I.M. and Vapnik V.N. (1992). A training algorithm for optimal margin classifiers. *COLT*, 144–152.

Chang, C.C. and Lin, C.J. (2001). Training ν support vector classifiers: theory and algorithms. *Neural Computation*, 13 (9), 2119–2147.

Cortes, C. and Vapnik, V. (1995). Support-vector networks. *Machine Learning*, 20 273–297.

Crisp, D.J. and Burges, C.J.C. (2000). A geometric interpretation of ν-SVM classifiers. In *Advances in Neural Information Processing Systems 12* (ed. Solla, S.A., Leen, T.K. and Müller, K.R.). Cambridge, MA: MIT Press, 244–250.

Csiszár, I. (1967). Information-type measures of divergence of probability distributions and indirect observations. *Studia Scientiarum Mathematicarum Hungarica*, 2, 299–318.

DeMiguel, V., Garlappi, L., Nogales, F.J. and Uppal, R. (2009). A generalized approach to portfolio optimization: improving performance by constraining portfolio norms. *Management Science*, 55 (5) 798–812.

Drucker, H., Burges C.J.C., Kaufman L., Smola A. and Vapnik V. (1997). Support vector regression machines. In *Advances in Neural Information Processing Systems 9*. Cambridge, MA: MIT Press, 155–161.

El Karoui, N., Lim, A.E.B. and Vahn, G.Y. (2012). Performance-based regularization in mean-CVaR portfolio optimization. arXiv:1111.2091v2.

Gotoh, J. and Takeda, A. (2012). Minimizing loss probability bounds for portfolio selection. *European Journal of Operational Research*, 217 (2), 371–380.

Gotoh, J. and Takeda, A. (2011). On the role of norm constraints in portfolio selection. *Computational Management Science*, 8 (4), 323–353.

Gotoh, J. and Takeda, A. (2005). A linear classification model based on conditional geometric score. *Pacific Journal of Optimization*, 1 (2), 277–296.

Gotoh, J., Takeda, A. and Yamamoto, R. (2014). Interaction between financial risk measures and machine learning methods. *Computational Management Science*, 11 (4), 365–402.

Gotoh, J. and Uryasev, S. (2013). *Support vector machines based on convex risk functionals and general norms.* Research report #2013-3. Gainesville, FL: University of Florida, Department of Industrial and Systems Engineering. www.ise.ufl.edu/uryasev/publications

Hoerl, E. and Kennard, R.W. (1970). Ridge regression: biased estimation for nonorthogonal problems. *Technometrics*, 12, 55–67.

Kanamori, T., Takeda, A and Suzuki, T. (2013). Conjugate relation between loss functions and uncertainty sets in classification problems. *Journal of Machine Learning Research*, 14, 1461–1504.

Koenker, R. and Bassett, G. (1978). Regression quantiles. *Econometrica*, 46, 33–50.

Lim, A.E.B., Shanthikumar, J.G. and Vahn, G.Y. (2011). Conditional value-at-risk in portfolio optimization: coherent but fragile. *Operations Research Letters*, 39 (3), 163–171.

Markowitz, H. (1952). Portfolio selection. *Journal of Finance*, 7, 77–91.

Perez-Cruz, F., Weston, J., Hermann, D.J.L. and Schölkopf, B. (2003). Extension of the ν-SVM range for classification. In *Advances in learning theory: methods, models and applications 190* (ed. Suykens, J.A.K., Horvath, G., Basu, S., Micchelli, C. and Vandewalle, J.). Amsterdam: IOS Press, 179–196.

Pflug, G.C. (2000). Some remarks on the value-at-risk and the conditional value-at-risk. In *Probabilistic constrained optimization: methodology and applications* (ed. Uryasev, S.). Berlin: Springer, 278–287.

Rockafellar, R.T. and Uryasev, S. (2000). Optimization of conditional value-at-risk. *The Journal of Risk*, 2 (3), 21–41.

Rockafellar, R.T. and Uryasev, S. (2002). Conditional value-at-risk for general loss distributions. *Journal of Banking and Finance*, 26 (7), 1443–1471.

Rockafellar, R.T. and Uryasev, S. (2013). The fundamental risk quadrangle in risk management, optimization and statistical estimation. *Surveys in Operations Research and Management Science*, 18 (1–2), 33–53.

Ruszczyński, A. and Shapiro, A. (2006). Optimization of convex risk functions. *Mathematics of Operations Research*, 31 (3), 433–452.

Schölkopf, B. and Smola, A.J. (2002). *Learning with kernels – support vector machines, regularization, optimization, and beyond.* Cambridge, MA: MIT Press.

Schölkopf, B., Smola, A.J., Williamson, R.C. and Bartlett, P.L. (2000). New support vector algorithms. *Neural Computation*, 12 (5), 1207–1245.

Takeda, A. and Sugiyama, M. (2008). ν-Support vector machine as conditional value-at-risk minimization. In *Proceedings of 25th Annual International Conference on Machine Learning*. New York: ACM Press, 1056–1063.

Tax, D.M.J. and Duin, R.P.W. (1999). Support vector domain description. *Pattern Recognition Letters*, 20, 1191–1199.

Tibshirani, R. (1996). Optimal reinsertion: regression shrinkage and selection via the lasso. *Journal of the Royal Statistical Society*, 58 (1), 267–288.

Tsyurmasto, P., Gotoh, J. and Uryasev, S. (2013). Support vector classification with positive homogeneous risk functionals. Research report #2013-4. Gainesville, FL: University of Florida, Department of Industrial and Systems Engineering. www.ise.ufl.edu/uryasev/publications/

Uryasev, S. (2016). Regression models in risk management. Chapter 11 of Ali N. Akansu, Sanjeev R. Kulkarni and Dmitry Malioutov ed. Financial Signal Processing and Machine Learning. Wiley.

Vanderbei, R.J. (2014). *Linear programming: foundations and extensions*, 4th ed. Berlin: Springer.

Vapnik, V.N. (1995). *The nature of statistical learning theory*. Berlin: Springer.

Wang, Y. (2012). Robust ν-support vector machine based on worst-case conditional value-at-risk minimization. *Optimization Methods and Software*, 27 (6), 1025–1038.

Wang, L., Zhu, J. and Zou, H. (2006). The doubly regularized support vector machine. *Statistica Sinica*, 16 (2), 589.

Xu, H., Caramanis, C. and Mannor, S. (2009). Robustness and regularization of support vector machines. *The Journal of Machine Learning Research*, 10, 1485–1510.

Zabarankin, M. and Uryasev, S. (2013). *Statistical decision problems: selected concepts and portfolio safeguard case studies*. Berlin: Springer.

Zhu, S. and Fukushima, M. (2009). Worst-case conditional value-at-risk with application to robust portfolio management. *Operations Research*, 57 (5), 1155–1168.

Zou, H. and Hastie, T. (2005). Regularization and variable selection via the elastic net. *Journal of the Royal Statistical Society*, 67 (2), 301–320.

11

Regression Models in Risk Management

Stan Uryasev
University of Florida, USA

This chapter discusses theory and application of generalized linear regression that minimizes a general error measure of regression residual subject to various constraints on regression coefficients and includes least-squares linear regression, median regression, quantile regression, mixed quantile regression, and robust regression as special cases. General error measures are nonnegative positively homogeneous convex functionals that generalize the notion of norm and, in general, are asymmetric with respect to ups and downs of a random variable, which allows one to treat gains and losses differently. Each nondegenerate error measure $\mathcal{E}$ yields the deviation measure $\mathcal{D}$ projected from $\mathcal{E}$ and the statistic $\mathcal{S}$ associated with $\mathcal{E}$. General deviation measures are also nonnegative positively homogeneous convex functionals, which, in contrast to error measures, are insensitive to a constant shift. They generalize the notion of standard deviation, but are not required to be symmetric. General deviation measures admit dual characterization in terms of risk envelopes, which is instrumental in devising efficient optimization formulations for minimization of deviation measures. The central theoretical result in generalized linear regression is the error decomposition theorem stating that minimization of an error measure of the regression residual can be decomposed into minimizing the projected deviation measure of the residual without the intercept and into setting the intercept to the statistic associated with the error measure. The value of this theorem is that minimization of deviation measures admits dual formulation in terms of risk envelopes and, as a result, yields efficient optimization formulations. Application of generalized linear regression includes examples of financial index tracking, sparse signal reconstruction, therapy treatment planning, collateralized debt obligation, mutual fund return-based style classification, and mortgage pipeline hedging. The examples also provide linear program formulations of the corresponding regressions.

Financial Signal Processing and Machine Learning, First Edition.
Edited by Ali N. Akansu, Sanjeev R. Kulkarni and Dmitry Malioutov.
© 2016 John Wiley & Sons, Ltd. Published 2016 by John Wiley & Sons, Ltd.

The chapter is organized as follows. The introduction discusses a general setup of linear regression. Section 11.2 introduces general error and deviation measures, whereas Section 11.3 introduces risk envelopes and risk identifiers. Section 11.4 states the error decomposition theorem. Sections 11.5, 11.6, and 11.7 formulate least-squares linear regression, median regression, and quantile regression, respectively, and present application of these regressions in financial engineering and signal processing. Section 11.7 also formulates mixed quantile regression as a generalization of quantile regression. Section 11.8 introduces unbiased linear regression and risk acceptable linear regression with application to financial index tracking. Section 11.9 discusses robust regression with application to mortgage pipeline hedging.

11.1 Introduction

In statistics, regression analysis aims to find the best relationship between a *response* random variable Y (*regressant*) and n independent variables $x_1, \dots, x_n$ (*regressors*) in the form:

$$Y = f(x_1, \dots, x_n) + \epsilon,$$

based on m available simultaneous observations of $x_1, \dots, x_n$ and Y (regression data): $x_{1j}, \dots, x_{nj}, y_j, j = 1, \dots, m$, where ϵ is the *approximation error*.

There are two main classes of regression: *parametric* and *nonparametric*. If the function f is determined by a finite set of parameters, regression is called parametric; otherwise, it is called nonparametric. The class of parametric regressions is further divided into *linear* and *nonlinear*. In linear regression, f is a linear function with respect to unknown parameters, whereas $x_1, \dots, x_n$ can be involved nonlinearly (though, in some definitions, linear regression is assumed to be linear with respect to $x_1, \dots, x_n$; see, e.g., Hastie *et al.*, 2008). Typically, in linear regression, f has the form

$$f(x_1, \dots, x_n) = c_0 + c_1 x_1 + \dots + c_n x_n = c_0 + \sum_{k=1}^{n} c_k x_k, \tag{11.1}$$

where $c_k \in I\!R$, $k = 0, 1, \dots, n$, are unknown regression parameters with c_0 called *intercept* or *bias*. (Estimates of $c_0, c_1, \dots, c_n$ found from the regression data are denoted by $\widehat{c}_0, \widehat{c}_1, \dots, \widehat{c}_n$, respectively.) In nonlinear regression, f is a nonlinear function of specified unknown parameters, which are usually found iteratively.

One of the main approaches for finding estimates of regression parameters is to maximize the likelihood of the observations of $y_1, \dots, y_m$ under the assumption that the residuals $e_j = y_j - f(x_{1j}, \dots, x_{nj}), j = 1, \dots, m$, are realizations of *independent and identically distributed* (i.i.d.) random variables $\epsilon_1, \dots, \epsilon_m$ with zero mean; see van der Waerden (1957). For example, if $\epsilon_1, \dots, \epsilon_m$ are i.i.d. and have the normal distribution $N(0, \sigma^2)$, then the likelihood of observing $y_1, \dots, y_m$ is given by

$$\frac{1}{(\sqrt{2\pi}\sigma)^m} \prod_{j=1}^{m} \exp\left(-\frac{1}{2\sigma^2} (y_j - f(x_{1j}, \dots, x_{nj}))^2 \right),$$

and its maximization simplifies to

$$\min \sum_{j=1}^{m} (y_j - f(x_{1j}, \dots, x_{nj}))^2,$$

which is called the *least-squares method*. With $f(x_1,\dots,x_n)$ in the form of (11.1), this minimization problem yields a system of linear equations for estimates $\hat{c}_1,\dots,\hat{c}_n$:

$$\sum_{k=1}^{n}\hat{c}_k\sum_{j=1}^{n}(x_{ij}-\tilde{x}_i)(x_{kj}-\tilde{x}_k)=\sum_{j=1}^{m}(x_{ij}-\tilde{x}_i)(y_j-\tilde{y}),\quad i=1,\dots,n, \tag{11.2}$$

with $\hat{c}_0=\tilde{y}-\sum_{k=1}^{n}c_k\tilde{x}_k$, where $\tilde{x}_i=\frac{1}{n}\sum_{j=1}^{m}x_{ij}$ for $i=1,\dots,n$ and $\tilde{y}=\frac{1}{n}\sum_{j=1}^{m}y_j$.

Even if $\epsilon_1,\dots,\epsilon_m$ are only uncorrelated (not necessarily independent) with zero mean and same variance, the *Gauss–Markov theorem* states that the *best linear unbiased estimator* (BLUE) of the form (11.1) is determined by *least-squares linear regression*. If $\epsilon_1,\dots,\epsilon_m$ are correlated and/or not identically distributed random variables, then the least-squares regression may not be appropriate.

Statistical approximation theory takes a different perspective on regression: when the response random variable Y is not understood completely and is better to be treated as a function $f(X_1,\dots,X_n)$ of *random* variables $X_1,\dots,X_n$, the error $Y-f(X_1,\dots,X_n)$ is sought to minimize some loss function or *error measure* with respect to unknown regression parameters; see Rockafellar *et al.* (2008). In this approach, central to regression analysis is the choice of error measure that should conform to *risk preferences* of an analyst. For example, if the problem is to track a stock market index by a portfolio of selected financial instruments, whose returns are random variables $X_1,\dots,X_n$, the analyst may penalize only underperformance of the portfolio return $f(X_1,\dots,X_n)$ with respect to the index return Y, so that symmetric measures like $\|\cdot\|_2$ are not appropriate.

This chapter pursues the statistical approximation approach to regression. It focuses on a general theory of approximating an output random variable Y by a linear combination of input random variables $X_1,\dots,X_n$:

$$f(X_1,\dots,X_n)=c_0+c_1X_1+\dots+c_nX_n=c_0+\sum_{k=1}^{n}c_kX_k$$

with an arbitrary error measure $\mathcal{E}$ under additional constraints on regression coefficients.

11.2 Error and Deviation Measures

Let $(\Omega,\mathcal{M},\mathbb{P})$ be a probability space of elementary events Ω with the sigma-algebra $\mathcal{M}$ over Ω and with a probability measure $\mathbb{P}$ on $(\Omega,\mathcal{M})$. Random variables are assumed to be measurable real-valued functions from $\mathcal{L}^2(\Omega)=\mathcal{L}^2(\Omega,\mathcal{M},\mathbb{P})$ unless otherwise specified,[1] and the relationships between random variables X and Y (e.g., $X\le Y$ and $X=Y$) are understood to hold in the almost-sure sense (i.e., $\mathbb{P}[X\le Y]=1$ and $\mathbb{P}[X=Y]=1$), respectively. Also, $\inf X$ and $\sup X$ mean *essential infimum* and *essential supremum* of X (i.e., $\operatorname{ess\,inf} X$ and $\operatorname{ess\,sup} X$), respectively. Two important integral characteristics of a random variable X are its *mean* and *variance*, defined by

$$\mu(X)=\int_\Omega X(\omega)\,d\mathbb{P}[\omega],\quad \sigma^2(X)=\int_\Omega (X(\omega)-\mu(X))^2\,d\mathbb{P}[\omega],$$

[1] $\mathcal{L}^2(\Omega)$ is the Lebesgue space of measurable square-integrable functions on Ω: $X\in\mathcal{L}^2(\Omega)$ is equivalent to $\int_\Omega|X(\omega)|^2d\mathbb{P}[\omega]<\infty$.

respectively, and $\sigma(X)$ is called the *standard deviation* of X. If $X \in \mathcal{L}^2(\Omega)$, then $\mu(X)$ and $\sigma^2(X)$ are well defined (bounded), which explains the choice of $\mathcal{L}^2(\Omega)$.

Rockafellar *et al.* (2002, 2006a, 2008) introduced *error measures* as functionals $\mathcal{E} : \mathcal{L}^2(\Omega) \to [0, \infty]$ satisfying

(E1) *Nonnegativity*: $\mathcal{E}(0) = 0$, but $\mathcal{E}(X) > 0$ for $X \neq 0$; also, $\mathcal{E}(c) < \infty$ for constants c.
(E2) *Positive homogeneity*: $\mathcal{E}(\lambda X) = \lambda \mathcal{E}(X)$ when $\lambda > 0$.
(E3) *Subadditivity*: $\mathcal{E}(X + Y) \leq \mathcal{E}(X) + \mathcal{E}(Y)$ for all X and Y.
(E4) *Lower semicontinuity*: set $\{X \in \mathcal{L}^2(\Omega) | \mathcal{E}(X) \leq c\}$ is closed for all $c < \infty$.

For example, $\mathcal{L}^p$ norms $\|X\|_p$ with $p \geq 1$ are error measures. However, error measures are not required to be symmetric; that is, in general, $\mathcal{E}(-X) \neq \mathcal{E}(X)$. An example of an *asymmetric* error measure is given by

$$\mathcal{E}_{a,b,p}(X) = \|a\, X_+ + b\, X_-\|_p, \quad a > 0, \quad b > 0, \quad 1 \leq p \leq \infty. \tag{11.3}$$

Another one is the asymmetric mean absolute error:

$$\mathcal{E}_\alpha(X) = \frac{1}{\alpha} E[\alpha\, X_+ + (1 - \alpha)\, X_-], \quad \alpha \in (0, 1), \tag{11.4}$$

where $X_\pm = \max\{0, \pm X\}$. For $\alpha = 1/2$, $\mathcal{E}_\alpha(X)$ simplifies to $\|X\|_1$. Observe that for $a = 1$ and $b = 1$, (11.3) simplifies to $\|X\|_p$, whereas for $p = 1$, $a = 1$, and $b = 1/\alpha - 1$, it reduces to (11.4).

An error measure $\mathcal{E}$ is *nondegenerate* if there exists $\delta > 0$ such that $\mathcal{E}(X) \geq \delta\, |E[X]|$ for all X. For example, (11.3) and (11.4) are both nondegenerate error measures with $\delta = \min\{a, b\}$ and $\delta = \min\{1, 1/\alpha - 1\}$, respectively; see Rockafellar *et al.* (2008).

Similar to error measures, Rockafellar *et al.* (2002, 2006a) introduced deviation measures as functionals $\mathcal{D} : \mathcal{L}^2(\Omega) \to [0, \infty]$ satisfying

(D1) *Nonnegativity*: $\mathcal{D}(X) = 0$ for constant X, but $\mathcal{D}(X) > 0$ otherwise.
(D2) *Positive homogeneity*: $\mathcal{D}(\lambda X) = \lambda \mathcal{D}(X)$ when $\lambda > 0$.
(D3) *Subadditivity*: $\mathcal{D}(X + Y) \leq \mathcal{D}(X) + \mathcal{D}(Y)$ for all X and Y.
(D4) *Lower semicontinuity*: set $\{X \in \mathcal{L}^2(\Omega) | \mathcal{D}(X) \leq c\}$ is closed for all $c < \infty$.

It follows from D1 and D3 that

$$\mathcal{D}(X - c) = \mathcal{D}(X) \qquad \text{for all constants } c,$$

which is known as *insensitivity to constant shift* (see Rockafellar *et al.*, 2006a). Axioms D1–D4 generalize well-known properties of the standard deviation; however, they do not imply symmetry, so that in general, $\mathcal{D}(-X) \neq \mathcal{D}(X)$.

Each error measure $\mathcal{E}$ yields a deviation measure through *penalties relative to expectation*

$$\mathcal{D}(X) = \mathcal{E}(X - E[X]), \tag{11.5}$$

and if $\mathcal{E}$ is nondegenerate, it furnishes another deviation through *error projection*

$$\mathcal{D}(X) = \inf_{c \in I\!R} \mathcal{E}(X - c), \tag{11.6}$$

which is called *the deviation of X projected from* $\mathcal{E}$; see Theorem 2.1 in Rockafellar *et al.* (2008). A solution to (11.6) is *the statistic of X associated with* $\mathcal{E}$

$$\mathcal{S}(X) = \arg\min_{c \in \mathbb{R}} \mathcal{E}(X - c), \tag{11.7}$$

which, in general, is an interval $[\mathcal{S}^-(X), \mathcal{S}^+(X)]$ of constants with $\mathcal{S}^-(X) = \min\{c | c \in \mathcal{S}(X)\}$ and $\mathcal{S}^+(X) = \max\{c | c \in \mathcal{S}(X)\}$, and has the following properties:

$$\mathcal{S}(X - c) = \mathcal{S}(X) - c \quad \text{for any constant } c,$$

$$\mathcal{S}(\lambda\, X) = \lambda\, \mathcal{S}(X) \quad \text{for any constant } \lambda > 0.$$

Well-known examples of the relationships (11.6) and (11.7) are given in the following table:

$\mathcal{E}(X)$	$\mathcal{D}(X)$	$\mathcal{S}(X)$
$\|X\|_2$	$\sigma(X)$	$E[X]$
$\|X\|_1$	$\|X - \text{med}\,(X)\|_1$	med (X)
$\frac{1}{\alpha}E[\alpha\, X_+ + (1-\alpha)\, X_-]$	$\text{CVaR}_\alpha^\Delta(X)$	$q_X(\alpha) = [q_X^-(\alpha), q_X^+(\alpha)]$

where med (X) is the median of X (possibly an interval),

$$q_X^-(\alpha) = \inf\{t \mid F_X(t) \geq \alpha\} \quad \text{and} \quad q_X^+(\alpha) = \sup\{t \mid F_X(t) \leq \alpha\}$$

are *lower* and *upper* α*-quantiles*, respectively, and $\text{CVaR}_\alpha^\Delta(X)$ is conditional value at risk (CVaR) deviation defined by

$$\text{CVaR}_\alpha^\Delta(X) = E[X] - \frac{1}{\alpha}\int_0^\alpha q_X^+(s)\, ds. \tag{11.8}$$

Observe that for $\mathcal{E}(X) = \|X\|_2$, deviations (11.5) and (11.6) coincide, whereas for $\mathcal{E}(X) = \|X\|_1$, they are different.

For a given deviation measure $\mathcal{D}$, a nondegenerate error measure can be obtained by *inverse projection*

$$\mathcal{E}(X) = \mathcal{D}(X) + |E[X]|,$$

which through (11.6) projects back to $\mathcal{D}$ with the associated statistic $\mathcal{S}(X) = E[X]$; see Rockafellar *et al.* (2008, Example 2.5).

If $\mathcal{E}_1, \ldots, \mathcal{E}_l$ are nondegenerate error measures that project to deviations $\mathcal{D}_1, \ldots, \mathcal{D}_l$, respectively, then, for any weights $\lambda_1 > 0, \ldots, \lambda_l > 0$ with $\sum_{k=1}^l \lambda_k = 1$,

$$\mathcal{E}(X) = \inf_{\substack{c_1, \ldots, c_l \\ \lambda_1 c_1 + \ldots + \lambda_l c_l = 0}} \sum_{k=1}^l \lambda_k\, \mathcal{E}_k(X - c_k)$$

is a nondegenerate error measure, which projects to the deviation measure

$$\mathcal{D}(X) = \sum_{k=1}^{l} \lambda_k \mathcal{D}_k(X)$$

with the associated statistic

$$\mathcal{S}(X) = \sum_{k=1}^{l} \lambda_k \mathcal{S}_k(X).$$

See Theorem 2.2 in Rockafellar *et al.* (2008). As an immediate consequence of this result, we restate Example 2.6 from Rockafellar *et al.* (2008).

Example 11.1 (mixed quantiles and mixed-CVaR deviation) *For any choice of probability thresholds* $\alpha_k \in (0,1)$ *and weights* $\lambda_1 > 0, \ldots, \lambda_l > 0$ *with* $\sum_{k=1}^{l} \lambda_k = 1$,

$$\mathcal{E}(X) = E[X] + \inf_{\substack{c_1,\ldots,c_l \\ \lambda_1 c_1 + \ldots + \lambda_l c_l = 0}} \sum_{k=1}^{l} \frac{\lambda_k}{\alpha_k} E[\max\{0, c_k - X\}] \tag{11.9}$$

is a nondegenerate error measure called mixed quantile error measure, *which projects to the mixed CVaR deviation*

$$\mathcal{D}(X) = \sum_{k=1}^{m} \lambda_k \,\mathrm{CVaR}_{\alpha_k}^{\Delta}(X), \quad \sum_{k=1}^{m} \lambda_k = 1, \lambda_k > 0, \quad k = 1, \ldots, m, \tag{11.10}$$

with the associated statistic

$$\mathcal{S}(X) = \sum_{k=1}^{l} \lambda_k \, q_X(\alpha_k), \quad q_X(\alpha_k) = [q_X^-(\alpha_k), q_X^+(\alpha_k)]. \tag{11.11}$$

11.3 Risk Envelopes and Risk Identifiers

Deviation measures have dual characterization in terms of *risk envelopes* $\mathcal{Q} \subset \mathcal{L}^2(\Omega)$ defined by the following properties:

(Q1) $\mathcal{Q}$ is nonempty, closed, and convex;
(Q2) for every nonconstant X, there is some $Q \in \mathcal{Q}$ such that $E[XQ] < E[X]$; and
(Q3) $E[Q] = 1$ for all $Q \in \mathcal{Q}$.

There is a one-to-one correspondence between deviation measures and risk envelopes (Rockafellar *et al.* 2006a, Theorem 1):

$$\mathcal{D}(X) = E[X] - \inf_{Q \in \mathcal{Q}} E[XQ],$$
$$\mathcal{Q} = \{Q \in \mathcal{L}^2(\Omega) \mid \mathcal{D}(X) \geq E[X] - E[XQ] \text{ for all } X\}. \tag{11.12}$$

The elements of $\mathcal{Q}$ at which $E[XQ]$ attains infimum for a given X are called *risk identifiers* for X:

$$\mathcal{Q}(X) = \arg\min_{Q\in\mathcal{Q}} E[XQ].$$

They are those elements of $\mathcal{Q}$ that track the downside of X as closely as possible.

The second relationship in (11.12) implies that the set of risk identifiers for X with respect to a deviation measure $\mathcal{D}$ is determined by

$$\mathcal{Q}_{\mathcal{D}}(X) = \{Q \in \mathcal{Q} \mid \mathcal{D}(X) = E[(E[X] - X)Q] \equiv \mathrm{Cov}(-X, Q)\}.$$

From the optimization perspective, $\mathcal{Q}_{\mathcal{D}}(X)$ is closely related to *subdifferential* $\partial\mathcal{D}(X)$ of a deviation measure $\mathcal{D}$ at X, which is the set of *subgradients* $Z \in \mathcal{L}^2(\Omega)$ such that

$$\mathcal{D}(Y) \geq \mathcal{D}(X) + E[(Y - X)Z] \quad \text{for all} \quad Y \in \mathcal{L}^2(\Omega).$$

Proposition 1 in Rockafellar *et al.* (2006b) shows that

$$\partial\mathcal{D}(X) = 1 - \mathcal{Q}_{\mathcal{D}}(X).$$

11.3.1 Examples of Deviation Measures $\mathcal{D}$, Corresponding Risk Envelopes $\mathcal{Q}$, and Sets of Risk Identifiers $\mathcal{Q}_{\mathcal{D}}(X)$

1. standard deviation $\mathcal{D}(X) = \sigma(X) \equiv \|X - E[X]\|_2$:

$$\mathcal{Q} = \{Q \mid E[Q] = 1, \sigma(Q) \leq 1\}, \quad \mathcal{Q}_{\sigma}(X) = \left\{1 - \frac{X - E[X]}{\sigma(X)}\right\}$$

2. standard lower semideviation $\mathcal{D}(X) = \sigma_-(X) \equiv \|[X - E[X]]_-\|_2$:

$$\mathcal{Q} = \{Q \mid E[Q] = 1, \|Q - \inf Q\|_2 \leq 1\}, \quad \mathcal{Q}_{\sigma_-}(X) = \left\{1 - \frac{E[Y] - Y}{\sigma_-(X)}\right\},$$

where $Y = [E[X] - X]_+$

3. mean absolute deviation $\mathcal{D}(X) = \mathrm{MAD}(X) \equiv |X - E[X]|_1$:

$$\mathcal{Q} = \{Q \mid E[Q] = 1, \sup Q - \inf Q \leq 2\},$$
$$\mathcal{Q}_{\mathrm{MAD}}(X) = \{Q = 1 + E[Z] - Z \mid Z \in \mathrm{sign}[X - E[X]]\}$$

4. lower worst-case deviation $\mathcal{D}(X) = E[X] - \inf X$:

$$\mathcal{Q} = \{Q \mid E[Q] = 1, Q \geq 0\},$$
$$\mathcal{Q}_{\mathcal{D}}(X) = \{Q \mid E[Q] = 1, Q \geq 0, Q(\omega) = 0 \,\text{when}\, X(\omega) > \inf X\}$$

5. CVaR deviation $\mathcal{D}(X) = \mathrm{CVaR}_{\alpha}^{\Delta}(X)$:

$$\mathcal{Q} = \{Q \mid E[Q] = 1, 0 \leq Q \leq 1/\alpha\} \tag{11.13}$$

and $\mathcal{Q}_{\mathrm{CVaR}_\alpha^\Delta}(X)$ is the set of elements Q such that $E[Q]=1$ and

$$Q(\omega)\begin{cases} =\alpha^{-1} & \text{on}\,\{\omega\in\Omega\mid X(\omega)<-\mathrm{VaR}_\alpha(X)\},\\ \in[0,\alpha^{-1}] & \text{on}\,\{\omega\in\Omega\mid X(\omega)=-\mathrm{VaR}_\alpha(X)\},\\ =0 & \text{on}\,\{\omega\in\Omega\mid X(\omega)>-\mathrm{VaR}_\alpha(X)\}.\end{cases} \tag{11.14}$$

If $\mathcal{D}_1,\ldots,\mathcal{D}_m$ are deviation measures, then

$$\mathcal{D}(X)=\sum_{k=1}^m \lambda_k\,\mathcal{D}_k(X),\quad \sum_{k=1}^m \lambda_k=1,\quad \lambda_k>0,\quad k=1,\ldots,m, \tag{11.15}$$

and

$$\mathcal{D}(X)=\max\{\mathcal{D}_1(X),\ldots,\mathcal{D}_m(X)\}, \tag{11.16}$$

are deviation measures as well, for which the risk envelopes are given by Proposition 4 in Rockafellar *et al.* (2006a):

$$\mathcal{Q}=\begin{cases}\text{closure of } \sum_{k=1}^m \lambda_k\,\mathcal{Q}_k \text{ for (11.15)},\\ \text{closure of the convex hull of } \cup_{k=1}^m \mathcal{Q}_k \text{ for (11.16)},\end{cases}$$

where $\mathcal{Q}_1,\ldots,\mathcal{Q}_m$ are the risk envelopes for the deviation measures $\mathcal{D}_1,\ldots,\mathcal{D}_m$. This result and the formula (11.13) imply that the risk envelope for the mixed CVaR deviation (11.10) is determined by

$$\mathcal{Q}=\text{closure of}\sum_{k=1}^m \lambda_k\,\mathcal{Q}_k,\quad\text{where}\quad E[Q_k]=1,\quad 0\le Q_k\le 1/\alpha_k,\quad k=1,\ldots,m. \tag{11.17}$$

Risk identifiers along with risk envelopes are instrumental in formulating optimality conditions and devising optimization procedures in applications involving deviation measures. For example, if X is discretely distributed with $\mathbb{P}[X=x_k]=p_k$, $k=1,\ldots,n$, then with the risk envelope representation (11.13), the CVaR deviation (11.8) is readily restated as a linear program

$$\mathrm{CVaR}_\alpha^\Delta(X)=E[X]-\min_{q_1,\ldots,q_n}\left\{\sum_{k=1}^n q_kp_kx_k\;\middle|\; q_k\in[0,\alpha^{-1}],\ \sum_{k=1}^n q_kp_k=1\right\},$$

whereas for the same X, mixed CVaR deviation (11.10) with (11.17) can be represented by

$$\sum_{i=1}^m \lambda_i\,\mathrm{CVaR}_{\alpha_i}^\Delta(X)=E[X]-\min_{q_{ik}}\left\{\sum_{i,k=1}^{m,n}\lambda_i\,q_{ik}p_kx_k\;\middle|\; q_{ik}\in[0,\alpha_i^{-1}],\ \sum_{k=1}^n q_{ik}p_k=1\right\}.$$

11.4 Error Decomposition in Regression

An unconstrained generalized linear regression problem is formulated as follows: *approximate a random variable $Y\in\mathcal{L}^2(\Omega)$ by a linear combination $c_0+\sum_{k=1}^n c_kX_k$ of given random variables $X_k\in\mathcal{L}^2(\Omega)$, $k=1,\ldots,n$, and minimize an error measure $\mathcal{E}$ of the error*

$Z = Y - c_0 - \sum_{k=1}^{n} c_k X_k$ *with respect to* $c_0, c_1, \ldots, c_n$, *where* $\mathcal{E}$ *is assumed to be nondegenerate and finite everywhere on* $\mathcal{L}^2(\Omega)$, or, formally,

$$\min_{c_0, c_1, \ldots, c_n} \mathcal{E}(Z) \quad \text{with} \quad Z = Y - c_0 - \sum_{k=1}^{n} c_k X_k. \tag{11.18}$$

Observe that because of possible asymmetry of $\mathcal{E}$, $\mathcal{E}(-Z) \neq \mathcal{E}(Z)$.

Well-known particular cases of the linear regression (11.18) include:

1. Least-squares linear regression with $\mathcal{E}(Z) = \|Z\|_2$;
2. Median regression with $\mathcal{E}(Z) = \|Z\|_1$; and
3. Quantile regression with the asymmetric mean absolute error

$$\mathcal{E}(Z) = \frac{1}{\alpha} E[\alpha\, Z_+ + (1 - \alpha)\, Z_-],$$

where $Z_\pm = \max\{0, \pm Z\}$ and $\alpha \in (0, 1)$.

The choice of error measure to be used in a given regression problem is determined by the particular application and risk preferences of a decision maker.

Theorem 3.2 in Rockafellar *et al.* (2008) shows that the generalized linear regression (11.18) can be decomposed into minimizing the projected deviation measure of $Y - \sum_{k=1}^{n} c_k X_k$ with respect to $c_1, \ldots, c_n$ and into setting the intercept c_0 to the associated statistic of optimal $Y - \sum_{k=1}^{n} c_k X_k$. In other words, (11.18) is reduced to

$$\min_{c_1, \ldots, c_n} \mathcal{D}(\widetilde{Z}) \quad \text{and} \quad c_0 \in \mathcal{S}(\widetilde{Z}) \quad \text{with} \quad \widetilde{Z} = Y - \sum_{k=1}^{n} c_k X_k, \tag{11.19}$$

where $\mathcal{D}(\widetilde{Z}) = \inf_{c \in \mathbb{R}} \mathcal{E}(\widetilde{Z} - c)$ is the deviation projected from $\mathcal{E}$, and $\mathcal{S}(\widetilde{Z}) = \arg\min_{c \in \mathbb{R}} \mathcal{E}(\widetilde{Z} - c)$ is the statistic associated with $\mathcal{E}$; see Rockafellar *et al.* (2008). This result is known as *error decomposition*.

Furthermore, Theorem 4.1 in Rockafellar *et al.* (2008) states that $c_1, \ldots, c_n$ is a solution to (11.19) if and only if

$$\text{there exists } Q \in \mathcal{Q}_{\mathcal{D}}(\tilde{Z}) \text{ such that } E[(1 - Q)X_j] = 0 \text{ for } j = 1, \ldots, n, \tag{11.20}$$

where $\mathcal{Q}_{\mathcal{D}}(\widetilde{Z})$ is the risk identifier for $\widetilde{Z}$ with respect to deviation measure $\mathcal{D}$; see Rockafellar *et al.* (2008).

In many applications (e.g., factor models, index tracking, and replication problems), the coefficients $c_0, c_1, \ldots, c_n$ are often required to satisfy additional constraints. Let $\mathcal{C}$ be a feasible set of $n + 1$ dimensional vector $c = (c_0, c_1, \ldots, c_n)$. For example, the requirement of $c_0, c_1, \ldots, c_n$ to be nonnegative translates into having $\mathcal{C} = \{c \in \mathbb{R}^{n+1} \mid c \geq 0\}$. In this case, the generalized linear regression takes the form

$$\min_{c_0, c_1, \ldots, c_n} \mathcal{E}\left(Y - c_0 - \sum_{k=1}^{n} c_k X_k\right) \quad \text{subject to} \quad (c_0, c_1, \ldots, c_n) \in \mathcal{C}. \tag{11.21}$$

Sections 11.5, 11.6, and 11.7 discuss the problem (11.21) with different error measures $\mathcal{E}$ and feasible sets $\mathcal{C}$ frequently arising in various statistical decision applications.

11.5 Least-Squares Linear Regression

A least-squares linear regression is one of the basic and most widely used statistical tools that finds its applications in virtually all areas of science dealing with data analysis and statistics (e.g., physics, biology, medicine, finance, and economics).

Unconstrained least-squares linear regression is a particular case of (11.18) with $\mathcal{E}(\cdot) = \|\cdot\|_2$ and is given by

$$\min_{c_0,c_1,\ldots,c_n} \left\| Y - c_0 - \sum_{k=1}^{n} c_k X_k \right\|_2^2 . \tag{11.22}$$

The first-order necessary optimality conditions for the optimization problem (11.22) yield a system of linear equations for $c_0, c_1, \ldots, c_n$:

$$\begin{cases} \sum_{k=1}^{n} c_k \operatorname{Cov}(X_k, X_j) = \operatorname{Cov}(Y, X_j), j = 1, \ldots, n, \\ c_0 = E[Y] - \sum_{k=1}^{n} c_k \, E[X_k]. \end{cases} \tag{11.23}$$

If the covariance matrix Λ of $X_1, \ldots, X_n$ is nonsingular, then the system can be solved either numerically or in a closed form through the inverse Λ^{-1} :

$$(c_1, \ldots, c_n)^\top = \Lambda^{-1}(\operatorname{Cov}(Y, X_1), \ldots, \operatorname{Cov}(Y, X_n))^\top .$$

This is the main advantage of the least-squares linear regression.

The system (11.23) shows that the least-squares linear regression is solved in two steps: finding $c_1, \ldots, c_n$ and then determining c_0. In fact, for $\mathcal{D} = \sigma$, the error decomposition formulation (11.19) takes the form

$$\min_{c_1,\ldots,c_n} \sigma(\widetilde{Z}) \quad \text{and} \quad c_0 = E[\widetilde{Z}], \quad \text{where} \quad \widetilde{Z} = Y - \sum_{k=1}^{n} c_k X_k,$$

which states that the least-squares linear regression is equivalent to minimizing variance of $Y - \sum_{k=1}^{n} c_k X_k$ with respect to $c_1, \ldots, c_n$ and then setting intercept c_0 to the mean of $Y - \sum_{k=1}^{n} c_k X_k$. This fact is often taken for granted and may create the impression that the linear regression with another error measure $\mathcal{E}$ also leads to c_0 being $E[Y - \sum_{k=1}^{n} c_k X_k]$. However, this is possible only if the *deviation projected from* $\mathcal{E}$ coincides with the *deviation from the penalties relative to expectation*; see Rockafellar *et al.* (2008).

With the risk identifier corresponding to the standard deviation, that is,

$$\mathcal{Q}_\sigma(X) = \left\{ 1 - \frac{X - E[X]}{\sigma(X)} \right\},$$

the optimality conditions (11.20) can be recast in the form

$$E\left[\left(Y - \sum_{k=1}^{n} c_k X_k \right) (X_j - E[X_j]) \right] = 0, \quad j = 1, \ldots, n,$$

which with $c_0 = E\left[Y - \sum_{k=1}^{n} c_k X_k \right]$ are equivalent to the system (11.23).

In contrast to (11.2), the system (11.23) yields "true" $c_0, c_1, \dots, c_n$ (not estimates) provided that the expected values $E[Y]$ and $E[X_k]$ and the covariances $\mathrm{Cov}(X_k, X_j)$ and $\mathrm{Cov}(Y, X_j)$ are known. However, in real-life problems, this is almost never the case: we are only given simultaneous observations of $X_1, \dots, X_n$ and Y: $x_{1j}, \dots, x_{nj}, y_j, j = 1, \dots, m$, so that the expected values and covariances should be estimated through the given data.

In applications, least-squares linear regression is often solved subject to additional constraints on regression coefficients and, in general, can be formulated by

$$\min_{c_0, c_1, \dots, c_n} \left\| Y - c_0 - \sum_{k=1}^{n} c_k X_k \right\|_2 \quad \text{subject to} \quad (c_0, c_1, \dots, c_n) \in \mathcal{C}, \tag{11.24}$$

where $\mathcal{C}$ is some feasible set of $(c_0, c_1, \dots, c_n)$. This problem admits a closed-form solution only in a few simple cases, for example when $\mathcal{C}$ is determined by a set of linear equalities. In a general case, (11.24) is solved numerically.

Example 11.2 (index tracking with mean square error) *Let Y be the daily rate of return of a stock market index (e.g., S&P 500 and Nasdaq), and let $X_1, \dots, X_n$ be the daily rates of return of chosen financial instruments. Suppose a unit capital is to be allocated among these instruments with capital weights $c_1, \dots, c_n$ to replicate the index's rate of return by a linear combination of $X_1, \dots, X_n$ without shorting of the instruments. The imposed requirements on $c_1, \dots, c_n$ correspond to the feasible set*

$$\mathcal{C} = \left\{ (c_1, \dots, c_n) \in \mathbb{R}^n \;\middle|\; \sum_{k=1}^{n} c_k = 1,\ c_k \geq 0, k = 1, \dots, n \right\}. \tag{11.25}$$

In this case, optimal allocation positions $c_1, \dots, c_n$ can be found through the least-squares linear regression (11.24) with $c_0 = 0$ and $\mathcal{C}$ given by (11.25), which is a quadratic optimization problem.

Another application of constrained least-squares linear regression is sparse signal reconstruction, whose objective is to find a decision vector that has few nonzero components and satisfies certain linear constraints. The SPARCO toolbox offers a wide range of test problems for benchmarking of algorithms for sparse signal reconstruction; see http://www.cs.ubc.ca/labs/scl/sparco/. Typically, SPARCO toolbox problems are formulated in one of three closely related forms: *L1Relaxed*, *L1Relaxed D*, and *L2 D* (or *LASSO*). Both "L1Relaxed" and "L1Relaxed D" formulations minimize the $\mathcal{L}_1$-error of the regression residual subject to box constraints on decision variables and subject to a constraint on the $\mathcal{L}_1$-norm of the decision vector.[2] The difference in these two formulations is that "L1Relaxed D" splits each decision variable c_i into two nonnegative variables $c_i^+ = \max\{c_i, 0\}$ and $c_i^- = \max\{-c_i, 0\}$ ($c_i = c_i^+ - c_i^-$ and $|c_i| = c_i^+ + c_i^-$) and, as a result, has all decision variables nonnegative. Since "L1Relaxed D" doubles the number of the decision variables, in some problems, it may be less efficient than "L1Relaxed." The "L2 D" formulation minimizes the weighted sum of the squared $\mathcal{L}_2$-norm of the regression residual and the $\mathcal{L}_1$-norm of the vector of regression coefficients subject to box constraints on the coefficients. As the "L1Relaxed D" formulation, this one also splits each regression coefficient into two nonnegative parts.

[2] $\mathcal{L}_1$-norm of a vector is the sum of absolute values of vector components.

Example 11.3 (sparse reconstruction problem from the SPARCO toolbox) *Let $L(c, X)$ be an error function that linearly depends on a decision vector $c = (c_1, \dots, c_n)$ and on a given random vector $X = (X_1, \dots, X_n)$. The "L2 D" formulation of the sparse reconstruction problem from the SPARCO toolbox is a regression that minimizes a linear combination of $\|L(c, X)\|_2^2$ and the regularization part $\sum_{i=1}^n |c_i|$ subject to box constraints $l_i \le c_i \le u_i$, $i = 1, \dots, n$, where l_i and u_i are given bounds with $u_i \ge 0$ and $l_i \le 0$. Let $c^\pm = (c_1^\pm, \dots, c_n^\pm)$ with $c_i^\pm = \max\{\pm c_i, 0\}$, then $c_i = c_i^+ - c_i^-$ and $|c_i| = c_i^+ + c_i^-$, and the "L2 D" formulation takes the form*

$$\begin{aligned}&\min_{c^+, c^-} \ \|L(c^+ - c^-, X)\|_2^2 + \lambda \sum_{i=1}^n (c_i^+ + c_i^-)\\ &\text{subject to} \quad 0 \le c_i^+ \le u_i, \quad 0 \le c_i^- \le -l_i, i = 1, \dots, n,\end{aligned} \tag{11.26}$$

where λ is a given parameter.

Constrained least-squares linear regression is also used in an intensity-modulated radiation therapy (IMRT) treatment-planning problem formulated in Men *et al.* (2008). To penalize underdosing and overdosing with respect to a given threshold, the problem uses *quadratic one-sided penalties* or, equivalently, *second-order lower and upper partial moments.*

Example 11.4 (therapy treatment planning problem) *Let $[L(c, X)]_+ \equiv \max\{0, L(c, X)\}$ be a loss function, where $L(x, \theta)$ linearly depends on a decision vector $c = (c_1, \dots, c_n)$ and on a given random vector $X = (X_1, \dots, X_n)$. The regression problem, arising in intensity-modulated radiation therapy treatment, minimizes $\|[L(c, X)]_+\|_2^2$ subject to box constraints $l_i \le c_i \le u_i$ with given bounds l_i and u_i:*

$$\min_{c_1, \dots, c_n} \ \|[L(c, X)]_+\|_2^2 \quad \text{subject to} \quad l_i \le c_i \le u_i, \quad i = 1, \dots, n. \tag{11.27}$$

11.6 Median Regression

In the least-squares linear regression, large values of the error $Z = Y - c_0 - \sum_{k=1}^n c_k X_k$ are penalized heavier than small values, which makes the regression coefficients quite sensitive to outliers. In applications that require equal treatment of small and large errors, the median regression can be used instead.

Unconstrained median regression is a particular case of (11.18) with $\mathcal{E}(\cdot) = \|\cdot\|_1$:

$$\min_{c_0, c_1, \dots, c_n} \left\| Y - c_0 - \sum_{k=1}^n c_k X_k \right\|_1, \tag{11.28}$$

for which the error decomposition formulation (11.19) takes the form

$$\min_{c_1, \dots, c_n} E|\tilde{Z} - \operatorname{med} \tilde{Z}| \quad \text{and} \quad c_0 \in \operatorname{med} \tilde{Z} \quad \text{with} \quad \tilde{Z} = Y - \sum_{k=1}^n c_k X_k, \tag{11.29}$$

where $\operatorname{med} \tilde{Z}$ is the median of $\tilde{Z}$, which, in general, is any number in the closed interval $[q_{\tilde{Z}}^-(1/2), q_{\tilde{Z}}^+(1/2)]$. Observe that the median regression does not reduce to minimization of the mean-absolute deviation (MAD) and that c_0 is not the mean of $Y - \sum_{k=1}^n c_k X_k$.

Let $c_1, \dots, c_n$ be an optimal solution to the problem (11.29), and let the random variable $\tilde{Z} = Y - \sum_{k=1}^{n} c_k X_k$ have no probability "atom" at $q_{\tilde{Z}}^{+}(1/2)$; then, the interval med $\tilde{Z}$ is a singleton, and the optimality conditions (11.20) reduce to

$$E[X_j - E[X_j] | \tilde{Z} \leq \text{med } \tilde{Z}] = 0, j = 1, \dots, n.$$

These conditions are, however, rarely used in practice.

In applications, $X_1, \dots, X_n$ and Y are often assumed to be discretely distributed with joint probability distribution $\mathbb{P}[X_1 = x_{1j}, \dots, X_n = x_{nj}, Y = y_j] = p_j > 0,\ j = 1, \dots, m,$ with $\sum_{j=1}^{m} p_j = 1$. In this case,

$$\|Z\|_1 = \sum_{j=1}^{m} p_j \left| y_j - c_0 - \sum_{k=1}^{n} c_k x_{kj} \right|,$$

and the median regression (11.28) reduces to the linear program

$$\begin{aligned} \min_{\substack{c_0, c_1, \dots, c_n, \\ \zeta_1, \dots, \zeta_m}} \quad & \sum_{j=1}^{m} p_j \zeta_j \\ \text{subject to} \quad & \zeta_j \geq y_j - c_0 - \sum_{k=1}^{n} c_k x_{kj}, \quad j = 1, \dots, m, \\ & \zeta_j \geq c_0 + \sum_{k=1}^{n} c_k x_{kj} - y_j, \quad j = 1, \dots, m, \end{aligned} \tag{11.30}$$

where $\zeta_1, \dots, \zeta_m$ are auxiliary variables.

The median regression with constraints on regression coefficients is formulated by

$$\min_{c_0, c_1, \dots, c_n} \left\| Y - c_0 - \sum_{k=1}^{n} c_k X_k \right\|_1 \quad \text{subject to} \quad (c_0, c_1, \dots, c_n) \in \mathcal{C}, \tag{11.31}$$

where $\mathcal{C}$ is a given feasible set of $(c_0, c_1, \dots, c_n)$. For an arbitrary joint probability distribution of $X_1, \dots, X_n$ and Y, the necessary optimality conditions for (11.31) are given in Rockafellar *et al.* (2006a).

If $X_1, \dots, X_n$ and Y are discretely distributed, and $\mathcal{C}$ is determined by a set of linear constraints, the Equation (11.31) reduces to a linear program.

Example 11.5 (index tracking with mean absolute error) *The setting is identical to that in Example 11.2. But this time, the optimal allocation positions $c_1, \dots, c_n$ are found through the median regression (11.31) with $\mathcal{C}$ given by (11.25). If $X_1, \dots, X_n$ and Y are assumed to be discretely distributed with joint probability distribution $\mathbb{P}[X_1 = x_{1j}, \dots, X_n = x_{nj}, Y = y_j] = p_j > 0, j = 1, \dots, m$, where $\sum_{j=1}^{m} p_j = 1$, then this regression problem can be formulated as the linear program*

$$\min_{\substack{c_1, \dots, c_n \\ \zeta_1, \dots, \zeta_m}} \sum_{j=1}^{m} p_j \zeta_j$$

$$\begin{aligned} \textit{subject to} \quad & \zeta_j \geq y_j - \sum_{k=1}^{n} c_k x_{kj}, \quad j = 1, \dots, m, \\ & \zeta_j \geq \sum_{k=1}^{n} c_k x_{kj} - y_j, \quad j = 1, \dots, m, \\ & \sum_{k=1}^{n} c_k = 1, \; c_k \geq 0, \quad k = 1, \dots, n, \end{aligned}$$

where $\zeta_1, \dots, \zeta_m$ are auxiliary variables.

Constrained median regression is also used to design a portfolio of *credit default swaps* (CDS) and credit indices to hedge against changes in a *collateralized debt obligation* (CDO) book. A CDS provides insurance against the risk of default (credit event) of a particular company. A buyer of the CDS has the right to sell bonds issued by the company for their face value when the company is in default. The buyer makes periodic payments to the seller until the end of the life of the CDS or until a default occurs. The total amount paid per year, as a percentage of the notional principal, is known as the *CDS spread*, which is tracked by credit indices. A CDO is a credit derivative based on defaults of a pool of assets. Its common structure involves tranching or slicing the credit risk of the reference pool into different risk levels of increasing seniority. The losses first affect the *equity* (first loss) tranche, then the *mezzanine* tranche, and finally the *senior* and *super senior* tranches. *The hedging problem is to minimize risk of portfolio losses subject to budget and cardinality constraints on hedge positions.* The risk is measured by mean absolute deviation (MAD) and by $\mathcal{L}_1$-norm (mean absolute penalty).

Example 11.6 (median regression and CDO) *Let $L(c,X)$ be a loss function in hedging against changes in a collateralized debt obligation (CDO) book, where $L(c,X)$ linearly depends on a decision vector $c = (c_1, \dots, c_n)$ (positions in financial instruments) and on a given random vector $X = (X_1, \dots, X_n)$. A regression problem then minimizes the mean absolute error of $L(c,X)$ subject to the budget constraint $\sum_{i=1}^{n} a_i|c_i| \leq C$ with given C and $a_i > 0$, $i = 1, \dots, n$, and subject to a constraint on cardinality of the decision variables not to exceed a positive integer S:*

$$\begin{aligned} & \min_{c_1,\dots,c_n} \; \|L(c,X)\|_1 \\ \textit{subject to} \quad & \sum_{i=1}^{n} a_i|c_i| \leq C, \\ & \sum_{i=1}^{n} I_{\{a_i|c_i| \geq w\}} \leq S, \\ & |c_i| \leq k_i, \quad i = 1, \dots, n, \end{aligned} \tag{11.32}$$

where w is a given threshold; $I_{\{\cdot\}}$ is the indicator function equal to 1 if the condition in the curly brackets is true, and equal to 0 otherwise; and $|c_i| \leq k_i$, $i = 1, \dots, n$, are bounds on decision variables (positions).

The next three examples formulate regression problems arising in sparse signal reconstruction. In all of them, $L(c,X)$ is an error function that linearly depends on a decision vector $c = (c_1, \dots, c_n)$ and on a given random vector $X = (X_1, \dots, X_n)$, and $l_i \le c_i \le u_i$, $i = 1, \dots, n$, are box constraints with given bounds l_i and u_i ($l_i \le u_i$).

Example 11.7 (sparse signal reconstruction I: "L1Relaxed" formulation) *This regression problem minimizes the mean absolute error of $L(c,X)$ subject to a constraint on cardinality of c with given integer bound S and subject to box constraints on c:*

$$\begin{aligned} &\min_{c_1,\dots,c_n} \ \|L(c,X)\|_1 \\ &\text{subject to} \ \sum_{i=1}^{n} \left(I_{\{a_i c_i \ge w\}} + I_{\{b_i c_i \le -w\}} \right) \le S, \\ &\qquad l_i \le c_i \le u_i, \quad i = 1, \dots, n. \end{aligned} \tag{11.33}$$

Example 11.8 (sparse signal reconstruction II) *This regression problem minimizes the mean absolute error of $L(c,X)$ subject to a constraint on the $\mathcal{L}_1$-norm of c, that is, $\sum_{i=1}^{n} |c_i| \le U$ with given bound U, and subject to box constraints on c:*

$$\begin{aligned} &\min_{c_1,\dots,c_n} \ \|L(c,X)\|_1 \\ &\text{subject to} \ \sum_{i=1}^{n} |c_i| \le U, \\ &\qquad l_i \le c_i \le u_i, \quad i = 1, \dots, n. \end{aligned} \tag{11.34}$$

Example 11.9 (sparse signal reconstruction III) *This estimation problems minimizes the cardinality of c subject to constraints on the mean absolute error of $L(c,X)$ and on the $\mathcal{L}_1$-norm of c with given bounds ϵ and U, respectively, and subject to box constraints on c:*

$$\begin{aligned} &\min_{c_1,\dots,c_n} \ \sum_{i=1}^{n} \left(I_{\{a_i c_i \ge w\}} + I_{\{b_i c_i \le -w\}} \right) \\ &\text{subject to} \ \|L(c,X)\|_1 \le \epsilon, \\ &\qquad \sum_{i=1}^{n} |c_i| \le U, \\ &\qquad l_i \le c_i \le u_i, \quad i = 1, \dots, n. \end{aligned} \tag{11.35}$$

Example 11.10 presents a reformulation of the regression problem (11.34).

Example 11.10 (sparse signal reconstruction from SPARCO toolbox) *Suppose the random vector X is discretely distributed and takes on values $X^{(1)}, \dots, X^{(m)}$ with corresponding positive probabilities $p_1, \dots, p_m$ summing into 1, so that $\|L(c,X)\|_1 = \sum_{j=1}^{m} p_j |L(c, X^{(j)})|$. Let*

$c_i^{\pm} = \max\{\pm c_i, 0\}$, $i = 1, \dots, n$, then $c_i = c_i^+ - c_i^-$ and $|c_i| = c_i^+ + c_i^-$. Given that $L(c, X^{(j)})$ is linear with respect to c and that $u_i \geq 0$ and $l_i \leq 0$, $i = 1, \dots, n$, the problem (11.34) can be restated as the linear program

$$\min_{\substack{c^+, c^- \\ \zeta_1, \dots, \zeta_m}} \sum_{j=1}^{m} p_j \, \zeta_j$$

$$\text{subject to} \sum_{i=1}^{n} (c_i^+ + c_i^-) \leq U,$$

$$\zeta_j \geq L(c^+ - c^-, X^{(j)}), \quad i = 1, \dots, n,$$

$$\zeta_j \geq -L(c^+ - c^-, X^{(j)}), \quad i = 1, \dots, n,$$

$$0 \leq c_i^+ \leq u_i, \quad 0 \leq c_i^- \leq -l_i, \quad i = 1, \dots, n, \tag{11.36}$$

where $\zeta_1, \dots, \zeta_m$ are auxiliary variables.

11.7 Quantile Regression and Mixed Quantile Regression

Both the least-squares linear regression and median regression treat ups and downs of the regression error equally, which might not be desirable in some applications. For example, in the index tracking problem from Example 11.2, a decision maker (financial analyst) may use a quantile regression that minimizes the asymmetric mean absolute error $\mathcal{E}_\alpha(Z) = \alpha^{-1} E[\alpha \, Z_+ + (1 - \alpha) \, Z_-]$ of $Z = Y - c_0 - \sum_{k=1}^{n} c_k X_k$ for some $\alpha \in (0, 1)$.

Unconstrained quantile regression is a particular case of the generalized linear regression (11.18) with the asymmetric mean absolute error measure:

$$\min_{c_0, c_1, \dots, c_n} \; E[\alpha \, Z_+ + (1 - \alpha) Z_-] \quad \text{with} \quad Z = Y - c_0 - \sum_{k=1}^{n} c_k X_k, \tag{11.37}$$

where $Z_{\pm} = \max\{\pm Z, 0\}$; and the multiplier α^{-1} in the objective function is omitted. Observe that for $\alpha = 1/2$, (11.37) is equivalent to the median regression (11.28).

In this case, the error decomposition formulation (11.19) takes the form

$$\min_{c_1, \dots, c_n} \text{CVaR}_\alpha^\Delta(\widetilde{Z}) \quad \text{and} \quad c_0 \in [q_{\widetilde{Z}}^-(\alpha), q_{\widetilde{Z}}^+(\alpha)] \quad \text{with} \quad \widetilde{Z} = Y - \sum_{k=1}^{n} c_k X_k. \tag{11.38}$$

In other words, the quantile regression (11.37) reduces to minimizing *CVaR deviation* of $Y - \sum_{k=1}^{n} c_k X_k$ with respect to $c_1, \dots, c_n$ and to setting c_0 to any value from the α-quantile interval of $Y - \sum_{k=1}^{n} c_k X_k$.

Let $c_1, \dots, c_n$ be an optimal solution to the problem (11.38), and let the random variable $\widetilde{Z} = Y - \sum_{k=1}^{n} c_k X_k$ have no probability "atom" at $q_{\widetilde{Z}}^+(\alpha)$, then the interval $[q_{\widetilde{Z}}^-(\alpha), q_{\widetilde{Z}}^+(\alpha)]$ is a singleton, and the optimality conditions (11.20) simplify to

$$E\left[X_j - E[X_j] \,\middle|\, \widetilde{Z} \leq q_{\widetilde{Z}}^+(\alpha)\right] = 0, \quad j = 1, \dots, n.$$

However, in this form, they are rarely used in practice.

If $X_1,\dots,X_n$ and Y are discretely distributed with joint probability distribution $\mathbb{P}[X_1 = x_{1j},\dots, X_n = x_{nj}, Y = y_j] = p_j > 0$, $j = 1,\dots,m$, where $\sum_{j=1}^{m} p_j = 1$, then with the formula (5) in Rockafellar *et al.* (2006a), the quantile regression (11.37) can be restated as the linear program

$$\min_{\substack{c_0, c_1, \dots, c_n \\ \zeta_1,\dots,\zeta_m}} \quad \sum_{j=1}^{m} p_j \left(y_j - c_0 + \alpha^{-1}\zeta_j - \sum_{k=1}^{n} c_k x_{kj} \right)$$

$$\text{subject to} \quad \zeta_j \ge c_0 + \sum_{k=1}^{n} c_k x_{kj} - y_j, \quad \zeta_j \ge 0, \quad j = 1,\dots,m, \tag{11.39}$$

where $\zeta_1,\dots,\zeta_m$ are auxiliary variables.

The *return-based style classification* for a mutual fund is a regression of the fund return on several indices as explanatory variables, where regression coefficients represent the fund's style with respect to each of the indices. In contrast to the least-squares regression, the quantile regression can assess the impact of explanatory variables on various parts of the regressand distribution, for example on the 95th and 99th percentiles. Moreover, for a portfolio with exposure to derivatives, the mean and quantiles of the portfolio return distribution may have quite different regression coefficients for the same explanatory variables. For example, in most cases, the strategy of investing into naked deep out-of-the-money options behaves like a bond paying some interest; however, in some rare cases, this strategy may lose significant amounts of money. With the quantile regression, a fund manager can analyze the impact of a particular factor on any part of the return distribution. Example 11.11 presents an unconstrained quantile regression problem arising in the return-based style classification of a mutual fund.

Example 11.11 (quantile regression in style classification) *Let $L(c,X)$ be a loss function that linearly depends on a decision vector $c = (c_1,\dots,c_n)$ and on a random vector $X = (X_1,\dots,X_n)$ representing uncertain rates of return of n indices as explanatory variables. The quantile regression (11.37) with $L(c,X)$ in place of Z takes the form*

$$\min_{c_1,\dots,c_n} E[\alpha[L(c,X)]_+ + (1-\alpha)[L(c,X)]_-]. \tag{11.40}$$

A constrained quantile regression is formulated similarly to (11.37):

$$\min_{c_0,c_1,\dots,c_n} \quad E[\alpha\, Z_+ + (1-\alpha)\, Z_-] \quad \text{with} \quad Z = Y - c_0 - \sum_{k=1}^{n} c_k X_k$$

$$\text{subject to} \quad (c_0, c_1,\dots,c_n) \in \mathcal{C}, \tag{11.41}$$

where $\mathcal{C}$ is a given feasible set for regression coefficients $c_0, c_1,\dots,c_n$.

Example 11.12 (index tracking with asymmetric mean absolute error) *The setting is identical to that in Example 11.2. But this time, the allocation positions $c_1,\dots,c_n$ are found from the constrained quantile regression (11.41) with $\mathcal{C}$ given by (11.25). If $X_1,\dots,X_n$ and*

Y are assumed to be discretely distributed with joint probability distribution $\mathbb{P}[X_1 = x_{1j}, \ldots, X_n = x_{nj}, Y = y_j] = p_j > 0, j = 1, \ldots, m,$ *where* $\sum_{j=1}^m p_j = 1$*, then this regression problem can be formulated as the linear program*

$$\min_{\substack{c_1,\ldots,c_n\\ \zeta_1,\ldots,\zeta_m}} \sum_{j=1}^m p_j \left(y_j - \sum_{k=1}^n c_k x_{kj} + \alpha^{-1}\zeta_j \right)$$

$$\text{subject to} \quad \zeta_j \geq \sum_{k=1}^n c_k x_{kj} - y_j, \quad \zeta_j \geq 0, \quad j = 1, \ldots, m,$$

$$\sum_{k=1}^n c_k = 1, \; c_k \geq 0, \quad k = 1, \ldots, n,$$

where $\zeta_1, \ldots, \zeta_m$ *are auxiliary variables.*

The linear regression with the *mixed quantile error measure* (11.9) is called *mixed quantile regression*. It generalizes quantile regression and, through error decomoposition, takes the form

$$\min_{\substack{c_1,\ldots,c_n,\\ C_1,\ldots,C_l}} E\left[Y - \sum_{j=1}^n c_j X_j\right] + \sum_{k=1}^l \lambda_k \left(\frac{1}{\alpha_k} E\left[\max\left\{0, C_k - \sum_{j=1}^n c_j X_j\right\}\right] - C_k \right) \tag{11.42}$$

with the intercept c_0 determined by

$$c_0 = \sum_{k=1}^l \lambda_k \, C_k,$$

where $C_1, \ldots, C_l$ are a solution to (11.42); see Example 3.1 in Rockafellar *et al.* (2008).

The optimality conditions (11.20) for (11.42) are complicated. However, as the quantile regression, (11.42) can be reduced to a linear program.

11.8 Special Types of Linear Regression

This section discusses special types of unconstrained and constrained linear regressions encountered in statistical decision problems.

Often, it is required to find an unbiased linear approximation of an output random variable Y by a linear combination of input random variables $X_1, \ldots, X_n$, in which case the approximation error has zero expected value: $E[Y - c_0 - \sum_{k=1}^n c_k X_k] = 0$. A classical example of an *unbiased linear regression* is minimizing variance or, equivalently, standard deviation with the intercept c_0 set to $c_0 = E[Y - \sum_{k=1}^n c_k X_k]$. If, in this example, the standard deviation is replaced by a general deviation measure $\mathcal{D}$, we obtain a generalized unbiased linear regression:

$$\min_{c_1,\ldots,c_n} \mathcal{D}(\widetilde{Z}) \quad \text{and} \quad c_0 = E[\widetilde{Z}], \quad \text{where} \quad \widetilde{Z} = Y - \sum_{k=1}^n c_k X_k. \tag{11.43}$$

In fact, (11.43) is equivalent to minimizing the error measure $\mathcal{E}(Z) = \mathcal{D}(Z) + |E[Z]|$ of $Z = Y - c_0 - \sum_{k=1}^{n} c_k X_k$. Observe that in view of the error decomposition theorem (Rockafellar *et al.*, 2008, Theorem 3.2), the generalized linear regression (11.18) with a nondegenerate error measure $\mathcal{E}$ and the unbiased linear regression (11.43) with the deviation measure $\mathcal{D}$ projected from $\mathcal{E}$ yield the same $c_1, \dots, c_n$ but, in general, different intercepts c_0.

Rockafellar *et al.* (2008) introduced *risk acceptable linear regression* in which a deviation measure $\mathcal{D}$ of the approximation error $Z = Y - c_0 - \sum_{k=1}^{n} c_k X_k$ is minimized subject to a constraint on the averse measure of risk $\mathcal{R}$ related to $\mathcal{D}$ by $\mathcal{R}(X) = \mathcal{D}(X) - E[X]$:

$$\min_{c_1,\dots,c_n} \quad \mathcal{D}(Z) \quad \text{subject to} \quad \mathcal{R}(Z) = 0 \quad \text{with} \quad Z = Y - c_0 - \sum_{k=1}^{n} c_k X_k, \tag{11.44}$$

which is equivalent to

$$\min_{c_1,\dots,c_n} \mathcal{D}(\widetilde{Z}) \quad \text{and} \quad c_0 = E(\widetilde{Z}) - \mathcal{D}(\widetilde{Z}), \quad \text{where} \quad \widetilde{Z} = Y - \sum_{k=1}^{n} c_k X_k. \tag{11.45}$$

The unbiased linear regression (11.43) and risk acceptable linear regression (11.45) show that the intercept c_0 could be set based on different requirements.

In general, the risk acceptable regression may minimize either an error measure $\mathcal{E}$ or a deviation measure $\mathcal{D}$ of the error Z subject to a constraint on a risk measure $\mathcal{R}$ of Z not necessarily related to $\mathcal{E}$ or $\mathcal{D}$. Example 11.13 illustrates a risk acceptable regression arising in a portfolio replication problem with a constraint on CVaR.

Example 11.13 (risk acceptable regression) *Let $L(c, X)$ be a portfolio replication error (loss function) that linearly depends on a decision vector $c = (c_1, \dots, c_n)$ and on a random vector $X = (X_1, \dots, X_n)$ representing uncertain rates of return of n instruments in a portfolio replicating the S&P 100 index. The risk acceptable regression minimizes the mean absolute error of $L(c, X)$ subject to the budget constraint $\sum_{i=1}^{n} a_i\, c_i \le U$ with known $a_1, \dots, a_n$ and U and subject to a CVaR constraint on the underperformance of the portfolio compared to the index:*

$$\begin{aligned} &\min_{c_1,\dots,c_n} \ \|L(c, X)\|_1 \\ &\textit{subject to} \ \sum_{i=1}^{n} a_i\, c_i \le U, \\ &\qquad CVaR_\alpha(L(c, X)) \le w, \\ &\qquad c_i \ge 0, \quad i = 1, \dots, n, \end{aligned} \tag{11.46}$$

where α and w are given.

11.9 Robust Regression

Robust regression aims to reduce influence of sample outliers on regression parameters, especially when regression error has heavy tails.

In statistics, robustness of an estimator is a well-established notion and is assessed by the so-called *estimator's breakdown point*, which is the proportion of additional arbitrarily large observations (outliers) needed to make the estimator unbounded. For example, the sample mean requires just a single such observation, while the sample median would still be finite until the proportion of such observations reaches 50%. Consequently, the mean's breakdown point is 0%, whereas the median's breakdown point is 50%.

As in the previous regression setting, suppose Y is approximated by a linear combination of input random variables $X_1, \ldots, X_n$ with the regression error defined by $Z = Y - c_0 - \sum_{i=1}^n c_i X_i$, where $c_0, c_1, \ldots, c_n$ are unknown regression coefficients. A robust regression minimizes an error measure of Z that has a nonzero breakdown point. Thus, in this setting, the regression's breakdown point is that of the error measure.

Often, a robust regression relies on order statistics of Z and on "trimmed" error measures. Two popular robust regressions are the *least median of squares (LMS) regression*, which minimizes the median of Z^2 and has a 50% breakdown point:

$$\min_{c_0,c_1,\ldots,c_n} \operatorname{med}(Z^2) \quad \text{with} \quad Z = Y - c_0 - \sum_{i=1}^n c_i X_i. \tag{11.47}$$

and the *least-trimmed-squares (LTS) regression*, which minimizes the average α-quantile of Z^2 and has a $(1-\alpha)\cdot 100\%$ breakdown point:

$$\min_{c_0,c_1,\ldots,c_n} \bar{q}_{Z^2}(\alpha) \quad \text{with} \quad Z = Y - c_0 - \sum_{i=1}^n c_i X_i. \tag{11.48}$$

Rousseeuw and Driessen (2006) referred to (11.48) as a challenging optimization problem.

Typically, in the LTS regression, α is set to be slightly larger than 1/2. For $\alpha = 1$, $\bar{q}_{Z^2}(\alpha) = \|Z\|_2^2$, and (11.48) reduces to the standard least-squares regression. The LTS regression is reported to have advantages over the LMS regression or the one that minimizes the α-quantile of Z^2; see Rousseeuw and Driessen (2006), Rousseeuw and Leroy (1987), and Venables and Ripley (2002).

Let h be such that $h(t) > 0$ for $t \neq 0$ and $h(0) = 0$, but not necessarily symmetric (i.e., $h(-t) \neq h(t)$ in general). Then, the LMS and LTS regressions have the following generalization:

1. *Minimizing the upper α-quantile of $h(Z)$*:

$$\min_{c_0,c_1,\ldots,c_n} q^+_{h(Z)}(\alpha) \quad \text{with} \quad Z = Y - c_0 - \sum_{i=1}^n c_i X_i, \tag{11.49}$$

2. *Minimizing the average α-quantile of $h(Z)$*:

$$\min_{c_0,c_1,\ldots,c_n} \bar{q}_{h(Z)}(\alpha) \quad \text{with} \quad Z = Y - c_0 - \sum_{i=1}^n c_i X_i. \tag{11.50}$$

For example, in both (11.49) and (11.50), we may use $h(Z) = |Z|^p$, $p \geq 1$. In particular, for $h(Z) = Z^2$, (11.49) with $\alpha = 1/2$ corresponds to the LMS regression (11.47), whereas (11.50) reduces to the LTS regression (11.48).

When $h(-t) = h(t)$, (11.49) and (11.50) do not discriminate positive and negative errors. This, however, is unlikely to be appropriate for errors with significantly skewed distributions.

For example, instead of med (Z^2) and $\bar{q}_{Z^2}(\alpha)$, we can use *two-tailed α-value-at-risk (VaR) deviation* of the error Z defined by

$$\begin{aligned}\text{TwoTailVaR}^{\Delta}_{\alpha}(Z) &= \text{VaR}_{1-\alpha}(Z) + \text{VaR}_{1-\alpha}(-Z)\\ &\equiv q^{-}_{Z}(\alpha) - q^{+}_{Z}(1-\alpha), \quad \alpha \in (1/2, 1]. \end{aligned} \tag{11.51}$$

The definition (11.51) shows that the two-tailed α-VaR deviation is, in fact, the range between the upper and lower $(1-\alpha)$-tails of the error Z, which is equivalent to the support of the random variable Z with truncated $(1-\alpha)\cdot 100\%$ of the "outperformances" and $(1-\alpha)\cdot 100\%$ of "underperformances." Consequently, the two-tailed α-VaR deviation has the breakdown point of $(1-\alpha)\cdot 100\%$. Typically, α is chosen to be 0.75 and 0.9.

Robust regression is used in mortgage pipeline hedging. Usually, mortgage lenders sell mortgages in the secondary market. Alternatively, they can exchange mortgages for mortgage-backed securities (MBSs) and then sell MBSs in the secondary market. The mortgage-underwriting process is known as the "pipeline." Mortgage lenders commit to a mortgage interest rate while the loan is in process, typically for a period of 30–60 days. If the rate rises before the loan goes to closing, the value of the loan declines and the lender sells the loan at a lower price. The risk that mortgages in process will fall in value prior to their sale is known as *mortgage pipeline risk*. Lenders often hedge this exposure either by selling forward their expected closing volume or by shorting either US Treasury notes or futures contracts. *Fallout* refers to the percentage of loan commitments that do not go to closing. It affects the mortgage pipeline risk. As interest rates fall, the fallout rises because borrowers locked in a mortgage rate are more likely to find better rates with another lender. Conversely, as rates rise, the percentage of loans that close increases. So, the fallout alters the size of the pipeline position to be hedged against and, as a result, affects the required size of the hedging instrument: at lower rates, fewer rate loans will close and a smaller position in the hedging instrument is needed. To hedge against the fallout risk, lenders often use options on US Treasury note futures.

Suppose a hedging portfolio is formed out of n hedging instruments with random returns $X_1, \ldots, X_n$. A pipeline risk hedging problem is to *minimize a deviation measure $\mathcal{D}$ of the underperformance of the hedging portfolio with respect to a random hedging target Y*, where short sales are allowed and transaction costs are ignored. Example 11.14 formulates a robust regression with the two-tailed α-VaR deviation used in a mortgage pipeline hedging problem.

Example 11.14 (robust regression with two-tailed α-VaR deviation) *Let a target random variable Y be approximated by a linear combination of n random variables $X_1, \ldots, X_n$, then the robust regression minimizes the two-tailed α-VaR deviation of the error $Y - c_0 - \sum_{i=1}^{n} c_i X_i$:*

$$\min_{c_0, c_1, \ldots, c_n} \text{TwoTailVaR}^{\Delta}_{\alpha}\left(Y - c_0 - \sum_{i=1}^{n} c_i X_i\right). \tag{11.52}$$

It has a $(1-\alpha)\cdot 100\%$-breakdown point.

References, Further Reading, and Bibliography

Hastie, T., Tibshirani, R. and Friedman, J. (2008). *The elements of statistical learning: data mining, inference, and prediction*, 2nd ed. New York: Springer.

Koenker, R. and Bassett, G. (1978). Regression quantiles. *Econometrica*, 46, 33–50.

Men, C., Romeijn, E., Taskin, C. and Dempsey, J. (2008). An exact approach to direct aperture optimization in IMRT treatment planning. *Physics in Medicine and Biology*, 52, 7333–7352.

Rockafellar, R.T., Uryasev, S. and Zabarankin, M. (2002). Deviation measures in risk analysis and optimization. Technical Report 2002-7, ISE Department, University of Florida, Gainesville, FL.

Rockafellar, R.T., Uryasev, S. and Zabarankin, M. (2006a). Generalized deviations in risk analysis. *Finance and Stochastics*, 10, 51–74.

Rockafellar, R.T., Uryasev, S. and Zabarankin, M. (2006b). Optimality conditions in portfolio analysis with general deviation measures. *Mathematical Programming, Series B*, 108, 515–540.

Rockafellar, R.T., Uryasev, S. and Zabarankin, M. (2008). Risk tuning with generalized linear regression. *Mathematics of Operations Research*, 33, 712–729.

Rousseeuw, P.J. and Driessen, K. (2006). Computing LTS regression for large data sets. *Data Mining and Knowledge Discovery*, 12, 29–45.

Rousseeuw, P. and Leroy, A. (1987). *Robust regression and outlier detection*. New York: John Wiley & Sons.

van der Waerden, B. (1957). *Mathematische Statistik*. Berlin: Springer-Verlag.

Venables, W. and Ripley, B. (2002). *Modern applied statistics with S-PLUS*, 4th ed. New York: Springer.

Index

References to figures are given in **bold** type. References to tables are given in *italic* type.

Financial Signal Processing and Machine Learning, First Edition.
Edited by Ali N. Akansu, Sanjeev R. Kulkarni and Dmitry Malioutov.
© 2016 John Wiley & Sons, Ltd. Published 2016 by John Wiley & Sons, Ltd.

Printed and bound by CPI Group (UK) Ltd, Croydon, CR0 4YY
16/04/2025
14658550-0002